COMPUTER-AIDED INJECTION MOLD DESIGN AND MANUFACTURE

PLASTICS ENGINEERING

Founding Editor

Donald E. Hudgin

Professor
Clemson University
Clemson, South Carolina

1. Plastics Waste: Recovery of Economic Value, *Jacob Leidner*
2. Polyester Molding Compounds, *Robert Burns*
3. Carbon Black-Polymer Composites: The Physics of Electrically Conducting Composites, *edited by Enid Keil Sichel*
4. The Strength and Stiffness of Polymers, *edited by Anagnostis E. Zachariades and Roger S. Porter*
5. Selecting Thermoplastics for Engineering Applications, *Charles P. MacDermott*
6. Engineering with Rigid PVC: Processability and Applications, *edited by I. Luis Gomez*
7. Computer-Aided Design of Polymers and Composites, *D. H. Kaelble*
8. Engineering Thermoplastics: Properties and Applications, *edited by James M. Margolis*
9. Structural Foam: A Purchasing and Design Guide, *Bruce C. Wendle*
10. Plastics in Architecture: A Guide to Acrylic and Polycarbonate, *Ralph Montella*
11. Metal-Filled Polymers: Properties and Applications, *edited by Swapan K. Bhattacharya*
12. Plastics Technology Handbook, *Manas Chanda and Salil K. Roy*
13. Reaction Injection Molding Machinery and Processes, *F. Melvin Sweeney*
14. Practical Thermoforming: Principles and Applications, *John Florian*
15. Injection and Compression Molding Fundamentals, *edited by Avraam I. Isayev*
16. Polymer Mixing and Extrusion Technology, *Nicholas P. Cheremisinoff*
17. High Modulus Polymers: Approaches to Design and Development, *edited by Anagnostis E. Zachariades and Roger S. Porter*
18. Corrosion-Resistant Plastic Composites in Chemical Plant Design, *John H. Mallinson*
19. Handbook of Elastomers: New Developments and Technology, *edited by Anil K. Bhowmick and Howard L. Stephens*
20. Rubber Compounding: Principles, Materials, and Techniques, *Fred W. Barlow*
21. Thermoplastic Polymer Additives: Theory and Practice, *edited by John T. Lutz, Jr.*
22. Emulsion Polymer Technology, *Robert D. Athey, Jr.*
23. Mixing in Polymer Processing, *edited by Chris Rauwendaal*

24. Handbook of Polymer Synthesis, Parts A and B, *edited by Hans R. Kricheldorf*
25. Computational Modeling of Polymers, *edited by Jozef Bicerano*
26. Plastics Technology Handbook: Second Edition, Revised and Expanded, *Manas Chanda and Salil K. Roy*
27. Prediction of Polymer Properties, *Jozef Bicerano*
28. Ferroelectric Polymers: Chemistry, Physics, and Applications, *edited by Hari Singh Nalwa*
29. Degradable Polymers, Recycling, and Plastics Waste Management, *edited by Ann-Christine Albertsson and Samuel J. Huang*
30. Polymer Toughening, *edited by Charles B. Arends*
31. Handbook of Applied Polymer Processing Technology, *edited by Nicholas P. Cheremisinoff and Paul N. Cheremisinoff*
32. Diffusion in Polymers, *edited by P. Neogi*
33. Polymer Devolatilization, *edited by Ramon J. Albalak*
34. Anionic Polymerization: Principles and Practical Applications, *Henry L. Hsieh and Roderic P. Quirk*
35. Cationic Polymerizations: Mechanisms, Synthesis, and Applications, *edited by Krzysztof Matyjaszewski*
36. Polyimides: Fundamentals and Applications, *edited by Malay K. Ghosh and K. L. Mittal*
37. Thermoplastic Melt Rheology and Processing, *A. V. Shenoy and D. R. Saini*
38. Prediction of Polymer Properties: Second Edition, Revised and Expanded, *Jozef Bicerano*
39. Practical Thermoforming: Principles and Applications, Second Edition, Revised and Expanded, *John Florian*
40. Macromolecular Design of Polymeric Materials, *edited by Koichi Hatada, Tatsuki Kitayama, and Otto Vogl*
41. Handbook of Thermoplastics, *edited by Olagoke Olabisi*
42. Selecting Thermoplastics for Engineering Applications: Second Edition, Revised and Expanded, *Charles P. MacDermott and Aroon V. Shenoy*
43. Metallized Plastics: Fundamentals and Applications, *edited by K. L. Mittal*
44. Oligomer Technology and Applications, *Constantin V. Uglea*
45. Electrical and Optical Polymer Systems: Fundamentals, Methods, and Applications, *edited by Donald L. Wise, Gary E. Wnek, Debra J. Trantolo, Thomas M. Cooper, and Joseph D. Gresser*
46. Structure and Properties of Multiphase Polymeric Materials, *edited by Takeo Araki, Qui Tran-Cong, and Mitsuhiro Shibayama*
47. Plastics Technology Handbook: Third Edition, Revised and Expanded, *Manas Chanda and Salil K. Roy*
48. Handbook of Radical Vinyl Polymerization, *Munmaya K. Mishra and Yusuf Yagci*
49. Photonic Polymer Systems: Fundamentals, Methods, and Applications, *edited by Donald L. Wise, Gary E. Wnek, Debra J. Trantolo, Thomas M. Cooper, and Joseph D. Gresser*
50. Handbook of Polymer Testing: Physical Methods, *edited by Roger Brown*
51. Handbook of Polypropylene and Polypropylene Composites, *edited by Harutun G. Karian*
52. Polymer Blends and Alloys, *edited by Gabriel O. Shonaike and George P. Simon*

53. Star and Hyperbranched Polymers, *edited by Munmaya K. Mishra and Shiro Kobayashi*
54. Practical Extrusion Blow Molding, *edited by Samuel L. Belcher*
55. Polymer Viscoelasticity: Stress and Strain in Practice, *Evaristo Riande, Ricardo Díaz-Calleja, Margarita G. Prolongo, Rosa M. Masegosa, and Catalina Salom*
56. Handbook of Polycarbonate Science and Technology, *edited by Donald G. LeGrand and John T. Bendler*
57. Handbook of Polyethylene: Structures, Properties, and Applications, *Andrew J. Peacock*
58. Polymer and Composite Rheology: Second Edition, Revised and Expanded, *Rakesh K. Gupta*
59. Handbook of Polyolefins: Second Edition, Revised and Expanded, *edited by Cornelia Vasile*
60. Polymer Modification: Principles, Techniques, and Applications, *edited by John J. Meister*
61. Handbook of Elastomers: Second Edition, Revised and Expanded, *edited by Anil K. Bhowmick and Howard L. Stephens*
62. Polymer Modifiers and Additives, *edited by John T. Lutz, Jr., and Richard F. Grossman*
63. Practical Injection Molding, *Bernie A. Olmsted and Martin E. Davis*
64. Thermosetting Polymers, *Jean-Pierre Pascault, Henry Sautereau, Jacques Verdu, and Roberto J. J. Williams*
65. Prediction of Polymer Properties: Third Edition, Revised and Expanded, *Jozef Bicerano*
66. Fundamentals of Polymer Engineering: Second Edition, Revised and Expanded, *Anil Kumar and Rakesh K. Gupta*
67. Handbook of Polypropylene and Polypropylene Composites: Second Edition, Revised and Expanded, *edited by Harutun G. Karian*
68. Handbook of Plastics Analysis, *edited by Hubert Lobo and Jose V. Bonilla*
69. Computer-Aided Injection Mold Design and Manufacture, *J. Y. H. Fuh, Y. F. Zhang, A. Y. C. Nee, and M. W. Fu*
70. Handbook of Polymer Synthesis: Second Edition, Revised and Expanded, *edited by Hans R. Kricheldorf, Graham Swift, and Samuel J. Huang*

Additional Volumes in Preparation

Metallocene Catalysts and Plastics Technology, *Anand Kumar Kulshreshtha*

COMPUTER-AIDED INJECTION MOLD DESIGN AND MANUFACTURE

J. Y. H. Fuh
Y. F. Zhang
A. Y. C. Nee
National University of Singapore
Singapore

M. W. Fu
Singapore Institute of Manufacturing Technology
Singapore

Marcel Dekker, Inc.   New York • Basel

Although great care has been taken to provide accurate and current information, neither the author(s) nor the publisher, nor anyone else associated with this publication, shall be liable for any loss, damage, or liability directly or indirectly caused or alleged to be caused by this book. The material contained herein is not intended to provide specific advice or recommendations for any specific situation.

Trademark notice: Product or corporate names may be trademarks or registered trademarks and are used only for identification and explanation without intent to infringe.

Library of Congress Cataloging-in-Publication Data
A catalog record for this book is available from the Library of Congress.

ISBN: 0-8247-5314-3

This book is printed on acid-free paper.

Headquarters
Marcel Dekker, Inc., 270 Madison Avenue, New York, NY 10016, U.S.A.
tel: 212-696-9000; fax: 212-685-4540

Distribution and Customer Service
Marcel Dekker, Inc., Cimarron Road, Monticello, New York 12701, U.S.A.
tel: 800-228-1160; fax: 845-796-1772

Eastern Hemisphere Distribution
Marcel Dekker AG, Hutgasse 4, Postfach 812, CH-4001 Basel, Switzerland
tel: 41-61-260-6300; fax: 41-61-260-6333

World Wide Web
http://www.dekker.com

The publisher offers discounts on this book when ordered in bulk quantities. For more information, write to Special Sales/Professional Marketing at the headquarters address above.

Copyright © 2004 by Marcel Dekker, Inc. All Rights Reserved.

Neither this book nor any part may be reproduced or transmitted in any form or by any means, electronic or mechanical, including photocopying, microfilming, and recording, or by any information storage and retrieval system, without permission in writing from the publisher.

Current printing (last digit):

10 9 8 7 6 5 4 3 2 1

PRINTED IN THE UNITED STATES OF AMERICA

To our families

for their kind understanding and unfailing support during the long hours we have spent on this book

Preface

Mold making is an important sector in the precision engineering industry since molded parts represent more than 70% of the consumer products, ranging from computers, home appliances, medical devices, to automobiles, etc. The high demand for shorter design and manufacturing lead-time, good dimensional accuracy, overall quality and rapid design changes have become the main bottlenecks to the mold-making industry. To maintain the competitive edge, there is an urgent need to shorten the lead-time and reduce manufacturing costs by automating the design process. Computer-aided design and computer-aided manufacturing (CAD/CAM) technologies, which emerged during the past two decades, have helped to increase engineering productivity significantly. They have provided the total integration of design, analysis and manufacturing functions and have had a large impact on the engineering practices.

While CAD/CAM has found a wide range of engineering applications, its applications to mold design and manufacturing have been relatively limited. Most of the mold-making companies now use 3-D CAD software to design tooling for increasing their productivity; however, the lack of a semi-auto- or fully auto-design system makes the design tasks manual, daunting, time-consuming and error-prone. Thus, the development of computer-aided injection mold design systems (CADIMDS) has become a must and a research focus in both industry and academia since the 1990s.

This book aims to report the latest research and development achieved in automating plastic injection mold (for plastic) and die casting mold (for metal) design and manufacture. It hopes to promote the use of CADIMDS and stimulate greater R&D efforts in this critical area. While most of the major CAD/CAM vendors are actively developing mold application modules, there are still many technical issues which need to be addressed. Based on the authors' past eight years of research on intelligent mold design technologies at the National University of Singapore, many important findings are summarized and concluded in this book. In particular, the development and commercialization of an Intelligent Mold Design and Assembly System (IMOLD®) are thoroughly presented. The system architectures and detailed technologies described will be very useful for the development of applicable CADIMDS in the CAD/CAM markets.

Chapter 1 introduces the historical background of CAD/CAM technology for tooling design such as fixtures and injection molds, highlighting the importance of its R&D efforts and impact on industry. The bottlenecks and technical issues are described in detail. The main concepts of plastic injection mold and molding design, based on report literature, are described in Chapter 2. The design flow from the creation of a containing box, parting generation, runner and gating design, mold base selection, ejector design, cooling layout, etc., is presented. The issues of slider and lifter design for undercut features are also highlighted. The approaches to intelligent mold design and assembly that can lead to future fully automated systems are presented in Chapter 3. The algorithms for optimal parting directions, parting lines, and parting surfaces are given with illustrative examples. The undercut features recognition for sliders and lifters design and core and cavity generation are also covered. Several examples are used to illustrate the methodology. Chapter 4 describes a semi-automated die casting die (mold) design methodology that is similar to the plastic injection mold design approach but is applied to the injection of metals, e.g., aluminum, magnesium, etc. A unique method for a parametric and feature-based design approach is presented.

Detailed discussions on computer-aided engineering (CAE) and analyses for mold design are given in Chapter 5. This chapter fills the gap between the design and manufacturing of injection molds and thus brings up the possibility of future integrated CAD/CAE/CAM systems for mold design applications. In Chapter 6, mold manufacturing and machining (cavity and core, in particular) are discussed. The key topics of cutter selection to ensure gouge-free 3-axis machining as well as automated EDM electrode design are comprehensively described. These issues are useful for CAM users in programming and planning the NC tool paths for the machining of complex injection molds. Chapter 7 reports on a computer-aided process planning (CAPP) approach specially developed for the manufacture of mold components. The approach, implementation, and prototype system are reported. Early cost estimation of mold making, a critical activity in mold manufacture, is presented in Chapter 8. One of the promising cost estimation approaches based on neural-network modeling is introduced. The case studies on both Unix-based and Windows-based intelligent mold design systems (i.e., IMOLD® and IMOLDWorks) are given in Chapter 9.

Preface vii

The prototypes of Windows-based mold design system and parametric die casting die design system are used to demonstrate the previously presented methodologies. The implementation details are shown together with many industrial example parts.

The mold design and molding technologies are important to the material processing of metals and plastics that constitute most of the engineering materials used today. This book is intended to give the most comprehensive descriptions and thorough study on CAD/CAM and CAE of injection molds for researchers and developers in the industry, R&D organizations and universities, with particular focuses on the algorithms, implementations and system architecture that can eventually lead to a fully automated or semi-automated CADIMDS. It also provides valuable information to the designers and developers in this challenging research field and can be used as either a design handbook or a reference/text for a graduate course in understanding intelligent mold design technologies. We sincerely hope that through this publication, greater R&D efforts can be generated from both academia and the industry in critical CAD/CAM and CAE technologies that eventually will benefit the precision engineering industry as a whole.

J. Y. H. Fuh
Y. F. Zhang
A. Y. C. Nee
M. W. Fu

Acknowledgments

We are indebted to the following organizations and individuals who have helped make this book possible due to their generous support and contributions to the intelligent mold design and assembly research conducted at the National University of Singapore (NUS) over the past eight years:

- The National University of Singapore (NUS), Faculty of Engineering, and Department of Mechanical Engineering for providing various research scholarships, grants, and resources which were pertinent to establishing our mold research group;

- The National University of Singapore for the support of a start-up company, Manusoft Technology Pte. Ltd., to promote our research outcomes in the industry and the CAD/CAM communities;

- The former National Science and Technology Board (NSTB), now renamed the Agency for Science, Technology and Research (A*STAR), Singapore, for sponsoring a key research project that led to a technology spin-off of IMOLD®;

- Associate Professor K. S. Lee for co-supervising several research students and co-investigating various research projects, and for his contributions to the start-up of the NUS spin-off company;

- Associate Professor M. Rahman for his research co-supervision and contribution to group discussion;

- Mr. Kok Ann Yap, Mr. Victor Gan, Mr. Zhiqiang Zhao, Mr. Ying Wang, Mr. P. Sameer, Dr. Leshu Ling, and Dr. Xilin Liu from Manusoft Technology Pte. Ltd., for providing mold design case studies and pictures; and

- Dr. Xiangao Ye, Dr. Xiaoming Ding, Mr. Shenghui Wu, Mr. Li Kong, Ms. Liping Zhang, Dr. M. R. Alam, Mr. Yifen Sun, and Ms. Guohua Ma for contributing part of their research results.

Finally, we would like to express our special gratitude to our family members for supporting us throughout the publishing process. Their understanding and support have been invaluable for the completion of this book.

J. Y. H. Fuh
Y. F. Zhang
A. Y. C. Nee
M. W. Fu

Contents

Preface v
Acknowledgments ix

1. Introduction 1
   1.1 CAD/CAM Technology in Tooling Applications 1
      1.1.1 Fixture Design 2
      1.1.2 Die and Mold Design 4
   1.2 CAD/CAM of Injection Molds 6
      1.2.1 Plastic Injection Molds 8
      1.2.2 Die Casting Molds 9
   1.3 Summary 9
   References 10

2. Plastic Injection Mold Design and Assembly 13
   2.1 Introduction 13
   2.2 Plastic Injection Mold Design 15
      2.2.1 Injection Molding and Mold 15
      2.2.2 Injection Mold Design Process 16
      2.2.3 Detailed Mold Design 20
      2.2.4 Mold Assembly 32
   2.3 Mold Design Methodology 33
      2.3.1 The Mold Development Process 33
      2.3.2 Top-Down vs. Bottom-Up Approach 35

	2.4	Computer-Aided Injection Mold Design and Assembly	35
		2.4.1 Assembly Modeling of Injection Molds	36
	2.5	Summary	41
	References	42
3.	**Intelligent Mold Design and Assembly**	**45**
	3.1	Introduction	45
	3.2	Feature and Associativity-Based Injection Mold Design	46
		3.2.1 Feature Modeling	47
		3.2.2 Associativity Within Injection Molds	50
	3.3	Representation of Injection Mold Assemblies	54
		3.3.1 Concepts and Notations for Object-Oriented Modeling	54
		3.3.2 Object-Oriented Representation of Mold Assembly	55
	3.4	Optimal Parting Design for Core and Cavity Creation	58
		3.4.1 Optimal Parting Direction	59
		3.4.2 Generation of Parting Lines	64
		3.4.3 Determination of Parting Surfaces	75
		3.4.4 Automatic Generation of Core and Cavity	78
	3.5	Automatic Cavity Layout Design	82
		3.5.1 Cavity Layout and Number	82
		3.5.2 Automatic Layout of Multi-Cavity	87
	3.6	Recognition and Extraction of Undercut Features	91
		3.6.1 Definitions and Classifications of Undercut Features	91
		3.6.2 Undercut Features Recognition	92
		3.6.3 Draw Range and Direction of Undercut Features	94
		3.6.4 Graph Representation of Solid Models	97
		3.6.5 Recognition Algorithms	106
	3.7	Generation of Side-Cores for Sliders and Lifters	110
		3.7.1 Slider and Lifter Mechanism	110
		3.7.2 Designing Side-Cores	112
		3.7.3 Recognition of Undercuts from Core and Cavity	113
		3.7.4 Automatic Generation of Side-Cores	117
	3.8	System Implementations and Case Studies	120
		3.8.1 System Architecture	120
		3.8.2 Development Platforms and Programming Languages	120
		3.8.3 Functional Modules and Graphical User Interfaces	121
		3.8.4 Case Study	128
	3.9	Summary	133
	References	133
4.	**Semi-Automated Die Casting Die Design**	**137**
	4.1	Introduction	137
	4.2	Principles of Die Casting Die Design	139
		4.2.1 Die Casting Die Design	139

		4.2.2 Gating and Runner System Design	140
		4.2.3 Die-Base Design	142
	4.3	Computer-Aided Die Casting Die Design	143
		4.3.1 Automated Design of Die Casting Die	143
		4.3.2 Computer-Aided Design of Gating System	144
	4.4	Design of Cavity Layout and Gating System	145
		4.4.1 Determination of Cavity Number	145
		4.4.2 Automatic Creation of Cavity Layout	147
		4.4.3 Determining the Parameters of Gating System	148
		4.4.4 Design of Gating Features	154
		4.4.5 Conforming the Gate Geometry to the Die-Casting Part	159
		4.4.6 Design of Shot Sleeve, Sprue, and Spreader	159
	4.5	Die-Base Design	161
		4.5.1 Design of Die-Base	161
		4.5.2 Die-Base Structure and Variables	162
		4.5.3 Creating the Parametric Assembly Models of Die-Base	162
		4.5.4 Building the Die-Base Database	165
		4.5.5 Automatic Generation of Die-Base	165
	4.6	Generation of Core and Cavity	166
	4.7	Automatic Subtraction of Die Component	167
	4.8	System Implementation and Examples	169
		4.8.1 Development Platforms and Languages	169
		4.8.2 System Architecture—*DieWizard*	170
		4.8.3 Examples	175
	4.9	Summary	184
	References		184
5.	**CAE Applications in Mold Design**		**187**
	5.1	Introduction	187
	5.2	CAE Analysis Procedures and Functionalities	189
		5.2.1 Analysis Procedures	190
		5.2.2 CAE Functionalities	190
	5.3	CAE in the Mold Development Life Cycle	192
	5.4	CAE Details in Mold Development	193
		5.4.1 What CAE Reveals	195
		5.4.2 CAE in Part Design	198
		5.4.3 CAE in Mold Design	199
		5.4.4 CAE in Process Design	202
		5.4.5 CAE in Product Quality Assurance	202
	5.5	Application Examples	204
		5.5.1 Injection Mold Cooling Analysis	205
		5.5.2 Simulation of the Casting Process	210
	5.6	CAE Challenges in Mold Design	214
	5.7	Summary	217

		References	218
6.		**Computer-Aided Die and Mold Manufacture**	**221**
	6.1	Introduction	221
	6.2	Interference-Detection in Mold Machining	223
		6.2.1 Machining Interference	223
		6.2.2 Methods of Interference Detection	224
	6.3	3-Axis End-Mill Interference Detection	230
		6.3.1 Local and Global Interference	232
		6.3.2 Illustrative Example	245
	6.4	Optimal Cutter Selection	246
		6.4.1 Previous Work	246
		6.4.2 Criteria of Selection	248
		6.4.3 Machining Time and Machining Area	250
		6.4.4 Step-Over and Machining Time Estimation	252
		6.4.5 Cutter Selection Algorithms	253
	6.5	Computer-Aided Electrode Design and Machining	254
		6.5.1 Introduction	254
		6.5.2 Principles of EDM Electrode Design	257
		6.5.3 Electrode Tool Design	258
		6.5.4 Electrode Holder Design	265
		6.5.5 Sharp Corner Interference Detection	265
		6.5.6 Illustrative Examples	270
	6.6	Modification of Mold Design and Tool Path Regeneration	275
		6.6.1 Introduction	275
		6.6.2 Basic Concepts and Notations	275
		6.6.3 Proposed Algorithm	276
		6.6.4 Illustrative Examples	281
	6.7	Summary	282
		References	283
7.		**Computer-Aided Process Planning in Mold Making**	**287**
	7.1	Introduction	287
	7.2	An Optimization Modeling Approach to CAPP	290
	7.3	CAPP for Sliders and Lifters	292
		7.3.1 Design of Sliders and Lifters	292
		7.3.2 A Hybrid CAPP Approach	293
		7.3.3 Process Planning Problem Formulation	296
		7.3.4 Optimization Techniques for Process Planning	298
		7.3.5 Discussions	304
	7.4	System Implementation and an Example	305
		7.4.1 The IMOLD_CAPP System	305
		7.4.2 An Example	309
	7.5	Summary	311

Contents

	References	312
8.	**Early Cost Estimation of Injection Molds**	**315**
	8.1 Introduction	315
	8.2 Cost Function Approximation Using Neural Networks	317
	8.3 Cost-Related Factors for Injection Molds	320
	8.4 The Neural Network Training	325
	8.4.1 The Neural Network Architecture	325
	8.4.2 The Training Process	327
	8.4.3 Training and Validation Results	328
	8.4.4 Neural Networks for Different Cost Ranges	329
	8.5 Summary	331
	References	332
9.	**Case Studies: IMOLD® and IMOLD-Works for Mold Design**	**335**
	9.1 Intelligent Mold Design and Assembly Systems	335
	9.1.1 Knowledge-Based Mold Design Systems	335
	9.1.2 IMOLD® Overview	338
	9.1.3 Development Platforms	339
	9.1.4 Functional Modules	340
	9.2 A Windows-Based Mold Design and Assembly System	344
	9.2.1 3D Windows-Native CAD Systems	345
	9.2.2 System Implementations	351
	9.2.3 Graphical User Interfaces (GUIs)	352
	9.2.4 Windows-Based Die Casting Die Design Systems	357
	9.2.5 Illustrative Examples	358
	9.3 Summary	362
	References	362

Glossary   *365*
Index   *369*

1
Introduction

1.1 CAD/CAM TECHNOLOGY IN TOOLING APPLICATIONS

Tool engineering constitutes an important branch of manufacturing engineering as any direct improvement of existing tooling processes or an introduction of novel processes could significantly improve the production efficiency, product quality and reduce design and processing time. This would enhance the competitiveness of a manufacturing company and maintain its leading edge over its competitors. Tooling design is as an important topic as the manufacturing process itself. Without suitable tooling, manufacturing processes are often crippled or rendered totally inefficient. The trade of a tooling designer, however, has been traditionally linked to long years of apprenticeship and skilled craftsmanship. There appears to be more heuristic know-how and knowledge acquired through trial-and-error than deep scientific analysis and understanding. With the increasing use of computer tools and technology, this scenario has changed rapidly since the introduction of CAD/CAM/CAE tools in the early 1980s. It is to the authors' beliefs that CAD/CAM technologies will play more and more important roles in the tooling industry to shorten the design and manufacturing time.

This chapter introduces the background of CAD/CAM technology for tooling design and highlights some of the R&D efforts in this area. One area is in the design of tooling for molds and die casting dies which will be discussed in detail in the rest of the chapters of this book. Another area is tooling for supporting machining operations, i.e., jigs and fixtures for orienting, locating, and supporting parts in a machining center. All these areas used to rely heavily on skilled tool designers, and unfortunately, there is a worldwide shortage of such people due to long years of acquiring the necessary skill and the reluctance of the younger generation to enter into this trade.

1.1.1 Fixture Design

1.1.1.1 Introduction

Fixtures are generally mechanical devices used in assisting machining, assembly, inspection, and other manufacturing operations. The function of such devices is to establish and secure the desired position(s) and orientation(s) of workpieces in relation to one another and according to the design specifications in a predictable and repeatable manner. With the advent of CNC technology and the capability of multi-axis machines to perform several operations and reduce the number of set-ups, the fixture design task has been somewhat simplified in terms of the number of fixtures which would need to be designed. However, there is a need to address the faster response and shorter lead-time required in designing and constructing new fixtures. The rapid development and application of Flexible Manufacturing System (FMS) has added to the requirement for more flexible and cost-effective fixtures. Traditional fixtures (e.g., dedicated fixtures) which have been used for many years are not able to meet the requirements of modern manufacturing due to the lack of flexibility and low reusability. The replacement of dedicated fixtures by modular and flexible fixtures is eminent in automated manufacturing systems, due to much smaller batch sizes and shortened time-to-market.

Modular fixtures are constructed from standard fixturing elements such as base-plates, locators, supports and clamps. These elements can be assembled together without the need of additional machining operations and are designed for reuse after disassembly [1]. The main advantages of using modular fixtures are their flexibility and the reduction of time and cost required for the intended manufacturing operations. Automation in fixture design [2,3,4] is largely based on the concept of modular fixtures, especially the hole-based systems, due to the following characteristics: (a) predictable and finite number of locating and supporting positions which allow heuristic or mathematical search for the optimum positions, (b) ease in assembly and disassembly and the potential of automated assembly using robotic devices, (c) relative ease of applying design rules due to the finite number of element combinations.

1.1.1.2 Computer-Aided Fixture Research

Fixture research employing computer aids started in the late 1970s and early 1980s. In the initial years, interactive or semi-automated fixture design techniques (see Fig. 1.1) were built on top of commercial CAD/CAM systems and expert system tools. These approaches were mainly concerned with fixture configuration,

Introduction

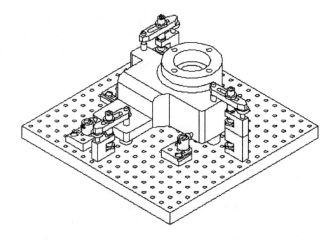

Fig. 1.1 A computer-aided modular fixture design [2].

and there was little analysis of the other aspects such as workpiece-fixture-cutting tool interactions.

A comprehensive fixture research plan should involve the analysis at different computational levels, *viz.*, geometric, kinematic, force and deformation analyses. The following sections will present brief overviews of the research activities in each of the above-mentioned areas, followed by the need to design an intelligent fixture which can be integrated with the machine tool.

(1) Geometric Analysis

Geometric analysis is closely associated with fixture planning and spatial reasoning. It determines the selection of the type and number of fixturing elements, support and locating elements, the order of datum planes, etc. The analysis also includes the checking of interference between workpiece and fixturing elements, as well as cutting tools.

Most of the early fixture research involved geometric analysis and synthesis of fixture construction with relatively little attention to kinematic and deformation analysis.

(2) Kinematic Analysis

Kinematic analysis is used to determine whether a fixture configuration is able to correctly locate and provide complete constraint to a workpiece.

Previous work on fixture design automation offers relatively little consideration in providing a comprehensive fixture-element database and effective assembly strategies for the generation and construction of modular fixtures. The assembly of modular fixtures is to configure the fixture elements such as locators, clamps and supports (in most cases, accessory elements are needed to generate fixture towers to fulfill the fixturing functions) on the base-plate according to a fixturing principle (e.g., 3-2-1 principle). The determination of the locating, supporting and clamping points for the assembly of modular fixtures is a key issue in fixture design automation.

(3) Force Analysis

In a machining fixture, different forces are experienced, *viz.*, inertial, gravitational, machining and clamping forces. While the first three categories of forces are usually more predictable, clamping force can be rather subjective in terms of magnitude, and point of application, as well as sequence of application.

It has been widely accepted that a thorough analysis of all the forces involved in a fixture is a formidable task since it is an indeterminate problem with a large number of fixturing elements. When friction is taken into account, the problem becomes even more complex because both the magnitude and the direction of the static friction forces are unknown. Recent effort in clamping force analysis can be found in [5,6].

(4) Deformation Analysis

Due to the complexity of force interaction, workpiece deformation can be attributed to a combination of factors. First, a workpiece would deform under high cutting and clamping forces. Second, a workpiece could also deform if the support and locating elements are not rigid enough to resist the above-mentioned forces. In reported literature, it is assumed that workpiece deformation is largely due to the first cause mentioned above. The most commonly used method in analysing workpiece deformation and fixturing forces is the finite-element method.

1.1.2 Die and Mold Design

Die and mold making is an important supporting industry since their related products represent more than 70% of the non-standard components in engineering consumer products. Their production runs are typically of small lot-size and with great varieties. The high demand for shorter design and manufacturing lead times, good dimensional and overall quality, and rapid design changes have become the bottlenecks in die and mold industry. For die and mold-making companies wishing to maintain the leading edge in local and international markets, they would attempt to shorten the manufacturing lead time by using advanced manufacturing equipment, automated manufacturing processes and improving the skill level of their employees.

In 1909, Baekeland and his associates developed a synthetic material, Bakelite, from phenol and formaldehyde, and this marked the early beginning of plastic

Introduction

materials [7]. Plastics are now commonly used to describe a group of synthetic organic materials which can be molded into complex shapes. Plastic components can be designed to have richer and more detailed features than cast parts and metal stampings. They can also be produced in large quantities inexpensively and in short lead-time. Virtually every household product has plastic components and hence the high demand for plastic injection molds.

In the plastic injection molding process, the plastic material, before injection, is made fluid-like by a combination of heat and mechanical action in a separate chamber outside the mold. It is then injected into the mold using high external pressure. The mold, made of the two halves locked together under pressure, is cooled and split to eject the part when the liquefied plastic has solidified. The entire cycle is then repeated. The accuracy of a part, the cooling efficiency and hence the cycle time, is very much dependent on the mold design and its operating parameters. Fig. 1.2 shows an assembled set of mold with its cavity and core inserts for injection molding.

Molding-making used to be largely experience-based and relies heavily on skilled craftsmen, typically trained under an apprenticeship scheme, and having acquired their expertise through years of practice. The supply of trained manpower, however, is rapidly diminishing as the younger generation is reluctant to go through the long training periods and would prefer other types of jobs. This situation was arrested to some extent with the introduction of CAD/CAM technologies in the late '70s. The mold-making scene has largely been transformed from the tacit knowledge acquired by the individuals to knowledge-based intelligent computer systems. Compared to the application of computer tools in other manufacturing areas, however, mold design has been relatively slow in progress. Quite recently, commercial software systems have begun to appear and are being adopted in the tool-making and molding industry.

According to CIMdata [8] a recent study of the mold-making industry revealed that mold design accounts for some 20% of the total work effort in mold shops and CAM programming accounts for about 8%. Approximately half of the 20% design work is associated with core and cavity design, with the remaining activities associated with mold base design/selection and the preparation of a design model for manufacturing. From this analysis, it can be seen that mold design is a major function in the mold shops. This high concentration of design activities has led to many software systems focusing their effort on providing solutions to the mold design aspects. These software systems are able to assist the mold designer in the various aspects of design such as automated parting line and surface determination, core and cavity design, runner and gate selection, analysis of temperature distribution and flow of plastic material in the mold, and the effective interfacing with different mold bases.

Fig. 1.2 A set of injection mold, core and cavity inserts.

1.2 CAD/CAM OF INJECTION MOLDS

Automating the design process is an important step to shorten design lead-time. The development of a Computer-Aided Injection Mold Design System, or CADIMDS (see Fig. 1.3) and the methodologies of CADIMDS have become the research focus in both industry and academia. CAD, CAE and CAM tools have been developed in the last two decades to assist mold designers in mold design configuration, analysis and machining. More specifically, they can be divided into the following categories.

(1) More plastic part designs are presented to the mold shops in electronic formats and this is expected to increase from the present 53% to 90% in the next few years. Mold designers would need to use compatible CAD software to retrieve the part design information, edit the design and make necessary alterations to prepare a final design model that meets all the manufacturing specifications.

(2) Computer-aided engineering (CAE) software such as temperature and flow analysis programs is used to simulate the plastic injection process to check for possible defects in the molded parts, and for the suitable design of runners and gates. Such feedbacks will be used for modification of the molded part and/or the mold so optimum cooling and material flow can be achieved.

(3) CAM software for the direct generation of NC tool-paths and programs for machining complex 3D surfaces (see Fig. 1.4) based on the number of axes and configuration of an NC machine is available from several vendors. With the advent of high speed machining technology and the rapid adoption by the mold-makers, a new breed of CAM software has now been designed to cater for the high machining speeds and feed rates.

Introduction

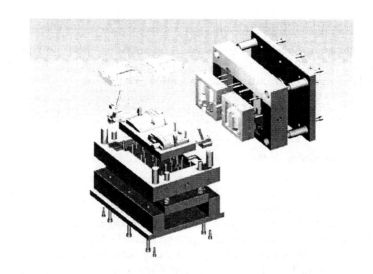

Fig. 1.3 An injection mold designed by IMOLD[R] system [9].
(Courtesy of *IMOLDWorks*, Manusoft Technology Pte. Ltd., Singapore.)

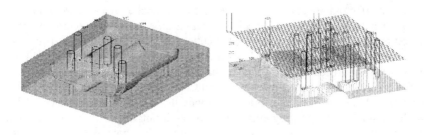

Fig. 1.4 Computer-aided mold machining.

(4) Although high speed machining of hardened die blocks has reduced the amount of EDM work, it is still necessary for EDM to handle the finer details of a mold. CAD/CAM software is able to extract features from cavities and design electrodes for EDM machining of complex 3D surfaces.

(5) Many mold-shops would still prefer to document their designs and assembly details with 2D drawings. CAD software for the automatic generation of 2D drawings with full dimensional and annotation capabilities from a final 3D design becomes easily available.

1.2.1 Plastic Injection Molds

As this book is dedicated to the details of computer-aided injection and die casting die (or mold) design and manufacture, some of the design issues such as parting direction and surface determination, mold cooling analysis, EDM electrode design, mold machining algorithms, etc. will be detailed in the later chapters. Some of the latest trends in design of injection molds are however briefly discussed here.

1.2.1.1 Automated Mold Design Using Knowledge-Based Systems

Automated mold design and manufacturing functions can provide great productivity to the tooling shops. Knowledge-based systems, such as those produced by Cimatron Technologies (Israel), Vero International (UK), Bentley Systems (US), IMOLD® (Singapore) are some of such systems that have added modules that automate many routine CAD functions for mold design.

In a typical knowledge-based system, mold components are pulled from a list of standard parts from mold base libraries. Once a component is indicated by a designer, e.g., an ejector, the system will automatically add a corresponding hole in the cavity block and the pin to the assembly drawing. Designers, based on the intelligent feature associativity built into the system, do not have to worry about missing out or mismatching of details in the final assembly. CAM systems are also getting smarter and are able to recognize where the stock remains at any time during the machining process [10]. Such a system is able to recognize corner materials left from a tool and define subsequent cutter paths with a smaller diameter tool to the unmachined material. Another feature is automatically recognize step-over areas between two cutter paths and re-machining of those areas to ensure good surface finish. The ultimate goal is to have a CAM system that is able to scrutinize the entire 3D profile and automatically provide the best machining strategy. This would require the CAM programs to employ both automatic feature recognition and knowledge-based system to create a system which is both experience and geometry reasoning-based.

1.2.1.2 Internet Technologies and Their Applications to Mold Design and Manufacture

With the advent of Internet technologies and much higher communication speed, web-based activities have increased manifold over the last five years. Mold design and fabrication have also taken advantage of this new technology, though at a slower pace than other forms of e-commerce and e-business activities. Web-based systems are able to disseminate information much faster and provide good support in distributed operations where design and fabrication are seldom concentrated in one location in order to take advantage of differences in pricing and availability of resources.

A relatively new term, "collaborative product definition management" or cPDM, has been in use to define product lifecycle throughout an entire enterprise

Introduction

[11]. The traditional PDM approach usually has its roots in design engineering, while cPDM is to establish enterprise-wide infrastructure to support management of product definition throughout its complete lifecycle, including manufacturing, marketing, field support, retailing and distribution, servicing and maintenance, end-of-product-life considerations, etc. Web-based cPDM systems provide enormous opportunities for a company to realize a full management of the product definition lifecycle. This approach has been rapidly adopted by the mold design and fabrication companies.

Recently, Lau et al. [12] described the development of an Internet-based integrated CAE system to support an agile plastic mold design and development (APMD) process. The system consists of four characteristics: (a) CAE driven collaborative design intelligence, (b) open and extensible web-centric architectural environment for integration of multi-engineering tools, (c) customer satisfactory requirement response driven design, and (d) knowledge capturing and reuse.

1.2.2 Die Casting Molds

Die casting is a manufacturing process for producing precisely dimensioned simple or complex-shaped parts from non-ferrous metals such as aluminum, magnesium, zinc, copper, and others [13]. It has a very strong similarity with the plastic injection molding process. The design process and mechanism used are quite closely related. Die casting is chosen over other manufacturing methods mainly because of its good part strength and better dimensional accuracy. Compared to plastic injection molding, in this process, molten metal is injected into the die cavity under high pressure, and the die-casting is ejected from the die halves after it has solidified. A die casting die not only consists of the main components such as the die base, but also has other variables that control metal flow, heat flow and the forces applied by the machine. In this book, we will call die casting molds as *die casting die* by following the industrial convention. Die design is a complex process that contains several critical design tasks. A complete die casting die design process involves the design of layout, gating system, die-base, parting, ejection system, cooling system, moving core mechanism, etc. To develop a design system that can accomplish all the above design tasks, the tasks are classified into different function groups, and corresponding modules are then established to realize them systematically. The details will be discussed in Chapter 4.

1.3 SUMMARY

This chapter summaries some of the historical developments and background in applying computer tools to tooling (i.e., fixture and plastic injection mold) research in the last two decades and projects some of the future trends.

Although fixture design and planning has traditionally been regarded as a science as much as an art, it is now possible to successfully design and plan fixtures using computer applications. Optimal clamping force control remains an essential issue to be addressed. With the development of intelligent fixtures incorporating dynamic clamping force control, workpiece accuracy can be controlled on-line, and this will further enhance the smooth operation of automated manufacturing systems.

Similarly, mold design and machining has taken a great step ahead as advanced software algorithms, machining, and computer technologies are now available to the mold makers. A trade that was used to be craft-based has now become totally revolutionized and largely IT-based. Customers can now expect to see faster delivery times as well as better quality products. An increasing trend is the use of Internet technologies, and this is expected to play a much bigger role in the next few years, in terms of distributed design as well as e-business activities.

Tooling design for manufacturing processes has been transformed from the "black art" into a science-based and IT savvy discipline. Readers can make reference to a Special Issue on Molds and Dies in the *IMechE Part B Proceedings* edited by one of the authors where researchers reported on the effective use of computer tools in the design, simulation and analysis of molds and dies [14].

The notable increasing trend is towards the greater use of computer tools and Internet technology. It is foreseeable that tooling design activities will now be drawn from various geographical locations and there will be reduced reliance on the diminishing expertise in tooling design.

REFERENCES

1. A.Y.C. Nee, K. Whybrew and A. Senthil Kumar, Advanced Fixture Design for FMS, Springer-Verlag, 1995.
2. J.R. Dai, A.Y.C Nee, J.Y.H. Fuh and A. Senthil Kumar, "An approach to automating modular fixture design and assembly", Proceedings of the Institution of Mechanical Engineers Part B, J of Engineering Manufacture, 211, Part B, pp. 509-521, 1997.
3. T.S. Kow, A. Senthil Kumar and J.Y.H. Fuh, "An Integrated approach to collision-free computer-aided modular fixture design", International J of Advanced Manufacturing Technology, 16, No. 4, pp. 233-242, 2000.
4. A. Senthil Kumar, J.Y.H. Fuh and T.S. Kow, "An automated design and assembly of interference-free modular fixture setup", Computer-Aided Design, 32:10, pp. 583-596, 2000.
5. Z.J. Tao, A. Senthil Kumar and A.Y.C. Nee, "Automatic generation of dynamic clamping forces for machining fixtures", International J of Production Research, 37:12, pp. 2755-2776, 1999.
6. A.Y.C. Nee, A. Senthil Kumar and Z.J. Tao, "An intelligent fixture with dynamic clamping scheme", Proceedings of the Institution of Mechanical Engineers Part B, J of Engineering Manufacture, Vol. 214, pp. 193-196, 2000.
7. M.L. Begeman and B.H. Amstead, Manufacturing Processes, John Wiley, 1968.
8. A. Christman, CIMdata – "Mold design – a critical and rapidly changing technology", MMS Online, Gardner Publication, Inc., 2001.

9. IMOLDWorks, Manusoft Technology Pte. Ltd., Singapore, http://www.manusoftcorp.com, 2003.
10. M. Albert and T. Beard, "Die/Mold machining on the march", MMS Online, Gardner Publication, Inc., 2001.
11. E. Miller, "Collaborative Product Definition Management for the 21st Century", Computer-Aided Engineering Magazine, March, 2000.
12. K.H. Lau, K.W. Chan and K.P. Cheng, "Internet based CAE in agile plastic mold design and development", Proceedings of 2002 International CIRP Design Seminar, 16-18 May 2002, Hong Kong.
13. NADCA, North American Die Casting Association, http://www.diecasting.org, 2001.
14. A.Y.C. Nee, Guest Editor, Special Issue on Molds and Dies, Proceedings of the Institution of Mechanical Engineers, Part B, J of Engineering Manufacture Vol. 216, No. B12, 2002.

2
Plastic Injection Mold Design and Assembly

2.1 INTRODUCTION

Injection molds are assemblies of parts in which two blocks, namely the *core* and *cavity* blocks, form an impression or a molding [1]. The mold core forms the interior of the molded parts (molding), and the mold cavity forms the exterior faces of the molding. Among all the components, the core and cavity are the main working parts. The molding is formed in the impression between the core and cavity and is ejected after the core and cavity are opened. The pair of opposite directions along which the core and cavity are opened are called the *parting direction*. For a given molding, one group of its surfaces is molded by the core and its inserts, and the other group of surfaces by the cavity and its inserts. The *parting lines* are therefore the intersection boundary between the surfaces molded by the core, the cavity and their inserts. The *parting surfaces* are the mating surfaces of the core and cavity. In a molding, the convex and concave portions of the molding are considered as *undercut features*. If the undercut features cannot be molded in by the core and cavity and their inserts, they will be the real *undercuts* and would require the incorporation of side-cores, side-cavities, form pins, or split cores in the mold structure. The side-cores, side-cavities, form pins, and split cores are called *local tools* since they are used to mold the local features in the molding. These local tools must be withdrawn by a mechanism prior to the ejection of the molding. Side-cores and side-cavities are normally removed by the sliders/lifters mechanism.

For an injection mold, the main components are the mold plates, core and cavity, sprue bush, register ring, ejection parts, side-cores, side-cavities, and actuating mechanisms such as sliders and lifters. The ancillary items include guide bushes, guide pillars, leader pins, and the fastening parts such as screws, bolts, and nuts. In addition, the cooling circuit and bubbler systems are important elements in an injection mold structure. The primary task of an injection mold is to accept the plastic melt from an injection unit and mold the molten material into the final shape of the plastic product. This task is fulfilled by a cavity system that consists of cores, cavities, and inserts. Besides the primary task of shaping the product, an injection mold also has to fulfill a number of tasks such as distribution of melt, cooling the molten material, ejection of a plastic product, transmitting motion, guiding, and aligning the mold halves. The component parts to fulfill these functional tasks are usually similar in the structure and geometrical shape for different injection molds. Therefore, an injection mold is a complex mechanical assembly of many functional parts to fulfill multiple functions, such as accepting plastic melt from an injection unit and cooling the plastic to the desired shape. Fig. 2.1 shows a typical injection mold (in cross section) which contains several subsystems for molding purposes. This chapter will detail the mechanical components of a mold and its main design methodology commonly adopted in the mold industry.

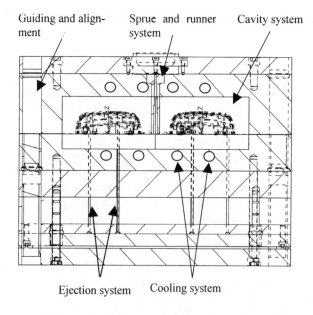

Fig. 2.1 A typical plastic injection mold.

2.2 PLASTIC INJECTION MOLD DESIGN

2.2.1 Injection Molding and Mold

With the broader and broader use of plastics parts in a wide range of products, from consumer products to machineries, cars and airplanes, the injection molding process has been recognized as an important manufacturing process, which depends on temperature changes in material properties to obtain the final shapes of the discrete parts. In this type of manufacturing process, liquid material is forced to fill and solidify inside the cavity of the mold.

Injection molding is a cyclic process of forming plastic into a desired shape by forcing the material under pressure into a cavity. The injection molding process for thermoplastics consists of three main stages: (1) injection or filling, (2) cooling, and (3) ejection and resetting. During the first stage of the process cycle, the material in the molten state is a highly viscous fluid. It flows through the complex mold passages and is subject to rapid cooling, which forms the mold wall on one hand and internal shear-heating on the other. The polymer melts and then solidifies under the high packing and holding pressure of the injection system. When the mold is finally opened, the part is ejected and the machine is reset for the next cycle to begin. The final shape is achieved by cooling (thermoplastics) or by a chemical reaction (thermosets). It is one of the most common and versatile operations for mass production of complex plastic parts with excellent dimensional tolerance. It requires minimal or no finishing operations. In addition to thermoplastics and thermosets, the process is being extended to other materials such as fibers, ceramics, and powdered metals, with polymers as binders. Approximately 32 percent by weight of all plastics processed go through injection molding machines [2]. Fig. 2.2 shows an injection mold mounted in the injection molding machine and ready for production.

An *injection mold* is an assembly of platens and molding plates typically made of tool steel. The mold shapes the plastics inside the mold cavity (or matrix of cavities) and ejects the molded part(s). The stationary platen is attached to the barrel side of the machine and is connected to the moving platen by the tie bars. The cavity plate is generally mounted on the stationary plate that houses the injection nozzle. The core plate moves with the moving plate guided by the tie bars. Occasionally, there are exceptions for which the cavity plate is mounted to the moving platen, and the core plate and a hydraulic knock-out (ejector) system is mounted to the stationary plate.

An injection mold system consists of tie bars, stationary and moving platens, as well as molding plates (bases) that house the cavity, sprue and runner systems, ejector pins, and cooling channels (see Fig. 2.1). An injection mold is essentially a heat exchanger in which the molten thermoplastic solidifies to the desired shape and dimensional details defined by the cavity. The mold consists of the moving mold half and fixed mold half. When a mold is closed, both halves are in contact, and molten plastic is injected into the impression formed by the core and cavity through a sprue. After injection, both mold halves are separated and the molded

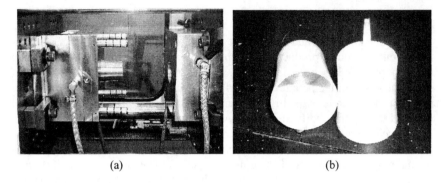

Fig. 2.2 Injection molding: (a) process, (b) molded or injected parts (molding).

part is ejected by the ejection system (including the ejector rod, ejector plate, tie rod, and stripper plate). In the subsequent molding cycle, the ejection system is pushed back to its original position.

Two-plate mold
　　The vast majority of molds essentially consist of two halves, as shown in Fig. 2.3. This kind of mold is used for parts that are typically gated on or around their edge, with the runner in the same mold plate as the cavity.

Three-plate mold
　　The three-plate mold is typically used for parts that are gated away from their edge. The runner is in two plates, separated from the cavity and core, as shown in Fig. 2.4.

2.2.2 Injection Mold Design Process

　　Due to the multi-functionality and varying configurations of injection molds, the design of an injection mold is a complicated task, whereby knowledge and skills from experienced designers are crucial to a successful design. A mold designer not only has to design the cavity and core that form the correct shape of a product, the runner system through which the impression can be filled properly and completely, but also position and orientate the various functional parts to ensure that the whole mold assembly works properly. Upon getting the complete information of a product design, a designer first decides on the type of injection machine and mold to be used primarily from the ordering quantity and manufacturing cost, then arranges the cavity layout and designs the runner system. After the initial mold design, the designer finally proceeds to the detailed design such as

Plastic Injection Mold Design and Assembly 17

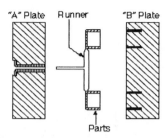

Fig. 2.3 A two-plate mold.

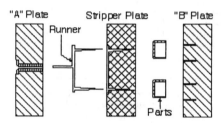

Fig. 2.4 A three-plate mold.

layout of ejection and venting, structures of sliders and lifters, etc.

In the designing process, designers would use both engineering rules and trial-and-error techniques in order to come up with a good mold design, which usually result in a long tooling lead time. However, the rapid changing and competitive markets require that premium quality molds be made within a shorter lead time (as short as 2-4 weeks). To solve this problem, CAD/CAM technologies have thus been introduced for the design and manufacturing of injection molds. Fig. 2.5 shows a systematic injection mold design process using CAD/CAM systems either in an interactive or semi-automated mode. The steps should be followed in the order presented with the exception of the last five steps. These five steps do not have to be carried out in any particular order and can even be performed in between any of the other steps.

Step 1 – Loading in a Part Model
The Data Preparation requires the designer to:

- load in the product model,
- assign the mold pull direction and the scale factor, and
- create the containing box.

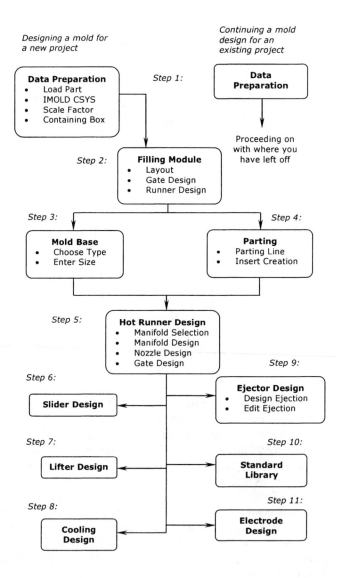

Fig. 2.5 Flow chart of a systematic injection mold design process.

To create a new mold design, the designer will load in the part and start the design process from step 1 through to step 10. For a design that has already been started, the designer will load in the top level assembly first, and continue from wherever the design process was left off.

Plastic Injection Mold Design and Assembly 19

Step 2 – Creating the Layout, Gates, and Runners
The creation of the Layout, Gates, and Runners is achieved in the Filling design. The designer is required to:
- define the cavitation of the mold,
- create gating for the part mode, and
- create the runners between the gate.

Step 3 – Loading in a Mold Base
To begin the design of the mold base, the designer will use the Mold Base Module to:
- define the style of the mold base,
- select the series of mold base, and
- enter the size or any other necessary customized specifications of the mold base.

Step 4 – Creating the Parting Line and Inserts
This step is carried out in the Parting design, where the designer is required to:
- define the parting line for the creation of core and cavity, and
- create any necessary inserts.

Step 5 – Designing the Hot Runner System (Optional)
For the design of hot runners using the Hot Runner Design Module, the designer will:
- define the style of the manifold, and
- design the nozzle and runner system.

Step 6 – Designing the Sliders
Using the Slider Design Module, the designer will create the slider and any other slider accessories that are needed for exterior undercuts.

Step 7 – Designing the Lifters
In the Lifter Design Module, the designer will create the lifter head and lifter body that are required in the design for interior undercuts.

Step 8 – Designing the Cooling System
This step is carried out in the Cooling Design Module where the designer will:
- define the cooling channels for the core, cavity, plates, slides, or lifters, and
- add standard cooling accessories to the cooling system.

Step 9 – Designing the Part Ejection System

This step is carried out in the Ejecting Design Module. In this module, the designer is required to:

- define the ejector pins needed to eject the part from the mold, and
- choose the different styles and sizes of ejector pins that are needed in the design.

Step 10 – Adding in Standard Parts
This step is carried out in the Standard Library Module. The designer will:

- define and locate standard parts like the socket head cap screws and locating rings to complete the mold design.

Step 11 – Design of Electrodes for CAM
This step is carried out in the Electrode Design Module. In this module, the designer is required to:

- define the boundaries of the geometry for the electrodes, and
- select an electrode tool holder for the electrode designed.

2.2.3 Detailed Mold Design

2.2.3.1 Parting

The location (either in lines or surfaces) at which the core and cavity components interface on mold closure is often referred to as the *parting* or *split* [3]. The parting location on the molded component is determined by the complexity of the core and cavity interface. Mold designers often locate the parting according to the product or customer requirement to simplify the tooling complexity, reduce the cost, and improve the aesthetic outlook of the injected parts. The parting design is indeed an important step in mold design. Fig. 2.6 shows the parted core and cavity blocks. The detailed parting method will be discussed in the next chapter in detail.

2.2.3.2 Feeding System - Gate, Runner, and Sprue

The feeding system consisting of the gate, runner, and sprue is of great importance to mold design. A *gate* in the injection mold is a small or thin passage for the molten material (plastic or metal) to travel via the runner to the mold cavity. Also, the entire injected component of a mold will therefore include the moldings, gates, runners, and sprue (i.e., the hole in a mold through which the molten material first enters). The position of the gate is important in determining how material flows into the mold impression. Other factors to be taken into consideration when choosing a gate position include [3]:

Plastic Injection Mold Design and Assembly

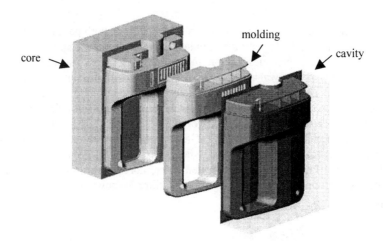

Fig. 2.6 Cavity and core mold inserts and the molded part.

(1) aesthetic considerations of the moldings;
(2) de-gating requirements of the molding;
(3) design complexity of the molding;
(4) mold temperature requirement;
(5) nature of the material to be processed;
(6) volume of the material to be fed through the gate and the feed rate; and
(7) significance, locations, and the effect of gas entrapments created as a result of the filling profile

There are usually more than 10 different types of available gatings within the gate design option. Some of these include the sprue gate, the side gate, the submarine gate, the fan gate (see Fig. 2.7), and so on. If there is a commonly used type of gating by the mold designer, it can be easily created into a gating library for easy retrieval either automatically or semi-automatically through CAD systems.

Fig. 2.7 Example of fan gates.

A *runner* is the channel in a mold joining one of or both parting surfaces of a molding to allow the injected molten material to flow from the sprue to the cavity. The runner design provides a venue for the mold designer to specify the shape of the runner that he/she wants to use as well as the positioning of it. This design option provides some very basic shapes like circle, semi-circle, U-shape, trapezoid, hexagon, etc. (see Fig. 2.8), as well as the circular and semi-circular shaped runner. The geometry, size, length, and volume of the feeding system has a strong influence on the quality of the molding. The shortest, smoothest and least winding flow path by which the molten material flows through will be a better solution. Fig. 2.9 shows a four-cavity feeding system. The more complicated examples for molding a single- and multi-cavity computer monitor is shown in Fig. 2.10.

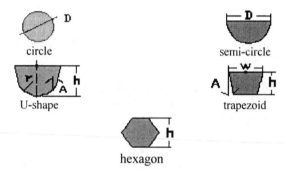

Fig. 2.8 Example of different runner cross sections.

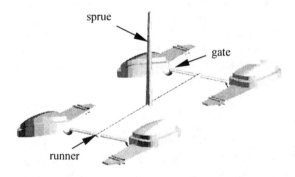

Fig. 2.9 A feeding system for a four-cavity mold.

Plastic Injection Mold Design and Assembly

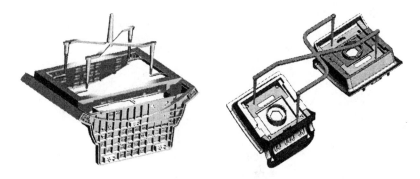

Fig. 2.10 Example of a single- and multi-cavity feeding system.

2.2.3.3 Mold Base

A mold base (Fig. 2.11) is a set of mold components that hold the major standard parts provided by mold makers and can be quickly customized according to users' needs. The size of the mold base is calculated based on the size of the layout. An allowance is added to the size so that there is a wall between the size of the mold base and the inserts. The mold designer will select the smallest mold base that matches these criteria. By default, the allowance is set to 100mm. The types of mold bases currently available in the market are HASCO (see Fig. 2.12), DME, FUTABA, LKM, HOPPT, STiEHL, NATIONAL, etc., either in metric unit or inch or both. Mold designers can choose any one of these vendors to satisfy the mold base requirements. The mold base database can be built adhering to the dimensions as stated in the vendors' catalogs. This is to ensure that there is absolute consistency between the mold base one needs and what the vendors have available in their standard catalogs.

2.2.3.4 Cooling

An adequate mold cooling is important to the quality of an injection molding. When the molten material (melt) flows through the mold cavity, heat must be removed from the molding for the material to solidify. To cool the molding, the heat generated from the molding to the mold needs to be removed through a designed cooling channel. This can be done by passing water through the cooling channels positioned in the mold. The cooling rate will, however, depend upon the thermal conductivity, melt temperature and the molding thickness [3]. To obtain the maximum cooling efficiency, the layout of the cooling channel (Fig. 2.13) is rather critical. The cooling channels must be positioned correctly and avoid interference with the other mold components such as ejector pins. The commonly used cooling designs include: flat plane cooling, spiral cooling, finger or bubbler cooling, baffle cooling, and so on. Fig. 2.14 shows a flat plane cooling system (in parallel or series) that is primarily used to cool large flat moldings of uniform cross-section.

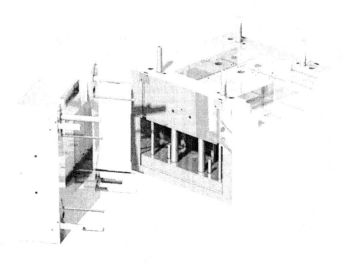

Fig. 2.11 A set of standard mold base.

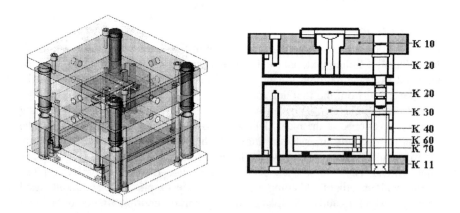

Fig. 2.12 HASCO type-A METRIC mold base [4].

Plastic Injection Mold Design and Assembly

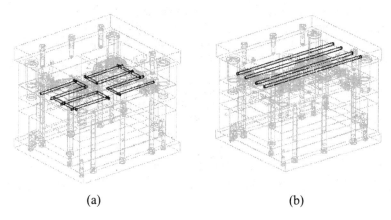

(a)   (b)

Fig. 2.13 Cooling layout (a) on core; (b) on cavity [5].

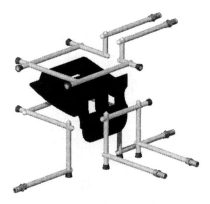

Fig. 2.14 A flat plane cooling with serial and parallel connections.

2.2.3.5 Ejectors

The injection molding cycle consists of several major stages. At the end of the injection cycle, the molded part has to be ejected when the injection mold opens. Since the ejection force from the ejector plate is usually large enough for the required releasing force, mold designers usually neglect the balance of the ejection force which is exerted on the sticking surface. Part deformation or damage might occur due to the unbalanced ejection force. Based on experience and with simple calculations, mold designers lay out an ejector system in which the size, number and location of ejectors are decided.

After the molding has solidified, cooled down and stuck to the core, it has to be removed by ejectors such as pins, sleeves, stripper rings, or stripper plates (Fig.

2.15), either singly or in combination. Ejectors are preferably positioned in areas where they can act on corners, side walls or ribs, which make de-molding difficult, but react more efficiently to the ejection force due to their rigidity. It will be ideal if each sticking surface is pushed by a force which is (1) larger than the required releasing force, and (2) in proportion to the required releasing force. Otherwise, either the molding is still stuck to the core or the local area of the molding is damaged or distorted. Fig. 2.16 shows an ejection system designed for a single- and 4-cavity mold.

Each ejector leaves a visible mark on the molding. This has to be considered when the designer determines the ejector location. Poor location of ejectors may cause flashing marks. After completing the ejector pins design, the contact pressure on the bottom surface has to be checked in order to avoid the deep marks and pins penetrating the thin surfaces.

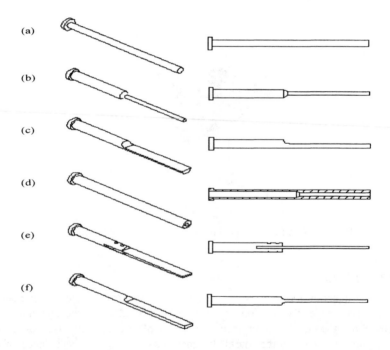

Fig. 2.15 Different forms of ejector pins: (a) plain; (b) stepped; (c) D-shape; (d) sleeve; (e) blade-1; and (f) blade-2 [6].

Plastic Injection Mold Design and Assembly

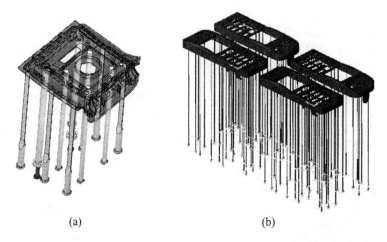

Fig. 2.16 Ejectors designed for (a) a single- and (b) 4-cavity mold.

Because $P = F / A$

and
$$P = \frac{4 \times S \times F}{\pi \times D^2} \quad (2.1)$$

where P: pressure of ejector pin,
D: ejector pin diameter,
F: pin force on the bottom surface, and
S: safety factor (e.g., 1.5).

The contact pressure has to be compared with the permissible pressure, which depends upon the plastic material. If the contact pressure is larger than the permissible pressure, the problem will arise. Another requirement is to check the buckling of small-diameter pins. Breakage of small ejector pins is common especially for ejecting some materials. It not only produces defective molded parts, but also causes the failure of the remaining ejector pins.

According to the Euler formula from the *Machinery's Handbook* [7]:

$$F = \frac{m \times \pi \times E \times A}{(L/k)^2} \quad (2.2)$$

where F: buckling load on the column,
L: length of the column,
E: modulus of elasticity, and
M: a constant depending on the end conditions of the column.

From the equation above, the maximum length of an ejector pin to withstand F can be calculated, and a suitable pin should be used to avoid buckling. The above procedure can be implemented and interfaced with CAD/CAM systems to automate the ejector design. A computer-aided ejector design system has been reported by Wang et al. [6]. The system applies the balance algorithm, i.e., incorporating heuristic knowledge, recognizing the molding feature and recommending optimum solution. It can also check the interference, one of the important criteria in designing ejection systems, with other mold components such as cooling channels automatically.

2.2.3.6 Sliders and Lifters

The slider and lifter are the mechanical components used to form the external and internal undercuts in molding respectively. For those undercut features that cannot be formed from the core and cavity blocks, slider (for external surfaces) and lifters (for internal surfaces) can be utilized. Before the mold opens, slider and lifter mechanisms must be activated and retracted first to avoid interference and damage to the molded parts. Depending on the shape of the molded surfaces, the mold designer first selects the slider/lifter head (Fig. 2.17) and the type of slider/lifter (slider/lifter body). Upon deciding, designers will model the slider body and the finger-cam and determine their default dimensions. After confirming with the parameters, the designer selects the face on the head where the slider body is attached to, and the slider body together with the finger-cam are incorporated into the mold assembly (Fig. 2.18). Next, the designer creates the accessory components for the slider, namely guides, wear plates, heel block, and the stop block. The selection of these component types is done according to the vendor catalogs. In a CAD environment, the designer simply confirms or changes the values of the parameters and the components are assembled to the slider body automatically. Pockets on the cavity and or core can also be created automatically. The lifer design works in the same way as in the slider design; however, the lifter head has to confirm to the shape of the inner surface of the molding compared to the slider head that applies to the external surfaces.

Fig. 2.17 Slider and lifter for molding undercut features.

Plastic Injection Mold Design and Assembly

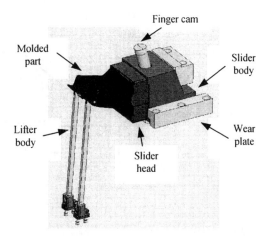

Fig. 2.18 A slider and lifter assembly.

2.2.3.7 Hot Runner

Hot-runner systems are often referred to as hot-manifold systems, or runnerless molding. In contrast to a cold-runner system, the material in the hot runners is maintained in a molten state and is not ejected with the molded part. The main advantages of employing a hot-runner system into a mold are [3]:
  (a) to reduce cycle times as a result of having the runner and feeding system in molten state above the quickly frozen gate;
  (b) to save material as there are no sprue or runner systems to dispose of;
  (c) to lower post-molding cost from finishing and de-gating of the moldings; and
  (d) to gain a better control over the mold filling and flow characteristics through achieving a locally altering temperature of the feeding system.

The main components of a typical hot-runner system is shown in Fig. 2.19. There are many types of hot-runner systems (either single-drop or multiple-drop manifolds) available in the market, and they tend to vary based on system designs and manufacturers. However, they can be categorized into three types (see Fig. 2.20): (a) the insulated hot runner, (b) the internally heated hot-runner system, and (c) the externally heated hot-runner system. The insulated hot runners are less complicated and costly to design compared to the other two but require a fast cycle to maintain a melt state and may create non-uniform mold filling. The internally heated runner systems can improve heat distribution but require more sophisticated heat control and balancing and thus increase the cost and complexity of the design. The externally heated runner systems have a better temperature control

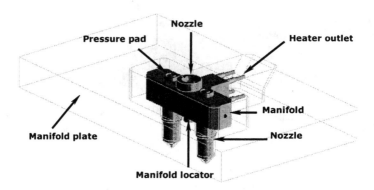

Fig. 2.19 Main components of hot-runner systems.

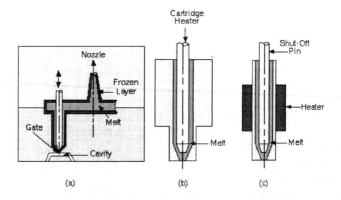

Fig. 2.20 Hot-runner system types: (a) insulated hot runner, (b) internally heated hot-runner, and (c) externally heated hot-runner systems [8].

than others but one needs to take into account the thermal expansion of various mold components and also the increased cost and design complexity. Fig. 2.21 shows a single-drop hot runner assembly built into an injection mold.

Plastic Injection Mold Design and Assembly

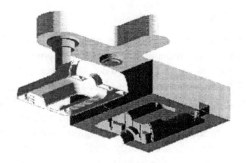

Fig. 2.21 A hot-runner assembly.

2.2.3.8 Standard Components

The standard components (Fig. 2.22) include connectors, cooling plugs, locating rings, sprue bushings, side locks, support pillars, side locks, etc. After creating miscellaneous components of a mold, designers can add them into the mold assembly. The standard components can also be added into the database as customization can be easily carried out by changing the dimension parameters and attributes in these parametric component models. It is advisable for mold designers to use standard components available from vendors for the two main reasons: (a) less manufacturing lead time, and (b) ease of tool maintenance. The cost required to incorporate standard components into a mold is usually lower than in-house designed and manufactured components. Other accessory components such as cooling components, ejection components, mold feeding systems and core pulling parts and accessories should also be considered to lower the cost of manufacturing a mold.

Fig. 2.22 Standard components of a mold.

2.2.4 Mold Assembly

An injection mold is a mechanical assembly that consists of product-dependent parts and product-independent parts. The geometrical shapes and sizes of a cavity system are determined directly by the plastic molded product. So the functional parts of a cavity system are product-independent. Besides the primary task of shaping the product, an injection mold has also to fulfill a number of tasks such as distribution of melt, cooling the molten material, ejection of the molded product, transmitting motion, guiding, and aligning the mold halves. The functional parts to fulfill these tasks are usually similar in the structure and geometrical shape for different injection molds. Their structures and geometrical shapes are independent of plastic molded products, while their sizes can be changed according to plastic products. It is important for mold designers to select critical components to work in an assembly domain that requires proper planning and checking on dimensions and tolerances.

Fig. 2.23 is a typical injection mold shown in closed and open conditions. When a mold is open, interference-free conditions must be ensured among all the components, especially when sliders and lifters are incorporated into a mold. An exploded view on the key components of a mold is displayed in Fig. 2.24. For a designer to work in a 3D assembly mode, the design task becomes more complicated. The number of components, accessories, and sub-assembly may be more than few hundreds or thousands depending on the molding complexity. The 3D mold design approach is now being adopted by most of the tool makers in the industry instead of 2D drafting design. However, the final assembly and checking is still very time-consuming and experience-dependent, and therefore a computer-aided assembly methodology as described in the later sections will be very beneficial to ease the mold assembly and thus enhance the quality of the injected molding.

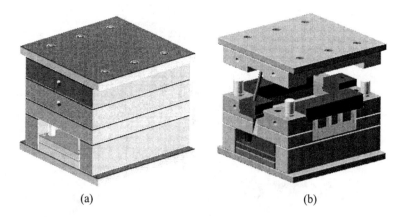

(a)   (b)

Fig. 2.23 An assembled injection mold: (a) mold closed and (b) open.

Plastic Injection Mold Design and Assembly

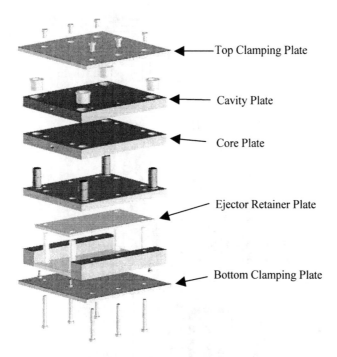

Fig 2.24 An exploded view of an injection mold.

2.3 MOLD DESIGN METHODOLOGY

2.3.1 The Mold Development Process

The design of molds for a new product is generally along the critical path of the production plan and thus the lead-time is essential in both design and subsequent manufacture. Injection mold design involves extensive empirical knowledge (heuristic knowledge) about the structure and the functions of the mold components. The typical process of a new mold development is previously shown in Fig. 2.5. The injection mold design phases can be organized as five major phases (see Fig. 2.25): product design, moldability assessment, detailed part design, insert/cavity design, and detailed mold design.

In Phase 0, a product concept is pulled together by a few people (usually a combination of marketing and engineering personnel). The primary focus of Phase 0 is to analyze the market opportunity and strategic fit. In Phase I the typical process-related manufacturing information is then added to the design to produce a detailed geometry. The conceptual design is transformed into a manufacturable one by using appropriate manufacturing information. In Phase II, the parting direc-

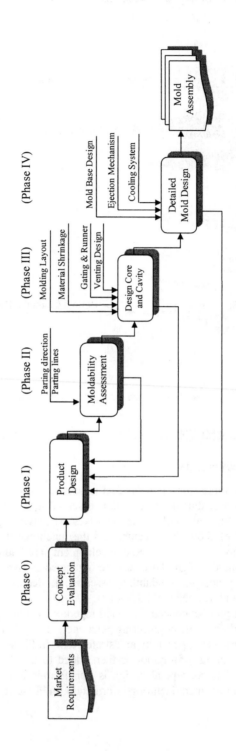

Fig. 2.25 Mold development process.

tion and parting lines location are added to inspect the moldability. Otherwise, the part shape may be modified. In Phase III, the part shape is used to establish the shape of the core and cavity that will be used to form the part. Shrinkage and expansion allowances are required so that the mold will be of the correct size and shape at the processing temperature to produce a specified part at its operating temperature. Gates, runners, and vents are added later. The association between geometric data and parting information is critical at this point. Phase IV is related to the overall mechanical structure of the mold, including the connection of mold to the machine, mechanisms for filling, cooling, ejection, and mold assembly.

2.3.2 Top-Down vs. Bottom-Up Approach

Top-down design is an approach for conceptual design of assemblies, where the focus is on the abstract specifications of the major component parts and the desired relationships. The natural way to design an injection mold assembly is the top-down approach where the designer begins with an abstract concept and recursively divides it into logical sub-assemblies until the level of part is reached. Unfortunately, most of the existing commercial CAD systems only support the bottom-up design approach that requires design of all the constituent parts and sub-assemblies before laying out the design for the assembly. To produce a compatible design paradigm for injection molds, a CAD system should support a top-down approach for assembly design.

2.4 COMPUTER-AIDED INJECTION MOLD DESIGN AND ASSEMBLY

The lack of specific facilities for injection mold design has blocked the widespread application of existing commercial CAD systems in the mold industry. Research into building up specific CAD systems for injection mold design has been carried out in recent years. Yuan et al. [9] developed an integrated CAD/CAE/CAM system for injection molding. Firstly, the drawings of a plastic product were transformed into the drawings of the mold impressions interactively, and then the mold design was carried out by using a group of design tools for injection molds. According to the user's needs, the system could analyze the runner balance, simulate a flow process, and generate the NC tapes for wire-cutting or milling machines. However, the design module of the system was based on AutoCAD platform. Thus, the design facilities provided by the system were actually a group of tools for editing and generating 2D drawings.

Kruth et al. [10] developed a design support system (IMES/DSS) for injection molds. The system was integrated within the AutoCAD systems and supports the design of injection molds through high-level functional mold objects, e.g., basic assemblies, components, and features. The system manages the low-level CAD entities and allows additional design information such as process planning information to be incorporated. The user could create or modify standard design objects and link them with a relational database. However, IMES/DSS could only support 2D design of injection molds. Lee et al. [11] developed a knowledge-based injection mold design system, which combines the facilities of Unigraphics [12] and ICAD [13] systems. Unigraphics was used as a geometric modeling package, while ICAD as a tool to process design knowledge. The system also contains the design libraries for mold bases and standard parts.

A mechanical assembly system consists of a number of component parts that, when assembled, performs a system function. When designing a mechanical assembly system, one first selects the constituent component parts and specifies their relationships in the assembly, then performs the detailed design of the individual component parts. Thus the design of a mechanical assembly is usually conducted in two phases: *conceptual design* and *detailed design*. During the conceptual design, the primary focus is on the selection of a collection of component parts and sub-assemblies, and the specification of their relations such that they deliver the desired function when assembled together. Detailed geometry becomes the focus in the detailed design phase, where the aim of the designer is to optimize the design under the constraints set by the desired performance of conceptual design, various engineering analysis, and the assembly feasibility and manufacturability of the parts. Common CAD systems are primarily geared towards the generation and utilization of the detailed geometric information, thus it is convenient to use these systems to perform detailed design. However, they do not offer much support for conceptual design, such as design with abstract concept and geometry, recursive decomposition of assemblies into sub-assemblies and parts, etc. Therefore, there is a need to enhance ordinary CAD systems to support conceptual design.

2.4.1 Assembly Modeling of Injection Molds

Assembly modeling has been the subject of research in diverse fields [14], such as kinematics, AI, and geometric modeling. Libardi et al. [15] compiled a review of research that had been done in assembly modeling. Their review reported that many researchers had used graph structures to model assembly topology. In this graph scheme, the components are represented by nodes, and transformation matrices are attached to arcs. However, the transformation matrices are not coupled together, which seriously affects the transformation procedure, i.e., if a sub-assembly is moved, all its constituent parts do not move correspondingly. Lee and Gossard [16] developed a system that supports a hierarchical assembly data structure containing more basic information about assemblies such as "mating feature" between the components. The transformation matrices were derived automatically from the associations of virtual links. However, this hierarchical topology model represents only "part-of" relations effectively.

Mantyla [17] developed a modeling system for top-down design of assembled parts. The system has the following characteristics not usually found in ordi-

Plastic Injection Mold Design and Assembly 37

nary CAD systems: structuring of product information in several layers according to the stage of the design process; representation of geometric information about components at several levels of details; and representation and maintenance of geometric relationships by means of a constraint-satisfaction mechanism.

An injection mold is a mechanical assembly that consists of a number of components. The task of designing an injection mold assembly involves the selection of these components and the specification of their relationships in the assembly. In addition to the functions for geometric modeling, a computer-aided injection mold design system should also provide a favorable design environment for assembly modeling of injection molds. This section will discuss a top-down design methodology for assembly modeling of injection molds, and then presents an approach to assembling an injection mold automatically.

2.4.1.1 Conceptual Design of a Mold Assembly

There exists two approaches to carry out the conceptual design of an assembly, namely the *bottom-up design* approach and the *top-down design* approach. The bottom-up design approach requires detailed design of all the constituent parts and sub-assemblies before laying out the design for the assembly. When designing an assembly by top-down design approach, the designer begins with an abstract concept and recursively divides it into logical sub-assemblies until the level of parts is reached. Hence this method is more natural for conceptual design. Fig. 2.25 compares the top-down and bottom-up design approaches. Fig. 2.25(a) shows the top-down procedure. A conceptual assembly of the slider is designed first. This is then decomposed into sub-assemblies and parts, whose detailed geometry is designed within the constraints set by top-level design. In Fig. 2.25(b) the opposite procedure is shown for packaging of a number of interconnected components of the slider system.

As observed by Libardi et al. [15], in the survey of computer-aided assembly design, most designers follow the top-down design approach while carrying out conceptual design. Thus, a CAD system for assembly design should support top-down design. At the initial stage of conceptual design, focus is on the abstract specifications of the major component parts and the desired relationships. These specifications may be expressed by means of numerical parameters and expressions; however, non-numeric, qualitative specifications are also needed. After the initial specifications, the design process proceeds from the abstract to the concrete. It becomes useful to use a geometric representation to express the desired relationships between the concepts introduced. The initial geometry can be mainly dimensionless, concentrating on the general geometric arrangement of the main component parts and their geometric interaction. Therefore, to support the top-down design process, a CAD system needs a number of capabilities; these include at least the following:

(1) The facility for modeling the designed objects on a purely abstract level: as a structure of model entities, their relationships, and their properties; and
(2) The abilities to support the creation of abstract geometry, where the designer can choose the level of details of the representation according to the particular requirements of the design task.

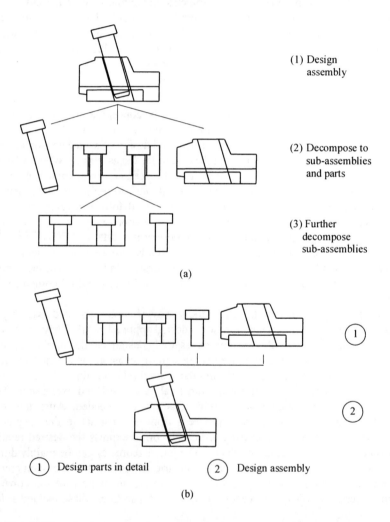

Fig. 2.25 Comparison of different assembly design paradigms, (a) top-down vs. (b) bottom-up approaches.

Plastic Injection Mold Design and Assembly 39

In addition to supporting the top-down design approach, a CAD system for assembly modeling should also provide the facility to infer the configuration of a component part in the assembly. During the conceptual design, the spatial relationships between parts are specified by the mating conditions, which implicitly describe the orientation and position of a part in the assembly. To exactly orientate and locate the part, a CAD system needs to calculate the numerical matrix of the part in the world coordinate system (WCS).

2.4.1.2 Top-down Design of Mold Assemblies

It can be seen from Fig. 2.1 that an injection mold mainly consists of a cavity system, a runner system, a guiding and locating system, a cooling system, and an ejection system. For different injection molds that have different cavity systems, there exists a structural similarity between their ejection systems, which usually consist of an ejector plate, a retainer plate, and a number of ejectors. This fact is also true for the guiding systems, cooling systems, and runner systems. The geometric shapes of their component parts are relatively independent of plastic products, while the sizes of their component parts can be changed according to the specific requirements. Thus, the assemblies of ejection systems, cooling systems, and guiding systems are called *product-independent* assemblies, while their component parts are called product-independent parts. Three important characteristics of the product-independent assemblies and parts are summarized as follows:

(1) The structures of product-independent assemblies are predictable.
(2) The geometric shapes of product-independent parts are predictable.
(3) The sizes of product-independent parts are pre-defined in a catalog.

For instance, DME mold base [18] is used to demonstrate the above characteristics. DME mold base is classified into five types: A, B, X, AX, and T. X type can be further classified into 5-plate series, 6-plate series, and 7-plate series. For type-*A* mold base, it consists of a top-clamping plate, an *A*-plate, a *B*-plate, a support plate, an ejector-retainer plate, an ejector plate, an ejector housing, guide pins, guide bushes, and screws. The dimensions of type-*A* mold bases have pre-defined series, such as 125×125, 125×156, 156×156, 196×246, and 246×296. Thus, when designing a mold base, what a designer needs to do is select the type of mold base and its dimension series.

Because of the characteristics of product-independent assemblies and parts, object-oriented models (to be described in the next chapter) can be established for product-independent assemblies, within which the geometric shapes of the product-independent parts are described by the various features, and the associativities between the consisting parts are represented by the properties of the objects. Based on these object-oriented models, a top-down browser is developed for the convenience of assembly modeling of injection molds. Fig. 2.26 shows the interface of the top-down browser. To design the product-independent assemblies by the top-down browser, one first selects the type and sub-type of the assembly, and then specifies the outline dimensions. Now the primary structure of the selected assembly is determined. Based on this primary structure, one can further add more optional parts and change some dimensions. Once the "apply" button is pressed, the program automatically instantiates the object-oriented model with data from the catalog database. The required product-independent assembly is displayed on the screen. Hence, the conceptual design of the product-independent assembly is considered complete. Fig. 2.27 shows the instantiation process of the mold base assemblies.

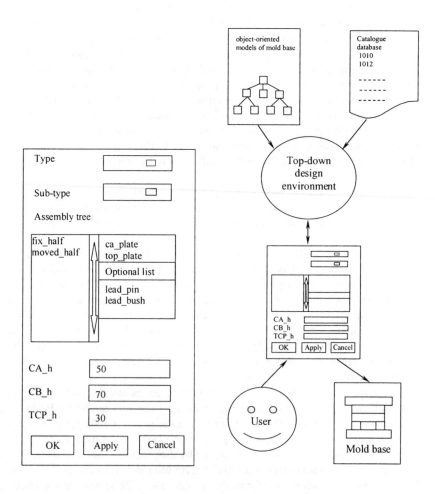

Fig. 2.26 Interface of the top-down browser.   **Fig. 2.27** Instantiation of mold base.

Plastic Injection Mold Design and Assembly 41

2.4.1.3 Automated Assembly Modeling of Injection Molds

It has been mentioned previously that any assembly of injection molds consists of product-independent parts and product-dependent parts. The design of individual product-dependent parts is actually an exaction of the geometry from the plastic part [19, 20]. Usually the product-dependent parts have the same orientation as that of the top-level assembly, while their positions are specified directly by the designer. As for the design of product-independent parts, conventionally, mold designers select the structures from the catalogs, build the geometric models for selected structures of product-independent parts, and then add the product-independent parts to the assembly of the injection mold. This design process is time-consuming and error-prone. In an automated or semi-automated assembly system, a database needs to be built for all the product-independent parts according to the assembly representation and object definition. This database not only contains the geometric shapes and sizes of the product-independent parts, but also includes spatial constraints between them. Moreover, some routine functions such as the interference check and pocketing are encapsulated in the database. Therefore, what the mold designer needs to do is to select the structure types of product-independent parts from user interfaces, and the system will automatically calculate the orientation and position matrices for these parts, and add them to the assembly.

As it can be seen from Fig. 2.28, the product-independent parts can be further classified as the mold base and standard parts. Mold base is the assembly of a group of plates, pins, guide bushes, etc. Besides shaping the product, a mold has to fulfill a number of functions such as clamping the mold, leading and aligning the mold halves, cooling, ejecting the product, etc. The fact that most molds have to incorporate the same functionality, result in a similarity of the structural build-up. Some form of standardization in mold construction needs to be adopted. Mold base is the result of this standardization.

2.5 SUMMARY

Injection molding is a technology predominantly used for processing thermoplastic polymers. An injection mold is a mechanical assembly that usually consists of more than hundreds of components. This chapter describes the basic mechanical components of an injection mold and its design practice commonly used in the mold industry. A systematic mold design process used in manual design is discussed in detail but the approach can be applied to a computer-aided mold design and assembly environment that uses advanced 3D CAD/CAM systems. The increasing complexity of a product design has resulted in a higher challenge to mold designers to create a more sophisticated mold design and assembly in a much shorter time. Therefore, an interactive or semi-automated design approach will facilitate the design tasks to a great deal compared to the manual approach. To develop a fully automated mold design system that does not rely heavily on ex-

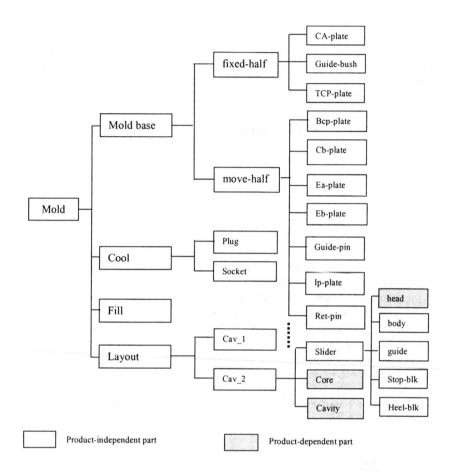

Fig. 2.28 Assembly structure of an injection mold [21].

perienced designers will require more research to be carried out. In the following few chapters, reported methodologies and prototypes will be discussed further.

REFERENCES

1. R.G.W. Pye, "Injection Mold Design: An introduction and design for the thermoplastics industry", 4th Ed., Essex, , UK: Longman Scientific and Technical Publishing, 1989.
2. Y.S. Yueh and R.A. Miller, "Systematic approach to support design for manufacturability in injection molding and die casting", Computers in Engineering ASME Database Symposium, ASME, New York, pp. 755-765, 1995.

3. P.S. Cracknell and R.W. Dyson, Handbook of thermoplastics injection mold design, Chapman & Hall, London, Chapter 5, pp. 42-55, 1993.
4. HASCO, mold base catalogue, Denmark, http://www.hasco.de, 2003.
5. Y.F. Zhang, J.Y.H. Fuh, K.S. Lee and A.Y.C. Nee, "IMOLD: An Intelligent Mold Design and Assembly System. In Computer Applications in Near Net-Shape Operations", edited by A.Y.C. Nee, S.K. Ong and Y.G. Wang. New York: Springer-Verlag, pp. 265-284, 1999.
6. Z. Wang, K.S. Lee, J.Y.H. Fuh, Z. Li, Y.F. Zhang, A.Y.C. Nee and D.C.H. Yang, "Optimum ejector system design for plastic injection mold" International J. of Computer Applications in Technology, Vol. 9, No. 4, pp. 211-218, 1996.
7. Amiss, John Milton, Machinery's handbook guide to the use of tables and formulas. New York: Industrial Press, 1984.
8. Hot runner systems, http://www.scudc.scu.edu/cmdoc/dg_doc/develop/design/runner/34000007.htm#208761, 2003.
9. Z.S. Yuan, D.Q. Li, X. Chen, and X.G. Ye, "Integrated CAD/CAE/CAM system for injection molding", Computing & Control Engineering Journal, Vol. 4, No. 6, pp. 277-279, 1993.
10. J. P. Kruth, R. Wilhelm, and D. Lecluse, "Design support system using high level mold objects", CIRP International Conference and Exhibition on Design and Production of Dies and Molds, Istanbul, Turkey, 1997, pp. 39-44.
11. K.S. Lee, Z. Li, J.Y.H. Fuh, Y.F. Zhang and A.Y.C. Nee, "Knowledge-based injection mold design system", Proc. of CIRP International Conference and Exhibition on Design and Production of Dies and Molds, Istanbul, Turkey, 1997, pp. 45-50.
12. UG-II, User Manual Version 10.5, Unigraphics, Electronic Data System, Cypress, CA., 1995.
13. ICAD is a knowledge-based CAD system from Knowledge Technologies International, Ltd, UK, 1997.
14. J.J. Shah and M.T. Rogers, "Assembly modeling as an extension of feature-based design", Research in Engineering Design, Vol. 5, No. 3&4, pp. 218-237, 1993.
15. E.C. Libardi, J.R. Dixon, and M.K. Simmon, "Computer environments for the design of mechanical assemblies: a research review". Engineering with Computers, Vol. 3, No. 3, pp. 121-136, 1988.
16. K. Lee and D.C. Gossard, "A hierarchical data structure for representing assemblies", Computer-Aided Design, Vol. 17, No. 1, pp. 15-19, 1985.
17. M.A. Mantyla, "Modeling system for top-down design of assembled products", IBM J. of Research and Development, Vol. 34, No. 5, pp. 636-658, 1990.
18. D-M-E Mold Base Catalogue, D-M-E Company, Madison Heights, MI: http://www.dme.net, 2003.
19. K.H. Shin and K. Lee, "Design of Side Cores of Injection Molds from Automatic Detection of Interference Faces", J. of Design and Manufacturing, 3(3), pp. 225-236, 1993.
20. Y. F. Zhang, K.S.Lee, Y. Wang, J.Y.H. Fuh and A.Y.C. Nee, "Automatic slider core creation for designing slider/lifter of injection molds". CIRP International Conference and Exhibition on Design and Production of Dies and Molds, Turkey, 1997, pp. 33-38.
21. X.G. Ye, J.Y.H. Fuh and K.S. Lee, Automated assembly modeling for plastic injection molds. International J of Advanced Manufacturing Technology, 16: pp. 739-747, 2000.

3
Intelligent Mold Design and Assembly

3.1 INTRODUCTION

Injection mold design in essence is a mental process. The mold designer starts with a mental outline, i.e., constructing the mold around the product, knowing exactly what components are necessary to accomplish a mold. Instead of low-level geometric entities provided by conventional CAD systems, the designer needs to concentrate on the desired functions and behavior of the design. To create a favorable environment for mold designers, a CAD system for injection mold design is needed to support design with high-level objects.

Injection mold design is also a complex and iterative process, during which changes are inevitable. In the geometry-based CAD systems, the low-level model does not preserve the design intent of the designers. The lack of associativity between the model entities means that a design change must be mapped to a set of changes of low-level geometric entities, and is therefore time-consuming. On the other hand, the construction of an injection mold is not a straightforward task even when using the high-level and powerful operations of a solid modeler. It is important that the modeling environment supports reuse of existing parts, such as standardized parts and part families in a new design. Reuse of existing parts requires a CAD system to be able to change the current design conveniently. So a useful CAD system for injection mold design should be able to automatically propagate changes throughout the whole model.

In the last two decades, CAD/CAM technologies have been widely used in the molding industry. Existing commercial CAD/CAM systems are able to help mold companies to enhance the mold quality and shorten the tooling lead time. However, these systems are usually generic mechanical CAD/CAM software and do not provide the specific facilities for injection mold design. For mold designers in many companies, the CAD/CAM systems are only a design tool to generate and edit the 2D drawings. Nowadays, the development of products is no longer a sequential process. Many companies start designing the injection molds with the end products still in the development phase. These require CAD systems to provide convenient functionalities in order to modify and fine-tune the injection mold designs. Therefore, there is a need to develop a computer-aided injection mold design system (CAIMDS) that not only provides specific functionalities for mold design, but also supports design with high-level objects and design change propagation.

Various research and development efforts and techniques used in automating the injection mold design are reviewed and discussed in this chapter. Most of the research work in computer-aided injection mold design has been concentrated on one of the following three topics: (1) determination of the optimal ejection direction for a plastic product, (2) automatic generation of cores and cavities through parting, and (3) development of interactive CAD systems for injection mold design. However, research and development on some key issues of computer-aided injection mold design, such as representing a three-dimensional (3D) injection mold in computers, automatically arranging multiple cavities, and assembly modeling of injection molds, has seldom been reported. An undercut region is a kind of features in the context of injection mold design. In-depth review in feature recognition shows that it is a challenge for all the existing feature recognition techniques to recognize undercut features from a plastic product with sculptured surfaces. Detailed overviews about assembly modeling are finally presented to address the significance of developing more effective solutions for assembly design of injection molds.

3.2 FEATURE AND ASSOCIATIVITY-BASED INJECTION MOLD DESIGN

Conventional CAD/CAM systems are essentially surface or solid modelers, which have been proven deficient for injection mold design. The following sections discuss the feature- and associativity-based methodology for computer-aided injection mold design, which can overcome the problems faced by the geometry-based CAD systems. After briefly introducing the feature concepts and feature creation techniques, in-depth discussions about features and their associativity within injection molds are presented. An object-oriented representation scheme for 3D injection molds, which can encapsulate both feature and associativity information into an assembly object will be described.

3.2.1 Feature Modeling

3.2.1.1 Feature Concepts

From an engineering viewpoint, *feature* means the generic shape or characteristics of a part or an assembly, with which engineers can associate certain attributes and knowledge useful for reasoning about that part or assembly. Features encapsulate the engineering significance of portions of the part geometry and, as such are applicable in part design, part definition, and reasoning about the part in a variety of applications. Hence features are related to some physical, geometric aspects of a part or an assembly. They are semantically significant and distinct entities in one or more engineering viewpoints. Features can be characterized as follows [1]:

- A feature is a physical constituent of a part
- A feature can be mapped to a generic shape
- A feature has engineering significance
- A feature has predictable properties.

A *feature model* is a data structure that represents a part or an assembly in terms of its constituent features. Each feature in the feature model is an identifiable entity that has some explicit representation. The shape of a feature may be expressed in terms of dimension parameters and enumeration of geometric and topological entities and relations. The engineering significance of a feature involves formalizing the function that the feature serves, or how it can be manufactured.

After the brief discussions of the fundamental concept of features, the attention is now turned to the contributions that features make in solving the problems such as low-level entities and lack of design intent, faced by geometric-based modeling. The main advantage of features is that a feature-based part model can be described using high-level primitives, defined as generic abstractions of recurring shapes. The open-ended nature of features makes it possible to include exactly those primitives that are natural for some application domains. For machined parts, slot, boss, hole, etc. can be defined as machined features; while for injection molds, undercuts can be defined as features. With the high-level entities ready in use, designers can concentrate on the desired functions and behaviors of their design. Thus, the initial design can be synthesized quickly from the high-level entities and their relations.

By letting a design be described in terms of application-oriented entities, feature models clearly provide a better record of the intent of a designer than mere geometric models. Features provide a basis for linking the design rationale with the model, hence support change propagation.

3.2.1.2 Feature Creation Techniques

The techniques for creating feature-based models can be divided into two main categories, i.e., feature recognition and design-by-features. The feature recognition techniques identify and extract features from geometric models. This class of techniques includes the interactive feature recognition and automatic feature recognition techniques. In interactive feature recognition, a geometric model is created first, then features are created by human users, such as picking entities in an image of the part; while in automatic feature recognition, portions of a geometric model are compared to predefined feature patterns. Important subtasks in automatic feature recognition are summarized as follows [2]:

- Constructing an auxiliary data structure from the geometric model to facilitate the search of features;
- Searching the geometric model to match topological and geometric patterns; and
- Extracting recognized features from the geometric model.

In all of the design-by-features techniques, parts are created directly using features and the geometric model is generated from the features. This requires that the CAD systems have generic feature definitions placed in a feature library from which features can be instanced by specifying the dimension and position parameters, the references on which the features are to be located, and various other attributes.

3.2.1.3 Features Within Injection Molds

Generally speaking, a feature represents the engineering meaning or significance of the geometry of a part. Features can be defined differently in many engineering applications, e.g., process planning, NC machining, inspection, and mold design. For a computer-aided injection mold design system, it is important to clearly define the features in the context of injection mold design.

The primary task of an injection mold is shaping the molten material into the final shape of a plastic product, and this task is fulfilled by the cavity system of an injection mold [3]. The geometric shape and size of the cavity system are directly determined by the plastic product. It is not practical to define form features that have stereotype shapes for design of cavity systems, because the plastic products are so versatile that large numbers of geometric shapes are required to model them. However, there still exist some special regions in plastic products that attract interests of mold designers. One of these regions is the undercut region that blocks the removal of a plastic product from the injection mold. The existence of undercut regions would substantially influence the design and manufacturing of injection molds. So it is worthwhile to define the undercut regions as a kind of design features. With the help of feature recognition techniques, the undercut features can be recognized automatically from plastic products, cores, and cavities.

Besides the cavity system, an injection mold consists of a group of functional parts to fulfill a number of secondary tasks such as distribution of melt, cooling the molten material, ejection of plastic products, etc. These functional parts are usually well standardized by mold base vendors or mold companies

Intelligent Mold Design and Assembly

themselves. What the mold designers need to do is select the structures and geometric shape of the functional parts from catalogues, and adjust their sizes according to plastic products. It is therefore natural to define a group of form features to allow mold designers to design the standardized functional parts in terms of high-level entities.

It can be seen from the above discussions that in the context of injection mold design, two classes of features can be defined, namely *undercut features* and *form features*.

(1) Undercut Features

A depression or protrusion region in a plastic product that blocks the removal of the plastic product from its injection mold is referred to as an *undercut feature*. The undercut features can be further classified as *depression undercut features* or *protrusion undercut features*. The depression undercut feature is a region that consists of concave face sets and is bounded by a convex edge loop, while the protrusion undercut feature is a region that consists of convex face sets and is bounded by a concave edge loop. To form the undercut features, side-cores are required, which incur a group of auxiliary parts to fulfill the side movement. It is obvious that for every undercut feature in a plastic product, there exists a corresponding region in the injection mold, which is also referred to as undercut feature on the injection mold. Fig. 3.1 shows the undercut features in a plastic product and its injection molds. Corresponding to the depression undercut feature in the plastic product, there exists a protrusion undercut feature in the injection mold.

(2) Form Features

After carefully analyzing the standard functional parts used in injection molds, a group of form features are defined to capture the design intents of mold designers. They are box, cylinder, simple hole, counter-bore hole, countersunk

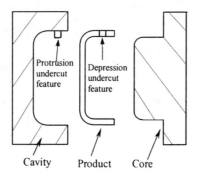

Fig. 3.1 Undercut feature.

hole, rectangular pocket, rectangular pad, and boss. When designing an injection mold, these form features are very useful to build the 3D models of product-independent parts, such as plates, guide bushes, pins, etc. Table 3.1 lists the parameters and positioning methods of these form features, and Fig. 3.2 illustrates their geometric shapes.

3.2.2 Associativity Within Injection Molds

3.2.2.1 Associativity on Feature Level

It is well known that a major function features serve is to create associativity between entities in a part definition. This associativity between entities makes it possible to encapsulate design or manufacturing constraints and to do geometric reasoning required in various engineering applications. The associativity encoded in a feature representation can further be divided into intra-feature associativity and inter-feature associativity.

Table 3.1 Form features for injection mold design [4]

Form features	Parameters	Positioning methods
Box	Length l, width w and thickness t	World coordinate system
Cylinder	Diameter d, height h	World coordinate system
Simple hole	Hole diameter d, hole depth h, tip angle α	Attach it to a planar face, and locate its axis
Counter-bore hole	Hole diameter d, hole depth h, counter-bore diameter D, counter-bore depth h_1, tip angle α	Attach it to a planar face, and locate its axis
Countersunk hole	Hole diameter d, hole depth h, countersunk diameter D, countersunk angle α_0, tip angle α.	Attach it to a planar face, and locate its axis
Rectangular pocket	Length X, Y, Z, corner radius r_c, floor radius r_f	Attach it to a planar face, and locate its edges
Boss	Diameter d, height h, and taper angle α	Attach it to a planar face, and locate its axis
Rectangular pad	Length X, Y, Z, corner radius r_c	Attach it to a planar face, and locate its edges

Intelligent Mold Design and Assembly

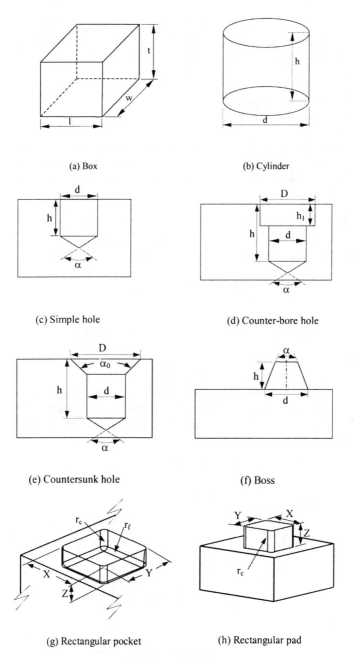

Fig. 3.2 Form features [4].

Intra-feature associativity. Intra-feature associativities are those attributes encoded in a feature representation that are independent of other features in the model and describe the relations between entities within the features. Two types of intra-feature associativities are as follows:

- Constraints of geometric shape
- Independent parameters.

Constraints of geometric shape describe the relationships among geometric entities within a feature, while independent parameters are the dimensions that are specified at the time when a feature is created. Parameters in Table 3.1 are all independent parameters.

Inter-feature associativity. Inter-feature associativities are those attributes encoded in feature representation that describe the relations between features. Four types of inter-feature associtivities are as follows:

- Derived feature parameters
- Feature location
- Feature orientation
- Constraints on feature size, location or orientation parameters.

When a new feature is added in a model, some of the parameters or dimensions of the new feature may be fixed by other features. These are termed *derived parameters* or *derived dimensions*. The diameter d of the boss in Fig. 3.3(a), for example is a derived feature parameter. Fig. 3.3(b) shows an example of size constraints. In the figure, t_1 of the counter-bore hole should be less than the t of the block feature.

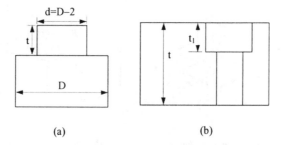

Fig. 3.3 Inter-feature associativity.

Feature location and orientation specify the information that is required to locate and orientate the feature. Usually, a new feature is positioned with respect to the existing entities or features. Positioning by means of entities is more than just a convenient way to specify where to put the feature, it enables relationships to be recorded and thus reused in other applications. In effect, the geometric entity used in positioning a feature becomes a kind of associativity encapsulated in the definition of the feature. Therefore, when one feature is modified or deleted, the change can be propagated to its adjacent and child features.

3.2.2.2 Associativity on Part Level

In addition to the intra-feature and inter-feature associativity, there exists associativity between various parts of an injection mold. Three kinds of associativities between parts that play important roles in injection mold design are parent-child relationships, spatial relationships, and geometric linkages between entities of parts.

Parent-child relationships. Parent-child relationships describe the affinity relations between parts of an injection mold assembly. An assembly can be composed of many sub-assemblies, while a sub-assembly can consist of many parts. The existence of a component may depend on the presence of its parent.

Spatial relationships. The spatial relationships between two parts describe the location and orientation of one part with respect to another part. Changes in the reference part can be propagated to the dependent part. Spatial relationships can be further classified into two categories, namely qualitative spatial relationships and numerical spatial relationships. The qualitative spatial relationships define geometrical and topological constraints between two objects, while numerical spatial relationships use matrices to describe the spatial relationships between two parts.

Geometric linkages. Geometric linkages between entities of parts describe the associations of geometrical entities within different parts. It is well known that the sizes and shape of the cavity system are directly determined by the plastic product, so any change in size or shape of the plastic product should be reflected in the cavity system. Fig. 3.4(a) shows a plastic product and its core and cavity. When the plastic product is changed, the geometric linkages between the product and core/cavity result in the changes of the shapes of the core and cavity accordingly (Fig. 3.4b). The geometric linkages between entities of plastic products and those of cavity systems provide the base for design change propagation.

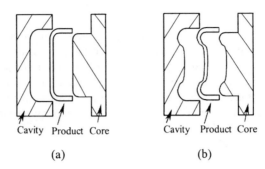

Fig. 3.4 Geometric linkages.

3.3 REPRESENTATION OF INJECTION MOLD ASSEMBLIES

One fundamental issue in developing a computer-aided injection mold design system is the representation of mold information, which includes not only the geometric and physical properties of the injection mold but also information about functions, constraints, and the design rationale. In this section, the object-oriented approach is used as the foundation for representing the information of an injection mold.

3.3.1 Concepts and Notations for Object-Oriented Modeling

Object-oriented modeling is a new way of thinking about problems using a model organized around real world concepts [5]. The fundamental entity is the object, which combines both data structures and behaviors in a single entity. Below are some basic concepts and notations for object-oriented modeling.

Object An *object* is defined as a concept, abstraction, or thing with crisp boundaries and meaning for the problem at hand. Objects serve two purposes: they promote the understanding of the real world and provide a practical basis for computer implementation. All objects have identities and are distinguishable.

Class A *class* describes a group of objects with similar properties (attributes), common behaviors (operations), common relationships to other objects, and common semantics. Classes are defined on objects as abstraction mechanisms to make common properties and semantics explicit.

Operation An *operation* is a function or transformation that may be applied to or by objects in a class. Each operation has a target object as an implicit argument. An object "knows" its class, and hence the right implementation of the operation.

Intelligent Mold Design and Assembly

Link and association A *link* is a physical or conceptual connection between object instances. An *association* describes a group of links with the common structure and common semantics. One important property of association is the multiplicity, which specifies how many instances of one class may relate to a single instance of an associated class. Multiplicity constrains the number of related objects. Usually association may be one-to-one, one-to-two, or one-to-many, so multiplicity is often described as being "one", "two", or "many". Fig. 3.5 shows the notations for multiplicity of associations. More details about notations for object-oriented modeling can be found in Rumbaugh's book [6].

3.3.2 Object-Oriented Representation of Mold Assembly

It is well known that every injection mold is an assembly that consists of sub-assemblies and component parts. A sub-assembly may contain other sub-assemblies and parts, while parts consist of features. In object-oriented representation of injection molds, mold assemblies, component parts, and features are all considered objects that provide a form of abstraction to model the different levels of functional abstraction and the actual physical objects. Associated with these objects are methods, which provide techniques to:
(1) perform procedural calculations;
(2) propagate implications of actions;
(3) aid in constraint management; and
(4) provide various accessing utilities.

Corresponding to the three kinds of objects, three generic classes can be defined, i.e., assembly class, and part class, feature class (Fig. 3.6). The relationships between the three classes are shown in Fig. 3.7.

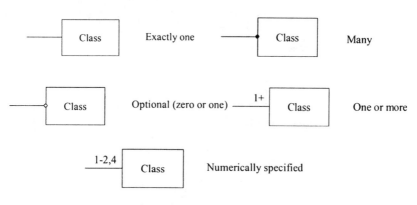

Fig. 3.5 Notations for multiplicity [6].

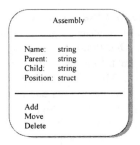

(a) Assembly class

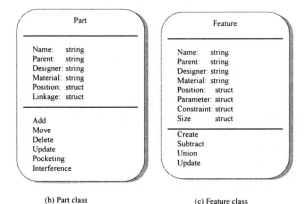

(b) Part class (c) Feature class

Fig. 3.6 Generic classes.

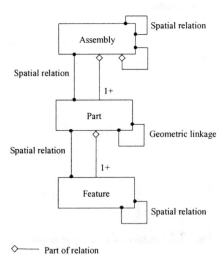

Fig. 3.7 Relationships between generic classes.

Intelligent Mold Design and Assembly

Based on the generic classes, more subclasses for injection molds, such as *plate*, *core*, *pin*, *screw*, etc., can be further derived. For example, the *plate* subclass is derived from the generic class *part*, which has three additional attributes: length, width, and thickness. Using these subclasses, the abstraction of injection molds can be represented by the diagram in Fig. 3.8.

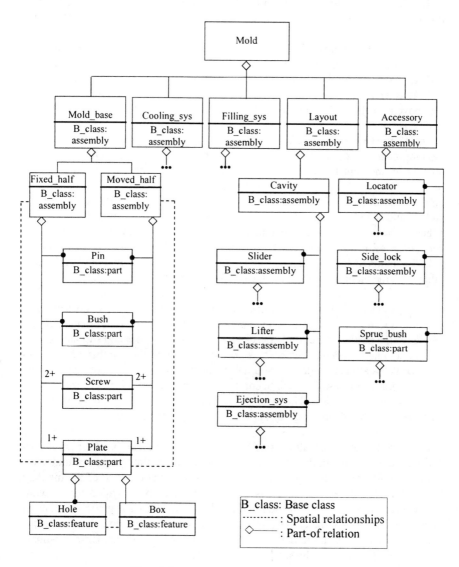

Fig. 3.8 Abstraction of an injection mold.

3.4 OPTIMAL PARTING DESIGN FOR CORE AND CAVITY CREATION

The literature on the recognition and extraction of undercut features in injection molded parts, automatic determination of optimal parting direction, parting lines and surfaces was quite limited until the 1990s. This is due to the complex geometry of plastic molded parts and the limited computing capabilities in the earlier days. Although some progress has been made in the 1990s, the work is limited to simple molded parts constituted by prismatic features and planar surfaces.

In injection molds, the presence of undercut features would affect molding cost and the entire structure of a mold since the parting directions, parting lines and surfaces, the number of local tools and their actuating mechanisms, the core and cavity and their inserts depend on them. The influence of undercut features on the above aspects can be summarized as follows:

(1) Parting directions In order to simplify the mold structure and reduce molding cost, the optimal parting direction should be in the direction where the number of undercut features and their corresponding volumes are maximum such that the number of the rest of the undercut features and their corresponding volumes are minimal. The rest of the undercut features will become the real undercuts and would need side-cores, side-cavities, or other local tools for molding. Whether the undercut features can be molded by the core or cavity depends on the parameters of undercut features. If the undercut directions are in the parting direction, the undercut features can be molded in by the core or cavity and their inserts, otherwise, the incorporation of the local tools is needed.

(2) Parting lines and parting surfaces Generally, the parting lines and surfaces are located where the projected area of the molding onto the plane perpendicular to the parting direction is maximum. The parting lines and surfaces should include as many undercut features as possible such that the number of the side-cores or side-cavities needed is minimal. If the parting lines and surfaces include most undercut features, they will become very complicated. There is a trade-off in dealing with this scenario. The principle is that the parting lines should include as many features, but the parting surfaces should be easy to manufacture.

(3) Side-cores and side-cavities A side-core is a local insert which is normally mounted at an angle to the parting direction for molding concave undercuts in the side face of a molding. The side-core prevents the removal of the molding from the parting direction, and thus some means must be provided for its withdrawal prior to the ejection of the molding. The side-cavity has a similar function to the side-core in that the convex undercuts can be molded in. In a molding, if the undercuts have the same undercut direction and appear in the suitable locations, they can be molded in by the same side-core or side-cavity.

3.4.1 Optimal Parting Direction

3.4.1.1   Parting Direction

Parting direction refers to a pair of opposite directions along which the core and cavity are opened, P_D is one of the parting directions, and $-P_D$ is the other. Chen, Chou, and Woo [7,8] determined the parting direction based on the visibility maps (V-map) of the pockets in a 3D molding. Fig. 3.9 shows the convex hull, sealed pockets and pockets of a 3D molding. Supposing that a set of pockets $P= [P_1, P_2, ..., P_m]$ are extracted from $CH(Q)-*Q$ and let $VM=[V_1(P_1), V_2(P_2),..., V_m(P_m)]$ denote the corresponding V-maps which are assumed to be non-empty. The parting direction is located at a pair of antipodal points p and $-p$ that maximize the number of V-map polygons in V-map sphere containing either p or $-p$. In other words, the parting direction includes as many undercut features as possible such that the number of the remaining undercut features is minimal. For the pockets shown in the figure, the parting direction is determined based on the V-maps (Fig. 3.10) of pockets P_1, P_2, P_3, P_4. The antipodal points of parting direction P_D maximize the number of V-map polygons V_1, V_2, and V_4. V_3, however, is not included, and a side-core is needed for molding the pocket P_3.

Mochizuki and Yuhara [8] used a similar principle as Chen [7,8]. All potential undercut directions and the parting directions are determined based on the potential undercut features. The parting direction should be in the direction in which the number of undercut features is maximum. Chen and Mochizuki did not consider the undercut volume when determining the parting direction. In practical mold design, the undercut volume is a very important factor affecting the parting direction, and this is the main drawback of their work. Hui and Tan [10] introduced a blocking factor to indicate the number of undercut features by measuring the relative amount of obstruction in a certain direction. The blocking factor is the ratio of the number of obscured points to the total number of test points. The obstructed point is a point which cannot be seen from a given direction. The blocking factor can be evaluated by first computing all the profile edges which are parallel to the given parting direction, then a number of test points are generated on these profile edges and all other edges of the molding. Each point in the edges is tested to see if it is obscured in the given direction. Another parameter is also introduced to determine the parting direction, namely, a preference value which is defined as the ratio of the projected area in the given direction to the maximum projected area of the molding among any direction. The high preference value reflects a larger projected area, and hence, the more preferred choice as the parting direction. After the blocking factor and preference value have been obtained, the parting direction can be determined. This is based on the evaluation of the preference value and the blocking factor of all the possible directions.

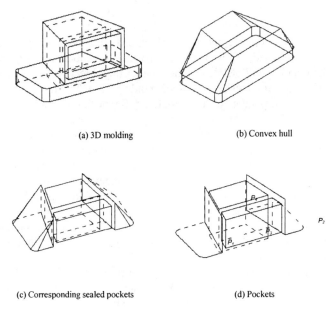

Fig. 3.9 Corresponding sealed pockets and pockets [7].

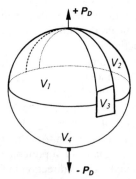

Fig. 3.10 V-map polygons and the parting direction [7].

Besides Hui and Tan, Weinstein and Manoochehri [11,12] presented their definitions of the convex and concave regions in a molding based on the part geometry for determining the parting direction. The convex region is a set of surfaces that cannot be seen from each other. The concave region, however, contains

Intelligent Mold Design and Assembly

those surfaces that are visible to each other. The allowable draw range of every concave region is determined based on the V-map of each surface of that region. The parting direction of the core and cavity is then determined by the intersection of the allowable draw range of the concave regions. The methodology is a three-step process. The first step is to find the allowable draw range for the subsets in each concave region. The second step is to determine the allowable range for the concave region from the union of the allowable draw ranges of the individual sets. The final step is to find the intersection of the allowable draw range of the concave regions and determine the allowable draw range for the molding. The methodology, however, does not consider the convex regions. The convex regions such as bosses, cylinders, cones, and spheres in a molding would become the real undercuts and affect the parting direction. Another weakness is that they did not mention how to recognize the concave regions in a molding, as the concave region can be made up by curved surfaces or free-formed surfaces. Sometimes, one single surface can form a concave region such as conical holes and spherical concave feature in the molding.

Automatic determination of parting directions is also one of the important issues in CAIMDS. There are some limitations in the previous work, and the methodologies reported did not include all the geometrical features. The geometry of the molded parts is limited to planar surfaces or simple curved surfaces in these approaches. All these approaches, however, are good exploratory and pioneering work in automated design of injection molds.

3.4.1.2   Criterion of Optimal Parting Direction

The optimal parting direction (O_P) is selected by maximizing the number of *undercut features* (see Section 3.6 for details) which can be molded in by the main core and cavity or minimizing the number of undercut features which can only be molded in by side-cores, etc. The approach described below [13,14], however, considers not only the number of undercut features, but their corresponding volumes as well. If direction x is chosen as the parting direction, the undercut features with the undercut directions not parallel to the parting direction will be the real undercuts and need the local tools for molding.

In order to determine the optimal parting direction, let all the undercut features be divided into different groups such that each group has the same undercut direction. The undercut grouping is illustrated in Fig. 3.11, which shows a molding with three undercut features. Undercut features 2 and 3 have the same undercut direction, and they are classified into one undercut feature group. The direction of undercut feature 1 is different from the direction of undercut features 2 and 3, so it belongs to its own group. Supposing that $V_{ij}(x_i)$ is the volume of the *j*th undercut feature in the *i*th group and x_i is the undercut feature direction of the group, then the optimal parting direction O_P is selected based on the following criterion [15]:

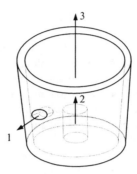

Fig. 3.11 Grouping of undercut features.

$$\left.\begin{array}{l} O_P = x_i \\ \textit{If and only if} \\ V_{imax} = \underset{i}{Max} [\sum_{j=1}^{m_i} V_{ij}(x_i)] \end{array}\right\} \quad (i=1,2,...n, j=1,2,...m_i) \quad (3.1)$$

where n is the undercut feature group number and m_i is the undercut feature number in the ith group. The undercut direction x_i which satisfies the above criterion will be the optimal parting direction O_P. Based on the above criterion, the sum of the undercut feature volumes (V_{imax}) in the optimal parting direction that can be molded by the main core and cavity should be greater than the ones in any other groups. In the optimal parting direction, the volume of undercut feature group will be maximized although it may not correspond to the largest number of undercut features. However, if there is more than one direction that satisfies Eq.(3.1), it means that there are more than two undercut feature groups with exactly the same group volume, then the optimal parting direction is chosen such that the undercut feature number is maximal in these undercut feature groups.

3.4.1.3 Implementation

In computer-aided design of injection molds, the product model can be input using the CAD part file in a wire-frame, surface, or 3D solid model. A 3D solid model can be directly used in CAD mold design, whereas wire-frame and surface models need to be converted into 3D solid models. If parts are designed without considering their moldability, part re-design for manufacturing is needed by taking into account of thermal expansion, drafting angle, structural, and moldablility analyses. Fig. 3.12 shows the flowchart for the determination of the optimal parting direction. Since the optimal parting direction is in the direction where the sum

Intelligent Mold Design and Assembly

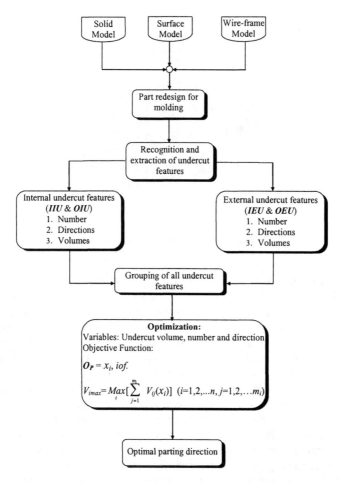

Fig. 3.12 Flowchart for the determination of optimal parting direction [15].

of undercut feature volumes in the same undercut feature group is maximal, the sum of the volumes of other undercut features not in the optimal parting direction will be minimal. After the optimal parting direction is determined, the undercut features with their directions in the optimal parting direction can be molded in by the core and cavity. The other undercut features which directions are not in the optimal parting direction will require the incorporation of side-cores, side-cavities, and so on.

3.4.1.4   Examples

Fig. 3.13 shows a thin-wall plastic molded part made up of various types of surfaces. All undercut features are recognized and extracted, and the optimal part-

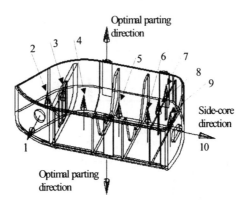

Fig. 3.13 Optimal parting direction: Case 1.

ing direction is generated based on the above methodology. According to the optimal parting direction shown in the figure, undercut features 2 to 9 can be molded in by the core, cavity, and their inserts due to their undercut feature directions appearing in the optimal parting direction. Undercut features 1 and 10 are not in the optimal parting direction, and thus two side-cores are needed for molding them.

Fig. 3.14 shows another plastic molded part. Since the undercut directions of undercut features 1, 2, 3, 4 are not in the parting direction, side-cores (for molding undercuts 1, 2, 4) and side-cavity (for molding undercut 3) are required to be incorporated in the mold structure for molding these undercut features. The other undercut features can be molded by the core. Another case is shown in Fig. 3.15. It is also a thin-wall plastic part made up of various types of surfaces. The optimal parting direction is chosen in the direction such that the sum of "undercut" feature volumes in the parting direction is maximal. Since the directions of undercut real features 1, 2, 3, 4 are not in the optimal parting direction, they can only be molded in by the side-cores. Due to different undercut directions, four different side-cores are needed for molding undercuts 1, 2, 3, 4. Undercut features 5, 6, 7, 8 can be molded in by the main core as their undercut feature directions are in the optimal parting direction.

3.4.2 Generation of Parting Lines

Parting lines are the intersection boundary of the surfaces of a part molded by the core and the cavity. The parting surfaces are the mating surfaces of the core and cavity in a mold. In CAIMDS, the parting lines and parting surfaces are generated after the parting direction is determined. For a given parting direction, the parting lines and parting surfaces can be determined based on the geometrical characteristics of a molded part.

Intelligent Mold Design and Assembly

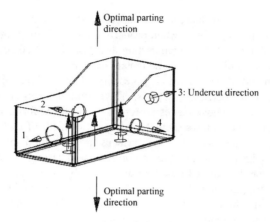

Fig. 3.14 Optimal parting direction: Case 2.

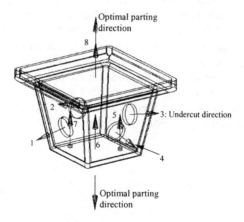

Fig. 3.15 Optimal parting direction: Case 3.

Tan et al. [16,17] classified all the parting surfaces into visible and invisible surfaces for a given parting direction based on the surface normal and the parting direction. If the surface normals contain positive vector components in the parting direction, the surfaces are visible. If the surface normals contain negative vector components in the parting direction, the surfaces are invisible. In the case where the surface normal of a surface is perpendicular to the parting direction, the surface is classified as visible if at least one of its adjacent surfaces is visible. The surface edges are classified into visible or invisible edges, if they are shared by two adjacent visible or invisible surfaces. When an edge is shared by a visible surface and an invisible surface, it is considered a tentative parting edge. A series of these tentative parting edges, when properly connected, may form the required parting lines with respect to a given parting direction. Fig. 3.16 shows a part with visible surfaces, invisible surfaces, and tentative parting edges.

Weinstein and Manoochehri [12] presented their methodology to determine the parting lines of molded and cast parts. The parting line sets are described in an in-order tree structure whose leaves represent the surfaces formed by half of the mold. The parting line follows the external edges of a set of surfaces in a given leaf. Based on the tree branches, the surfaces are classified into many surface groups, and the edges of each surface group represent parting lines. The optimum parting line can be determined based on the multi-objective function criteria considering the parting line complexity, draw depth, number of undercuts, number of side-cores, and machining complexity of a mold. The optimum parting lines minimize the objective function value with the above variables.

Considering the five parameters mentioned above, the parting lines are determined according to the external edges of the convex region of the molded part based on the multi-objective function criteria. It is a good exploratory work to investigate how the various parameters would affect the mold manufacturability and cost. This methodology to generate the parting lines has two drawbacks: on the one hand, it may underestimate the other possible cases since the parting lines are determined only based on the surfaces of the convex regions; on the other hand, it complicates the process in determining the parting lines since the multi-objective function criterion is over-elaborated. For a practical part, the parting lines should be at the place where there is a maximum projected area in the plane perpendicular to the parting direction.

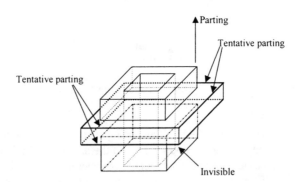

Fig. 3.16 Visible surfaces, invisible surfaces, and tentative parting edges.

3.4.2.1 Method to Auto-Generate Parting Lines

The parting lines in a molded part are the intersection boundary of the surfaces molded by the core and the surfaces molded by the cavity. In general, the types of parting lines to be dealt with are *flat, step,* and *complex partings*. If the parting lines are in the same plane, it is considered as a flat parting. If parting lines are not in the same plane, and surfaces generated by extruding the parting lines to the boundary of the core and cavity are planar surfaces, it is treated as a step parting. If the parting lines are not in the same plane and not all surfaces generated by extruding the parting lines are planar surfaces, this type is a complex parting. Since every surface which normal vector is not in the parting direction should have a certain draft angle, the core-molded and the cavity-molded surface groups will have the equal maximal projected area on the plane perpendicular to the parting direction. Based on these assumptions, it is found that parting lines are the common boundary of the two surface groups molded by the core and cavity, and they are the largest edge-loop in these two surface groups. Furthermore, the parting lines will have the maximum projected area onto the plane perpendicular to the parting direction. If the common boundary of the two surface groups can be found, it will be quite easy to obtain the parting lines. The two surface groups do not include the surface group molded by the side-cores and side-cavities, therefore, it is necessary to analyze this surface group in determining the parting lines.

(1)   Classification of Molding Surfaces

A 3D model of a molded part can be described in B-Rep in which the boundary is assumed to be partitioned into a finite number of bounded subsets called surfaces, where each surface is, in turn, represented by its boundary edges and each edge by its vertices. The solid model of a molding is therefore a closed body comprised of surfaces. Supposing that the parting direction is known and is represented by a pair of opposite vectors P_D and $-P_D$. Let L_i be the centre normal vector of the surface F_i (planar, curved or free-formed surfaces). It can be the axis directional vector of the cylindrical, conical, and revolved surfaces. All the molding surfaces as summarized in Table 3.2 can be classified into G_1, G_2, and G_3 surface groups based on their orientations with P_D. If P_D is the parting direction of the core, then most of the G_1 surfaces can be molded by the core and most of the G_3 surfaces by the cavity; some G_2 surfaces can be molded by the core and cavity or by the side-core or side-cavity.

Let the elements of G_1, G_2, and G_3 be denoted by G_1^i, G_2^i, and G_3^i and their edge-loops expressed as E_1^i, E_2^i and E_3^i, respectively. The surface re-classification is also listed in Table 3.2. In the table, m_1, m_2 and m_3 are the surface number of G_1, G_2 and G_3 respectively. For G_1 surfaces, if they are surrounded by G_3 surfaces and do not have any connectivity with other G_1 surfaces, they are re-classified into G_3. For G_3 surfaces, they are re-classified in the similar way. For G_2 surfaces, if they are totally surrounded by G_1 and do not have any connectivity with G_3 surfaces, they are re-classified into G_1. Similarly, if G_2 surfaces are totally surrounded

Table 3.2 Surface classification

Surface types	Classification criteria	Re-classification criteria
G_1	$L_i \bullet P_D > 0$	If $(E_1^i \cap E_1^j = 0) \wedge (E_1^i \cap E_3^k \neq 0)$ ($i=1, 2....m_1, j=1, 2... m_2$ and $k=1, 2,...m_3$; $i \neq j$), G_1^i is re-classified into G_3.
G_2	$L_i \bullet P_D = 0$	If $(E_2^i \cap E_1^j \neq 0) \wedge (E_2^i \cap E_3^k = 0)$, ($i=1, 2....m_2; j=1, 2... m_1$ and $k=1,2,...m_3$), G_2^i is re-classified into G_1; If $(E_2^i \cap E_1^j = 0) \wedge (E_2^i \cap E_3^k \neq 0)$, G_2^i is re-classified into G_3; If $(E_2^i \cap E_1^j \neq 0) \wedge (E_2^i \cap E_3^k \neq 0)$, G_2^i is re-classified into G'_2.
G_3	$L_i \bullet P_D < 0$	If $(E_3^k \cap E_3^j = 0) \wedge (E_3^k \cap E_1^i \neq 0)$, ($k=1, 2....m_3, j=1, 2... m_2$ and $i=1, 2,... m_1$; $k \neq j$), G_3^k is re-classified into G_1.

by G_3, they are re-classified into G_3. After such a re-classification, G_2 surfaces would usually appear in the undercut features and will not affect the parting line location. In G_2 surfaces, if they have connectivity with both G_1 and G_3 surfaces, they are re-classified into G'_2. G'_2 surfaces are the transitional surfaces of G_1 and G_3. They seldom occur because of the presence of the draft angle of the surfaces. If G'_2 surfaces exist, they are usually in the undercut features and do not affect the parting line location for the given P_D, except for those small surfaces without the draft angle.

In a molding, cylindrical, conical, and revolved surfaces are frequently used in blending a solid, creating bosses or hole features. The classification of these surfaces is different from the above method. If the edges of these surfaces are surrounded by G_1 surfaces, they belong to G_1 surface group. Similarly, if the edges of these surfaces are surrounded by G_3 surfaces, they belong to G_3 surface group. In solid blending, the blended surfaces are usually the cylindrical surfaces and have more than three bounded edges. If two of the edges are the common edges of G_1 or G_3 surfaces, they belong to G_1 or G_3 surface group. For spherical surfaces, they also abide by the above rules.

Fig. 3.17 shows the classification of all surfaces in a molding example based on the above methodology. Fig. 3.17(a) is a molding, (b) is a G_1 surface group since their center normal vectors meet the requirement of surface classification criteria for a given parting direction. Surfaces in (c) are classified into G_3 surface group. Surfaces in (d) are G_2 and G'_2 surfaces. Based on the surface re-

Intelligent Mold Design and Assembly

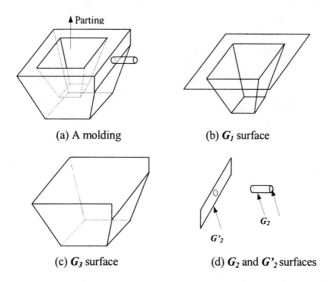

Fig. 3.17 Surface grouping in a molding.

classification criteria, some G_2 surfaces are re-classified into G'_2 surfaces. The cylindrical surface and its top adjacent surface belong to G_2 surfaces. The planar surface belongs to G'_2 surface since it is the transitional surface between G_1 and G_3 surfaces and meets the G_2 re-classification criteria. From the above classification, it can be found that the surface groups influencing the generation of parting lines are G_1, G_3, and G'_2 surface groups.

(2) Definition of Parting Lines

Assuming that P_D is the parting direction of the core, then G_1 surfaces are molded by the core and its inserts or by related local tools, and G_3 surfaces by the cavity and its inserts or by related local tools. Most G_2 surfaces form the undercut features and can be molded by the side-cores or side-cavities, etc. Only in special cases, some small G'_2 surfaces have no draft angles and they may affect the selection of parting lines. In most cases, the parting lines are the common boundary of the G_1 and G_3 surfaces. The parting lines P_L can be defined as:

$$P_L = (G_1 \cap G_3) \vee [(G_1 + G'_2) \cap G_3] \vee [G_1 \cap (G_3 + G'_2)] \qquad (3.2)$$

Since G'_2 group of surfaces may affect the determination of the parting lines, the definition of Eq.(3.2) thus includes G'_2 surface. In general, the parting lines are the common boundary of G_1 and G_3 surfaces.

(3) Generation of the Largest Edge-Loop

In G_1 and G_3 groups, surfaces are connected together by common edges. In each group, all surfaces are linked together to form the different sub-groups of surfaces. Among all the surface sub-groups, the largest sub-group is molded by the core or cavity and the other small sub-groups occurring in the undercut features are molded by local tools. In each surface group, the small sub-group surfaces do not affect the parting line location if the parting direction is known or fixed. Supposing that G_k^i and G_k^j are the two adjacent surfaces in the largest sub-group in G_1 or G_3 group and their external edge-loops are EL_k^i and EL_k^j, respectively, the following edge union operation is called the "edge-loop enlarging" process. The result is denoted by EL_k^{i-j}, and it can be named as the enlarged edge-loop.

$$EL_k^{i-j} = EL_k^i \cup EL_k^j \quad (\text{if } EL_k^i \cap EL_k^j \neq 0 \text{ and } k=1,3) \tag{3.3}$$

This process is illustrated in Fig. 3.18. In the implementation of above equation, all the common edges should be extracted first. To obtain the surface external edge-loops EL_k^i and EL_k^j (k is group number, $k=1,3$), the following criterion can be used to determine whether E_k^{i-j} is the common edge of the adjacent surfaces of G_k^i and G_k^j.

$$E_k^{i-j} = \{E_k^{i-j} \mid (E_k^{i-j} \subset EL_k^i) \wedge (E_k^{i-j} \subset EL_k^j)\} \tag{3.4}$$

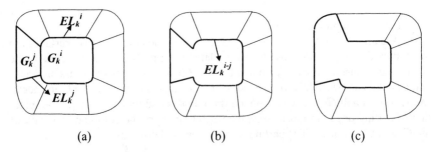

Fig. 3.18 Edge-loop enlarging process.

Intelligent Mold Design and Assembly

If the above criterion is met, E_k^{i-j} is considered as the common edge of the surfaces G_k^i and G_k^j. The search for common edges can be operated step by step. After all of the common edges are found, the rest of the edges will form the different edge-loops in which the edges are linked together in each group of surfaces. The largest edge-loop in G_1 and G_3 groups are denoted by EG_1 and EG_3, respectively. They can be the possible parting lines. The other smaller edge-loops in G_1 and G_3 groups usually appear in the undercut features.

(4) Criteria of Parting Lines

From the industry practice, parting lines should correspond to the edge-loop which has the maximal projected area in the plane perpendicular to the parting direction. Based on the present methodology, the largest edge-loop in the two surface groups should firstly have the maximal projected area in the plane perpendicular to P_D; secondly, the two largest edge-loops should be exactly the same. If the two largest edge-loops EG_1 and EG_3 in the two surface groups G_1 and G_3 are the parting lines, the following equations should be satisfied.

$$EG_1 = EG_3 \tag{3.5}$$
Constraint conditions:
$$Min\{|\ Project\ (Q, P_D) - Project\ (EG_1, P_D)\ |\}$$
$$Min\{|\ Project\ (Q, P_D) - Project\ (EG_3, P_D)\ |\}$$

where Q is a 3D molding. Eq.(3.5) can be regarded as the criteria of parting lines. The first equation implies that if the edge-loops EG_1 and EG_3 are the parting lines, they should be exactly the same. Since $Project\ (Q, P_D)$ and $Project\ (EG_1, P_D)$ refer to the projected areas of 3D molding and EG_1 projected onto the plane perpendicular to the parting direction P_D, respectively. The second equation means that the difference of the projected areas between Q and EG_1 should be minimal. Similarly, the third equation has the same meaning.

3.4.2.2   Implementation

Fig. 3.19 is the flowchart for generating parting lines. Based on the methodology, the parting lines can be determined from a given 3D molding and the optimal parting direction. The procedures to generate the parting lines of the molding are summarized as follows:

(1) Classify all the molding surfaces into G_1, G_2 and G_3 surface groups;
(2) Determine the principal surface in each surface group;
(3) Identify the common edges and the largest edge-loop in each surface group and then determine the largest edge-loops EG_1 and EG_3 in G_1 and G_3 surface groups, respectively;

(4) Calculate the projected areas of the molding, EG_1 and EG_3 in the plane perpendicular to the optimal parting direction; and
(5) Determine whether the parting line criteria are satisfied or not. If the largest edge-loop in each surface group meets the requirement of the parting line criteria, they can become the parting lines; otherwise, it is necessary to re-design the parting lines manually. In this case, the largest edge-loops in the two surface groups do not appear in the maximal contour of the molding or do not include the actual parting lines in the molding, and the methodology cannot determine the parting lines automatically. The limitations of the methodology will be discussed in later.

In this method, the parting surfaces are generated by extruding all the parting line edges (parting line edges referring to the edges in the largest edge-loop in the two surface groups) to the boundary of the core and cavity blocks to form the extruded surfaces. The output information of the parting lines includes the following items:

  b. Parting line edge ID, parting line edge number, and its sequence number in the edge set of the molding,
  c. Vertices of the parting line edge and their coordinates, and
  d. Topological relationships of the parting line edges with their adjacent edges and surfaces.

3.4.2.3   Examples

Fig. 3.20 and Fig. 3.21 are all thin-wall plastic parts. The parts are made up of various types of surfaces, and their respective optimal parting directions are shown. For the given parting direction, the parting lines can be generated based on the above methodology and algorithms. The parting lines are generated and shown in the figure. From the figure, it can be found that the parting lines are the intersection boundary of the two surface groups, namely G_1 and G_3 molded by the core and cavity, respectively. From Fig. 3.20(a), it can also be found that the parting lines are the largest edge-loop with the maximal projected area on the plane perpendicular to the parting direction P_D and $-P_D$. The projected area of the largest edge-loop on the plane perpendicular to P_D is the same as the projected area of the molding.

Fig. 3.21 shows a plastic molded part with a more complex parting. Based on the given parting direction, the parting lines are generated and shown in the figure. It can also be found that the parting lines have the largest contour. In the above two cases, the parting lines generated are consistent with the practical parting design in industry. The results show that the proposed methodology to automatically generate the parting lines is satisfactory.

Intelligent Mold Design and Assembly

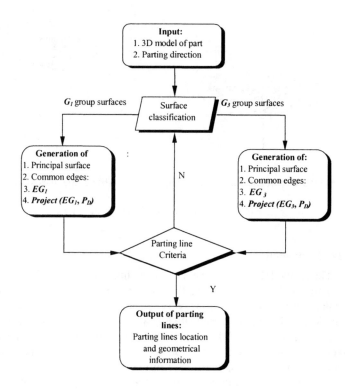

Fig. 3.19 Algorithm for the generation of parting lines [15].

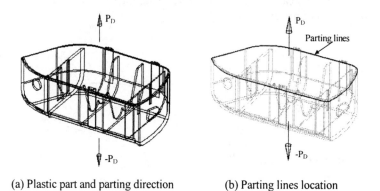

(a) Plastic part and parting direction   (b) Parting lines location

Fig. 3.20 Generation of parting lines: Case 1.

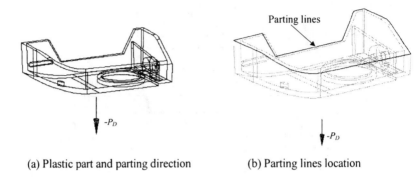

(a) Plastic part and parting direction   (b) Parting lines location

Fig. 3.21 Generation of parting lines: Case 2.

3.4.2.4 Limitations

The methodology to determine parting lines is based on the geometrical features, namely the edge-loops in the molding. If there is no edge-loop in the particular parting location, it is very difficult to determine the parting lines automatically. In this case, manual generation of parting lines is needed. This is not within the purview of the discussed methodology. It is the intrinsic geometric characteristics that make it impossible to generate the parting lines automatically. The above methodology is ineffective in the following two cases. (*i*) EG_1 or EG_3 does not appear in the maximal contour of the molding, namely, the actual parting lines location. This case is shown in Fig. 3.22(a). (*ii*) The largest edge-loop does not include the total geometric features where the practical parting lines are located, as shown in Fig. 3.22(b). In these cases, EG_1 and EG_3 do not include the practical parting lines. One possible solution is to create the edges manually in the location of the practical parting lines before using the developed programs or to create practical parting lines manually.

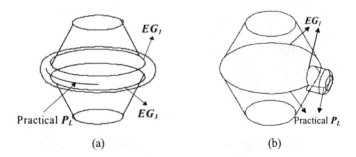

Fig. 3.22 Limitations of the methodology to generate the parting lines.

Intelligent Mold Design and Assembly

3.4.3 Determination of Parting Surfaces

In the previous sections, the algorithms to determine the parting lines have been proposed. In this section, the methodology to generate the parting surfaces for the given parting direction and parting lines are described. Using these parting surfaces as the splitting surfaces will split the containing box into two mold halves, namely the core and cavity.

The parting surfaces are the mating surfaces of the core and cavity. The parting surfaces are generated by extruding the parting lines outwards to the outside boundary of the core and cavity. Two methods [18] can be used to generate the parting surfaces:

(1) *Sweep method* : The parting surfaces are generated by sweeping the parting lines along their extruding direction outwards to the outside boundary of the core and cavity.

(2) *Radiate method* : The parting surfaces can also be created by radiating the parting lines with a specified radiate distance that is large enough to extend the surface beyond the outside faces of the containing box.

The parting surfaces can be used as the splitting surfaces to cut the containing box of a mold box into two halves.

3.4.3.1 Sweep Method for Complex Parts

Every parting edge has its own parting surface generated by sweeping it along its corresponding extruding direction. The swept surfaces can be denoted by P_l^i ($i = 1,2,...,n_e$). Besides the swept surfaces, there are still four corner surfaces P_v^j ($j = 1,2,3,4$) that need to be generated. After all the swept surfaces are generated, the final parting surfaces P_S can be defined as:

$$P_S = \bigcup_i^{n_e} P_l^i + \bigcup_j^4 P_v^j \qquad (3.6)$$

For the given parting direction and parting lines, the procedure for generating parting surfaces can be summarized as follows:

- Compute the sweep distance to confirm parting surfaces cut through the containing box and the sweep distance will be long enough;
- Divide the parting lines into four groups;
- Determine the extruding direction for each edge group (Fig. 3.23a);
- Sweep each edge-group along the corresponding extruding direction (Fig 3.23b);

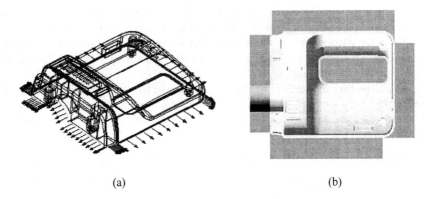

(a)   (b)

Fig. 3.23 Determination of extruding direction.

From the parting lines to the boundary of the core or cavity block, the extruding direction is the path that the parting lines will follow when sweeping. It is perpendicular to the parting direction but parallel to the surface normal of the side face of the mold box.

- Sweep the vertices at the four corners along the two extruding directions. The $P_v^{\,j}$ surfaces are shown and highlighted in Fig. 3.24; and
- All the swept surfaces are subtracted with the product. Fig. 3.25 shows the process to generate the parting surfaces.

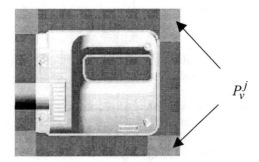

Fig. 3.24 $P_v^{\,j}$ surfaces.

Intelligent Mold Design and Assembly

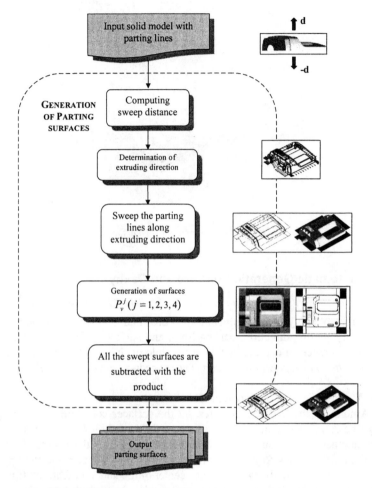

Fig. 3.25 Sweep algorithms to generate the parting surfaces.

3.4.3.2 Radiate Method

For some simple parts, the parting surfaces can also be created by radiating the parting lines with a specified radiate distance that is large enough to extend the surface beyond the outside faces of the containing box. The radiate surfaces are shown in Fig. 3.26. The radiate surfaces can be used as the parting surfaces to cut a mold into two halves. The procedures are:
- Input a solid model with parting lines;
- Compute the radiate distance;
- Determine the radiate direction;
- Radiate the parting lines along the radiate direction; and
- Subtract all the swept surfaces from the product.

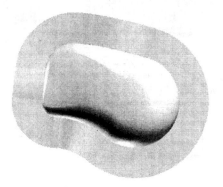

Fig. 3.26 Radiated parting surfaces.

3.4.4 Automatic Generation of Core and Cavity

3.4.4.1 Generation of Cores and Cavities by Boolean Operations

A methodology [15] to generate the cores and cavities is described below based on the parting direction, parting lines, and surfaces. As the operation is a Boolean difference operation in this methodology, it is called the *Regularized Boolean Difference Operation Method*. The procedures are outlined below.

(1) Identification of Undercut Hole Features

Among all the undercut features, let the surfaces constituting the undercut features be called undercut feature surfaces and denoted by U_j^i (i refers to the sequence number of undercut feature surfaces, j refers to the jth undercut feature in the molding). Two rays $\boldsymbol{Ray}_U$ are fired in the undercut direction and the opposite undercut direction from the target surface center of the undercut feature. If there is no intersection point of $\boldsymbol{Ray}_U$ with the surfaces U_j^i, the undercut features are defined as the hole undercut feature. This can be expressed as:

$$U_j^i \cap \boldsymbol{Ray}_U = 0 \tag{3.7}$$

For the undercut hole features, they must be patched for generation of cores and cavities. Otherwise, the core and cavity cannot be separated after the Boolean operation. Patching the undercut hole features is another issue in the generation of cores and cavities. One simple way is to generate a characteristic undercut solid based on the undercut feature geometric entities (surfaces, edges, and vertices). The undercut solid has exactly the same volume as the undercut hole feature. The undercut hole feature will be patched after uniting the characteristic undercut solid with the molding. The method proposed is shown in Fig. 3.27.

Intelligent Mold Design and Assembly

Fig. 3.27(a) shows a simple molded part. There is an undercut hole feature in the side surface; (b) is the characteristic undercut solid generated based on undercut geometric entities; and (c) is the part after patching the undercut hole feature. The undercut solid is actually the "head" of the side-core. The generation of the characteristic undercut solid based on undercut geometric entities is very important in the creation of local tools. After patching all the undercut hole features, the part can be used to generate the cores and cavities.

(2) Generation of Cores and Cavities
The cores and cavities can be generated by the following steps:

Step 1: Generation of the containing box which encloses the molded part with suitable dimensions;
Step 2: Subtracting the containing box with the patched part by the Regularized Boolean Difference Operation method; and
Step 3: Splitting the containing box using the parting surfaces generated.

In Step 1, the maximum dimensions of the part in x, y, z directions must be known. The size of the box can be determined based on the strength of the mold and the process parameters of the molding. In Step 2, the Regularized Boolean Difference Operation is carried out between the containing box and the patched part. After the operation, the box has an empty space inside. In Step 3, the parting surface generated previously is used as the splitting surface to split the box into two mold halves. One is the core block, the other is the cavity block. The procedures are illustrated in Fig. 3.28.

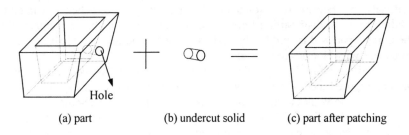

(a) part   (b) undercut solid   (c) part after patching

Fig. 3.27 Patching undercut hole feature.

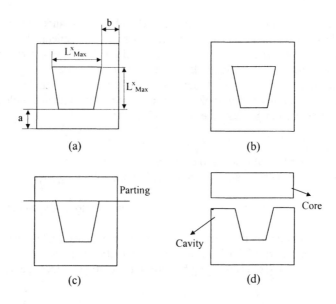

Fig. 3.28 Procedures to generate the core and cavity.

3.4.4.2 Examples

Fig. 3.29 shows a plastic molded part with four hole undercuts in the side-surfaces. The type of parting lines belongs to flat parting. First of all, the information of undercut geometrical entities is extracted for the generation of the "heads" of the local tools, and all the undercut hole features in the molded part are patched. Next, the containing box is generated. The dimensions of the containing box are determined based on the experience, which considers the strength of mold materials and the molding parameters. The parting surface is then generated according to the known parting lines. Finally, the containing box is subtracted by the molded part and is split by the parting surface. The two mold halves, *viz.*, the core and cavity blocks, are generated.

In Fig. 3.29(a), the highlighted surface is the parting surface generated based on the proposed methodology. The core block is also shown in Fig. 3.29(b). The four circles shown in the figure are the location of the four side-cores. The cavity block is shown in Fig. 3.29(c). Fig. 3.30(a) shows another plastic part. The parting surface is a complex parting. The procedures to generate the parting surface, core, and cavity are the same as in Case 1. The parting surfaces are the highlighted surfaces shown in the figure. After the hole undercuts are patched, the containing box is split into two mold halves and becomes the core and cavity blocks as shown in Fig. 3.30(b) and (c).

Intelligent Mold Design and Assembly

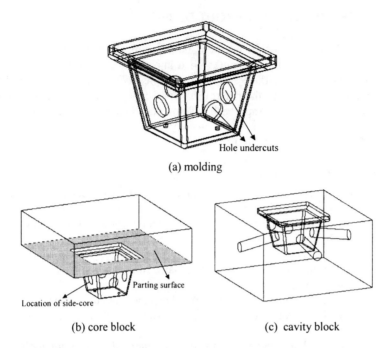

Fig. 3.29 Generation of parting surface, core and cavity: Case 1.

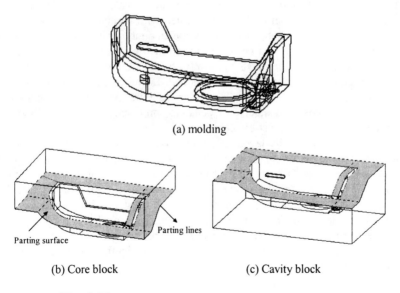

Fig. 3.30 Generation of parting surfaces, core and cavity: Case 2.

3.5 AUTOMATIC CAVITY LAYOUT DESIGN

At the initial stage of the injection mold design, a decision must be made as to the cavity layout. Thus a computer-aided injection mold design system should have functionalities to facilitate arranging multiple cavities in a single mold base. This section first discusses the determination of cavity number based on various criteria, and then presents an approach to automatically arrange the cavity layout.

3.5.1 Cavity Layout and Number

3.5.1.1 Layout Requirements

In modern injection molding machines, the barrel is usually positioned in the central axis of the stationary platen. This establishes the position of the sprue. The cavities have to be arranged relative to the central sprue in such a way that the following conditions are satisfied:

(1) All cavities should be filled at the same time with molten plastic of the same temperature;
(2) The flow length should be short to keep the scrap to a minimum; and
(3) The final layout should not result in extra cost of manufacturing.

The first condition makes sure the runner system of an injection mold is balanced. Runner balancing means that the distance the plastic material travels from the sprue to the gate should be the same for each plastic product. The balanced runner system ensures that each cavity is filled with molten material at the same time, and all cavities are subjected to the same amount of injection pressure so that identical plastic products can be produced for each cavity. The second condition requires that the runner length should always be kept minimum to reduce pressure losses and amount of runner volume scrapped. Lower pressure losses enable the production of plastic products on a smaller machine. As larger machines incur higher overhead cost and other cost penalties, producing products with smaller machines is always more cost-effective.

To arrange the cavity layout of an injection mold, mold designers must first determine the cavity number, then select the layout pattern and arrange the orientation of cavities. Cavity number depends primarily on three factors: (a) the quantity of products, (b) the production cost, and (c) machine parameters. The layout pattern substantially influences the runner system of an injection mold. Thus, when selecting layout patterns, mold designers must keep in mind that the runner system of the selected patterns should be balanced. The orientation of individual cavity in a mold base is mainly determined by the gate location and the shape of a product.

3.5.1.2 Layout Patterns

Three popular types of layout patterns are circular layout, series layout, and symmetrical layout. Table 3.3 summarizes the advantages and disadvantages of various layout patterns. The circular layout results in extra cost of manufacturing, thus this kind of layout would not be used unless otherwise specified by designers. The most favorable layout pattern is the symmetrical layout, because this layout has equal lengths and sizes for all flow channels. Cavity numbers for symmetrical layout are restricted to a 2^n sequence (i.e., 2, 4, 8, 16, 32, 64, 128, etc.).

Table 3.3 Various layout patterns

Layout patterns	Advantages	Disadvantages
Circular	(1) Equal flow lengths to all cavities (2) Easy demolding of parts that require unscrewing	Extra cost of manufacturing
Series	(1) Space for more cavities than with circular layout (2) Easy to manufacture	Unequal flow lengths to individual cavities
Symmetric	(1) Equal flow lengths to all cavities without gate correction (2) Easy to manufacture	Larger runner volume; more scrap than circular and series layout

3.5.1.3 Determination of Cavity Number

The use of a multi-cavity mold reduces the direct labor cost per molding and increases the output, but using a greater number of cavities in a larger mold may increase the size of injection machine necessary and hence involves higher press-depreciation charges in the cost. Further, the larger mold will obviously be more costly, although not necessarily in proportion. To determine the cavity number, technical and economical criteria must be considered comprehensively.

(1) Cavity Number Based on Delivery Date

The cavity number must ensure that the order can be fulfilled within the available time span. Based on the time for plastic molding, the minimum number of cavity for meeting the delivery date can be determined using the following equation [19]:

$$n_{deliv} = \frac{K_r \times t_{cycle} \times L}{3600 \times t_m} \qquad (3.8)$$

where

K_r is the reject factor (K_r=1.05),
t_{cycle} is the cycle time of every product (second),
L is the lot size of the product, and
t_m is the time for molding production (working hours).

(2) Cavity Number Based on Cost

Through the use of empirical relationships for molding machine, mold base, and cavity manufacturing costs, it is possible to predict the economically optimum number of cavities. Based on the regression analysis of experimental data and the assumption that the cost of mold base is a constant independent of the number of cavities machined, Dewhurst and Kuppurajam [20] proposed a formula to calculate the economically optimum number of cavities as follow:

$$n_c = \left(\frac{L(k_1 + C_s) \times t}{1.1 \times C_1} \right)^{0.6} \qquad (3.9)$$

where

L is lot size,
C_1 is the cost of one unique cavity ($),
C_s is the cost of machine supervision ($/s), and
t is the cycle time (s),

and

$$k_1 = \frac{j_1}{PK \times SH \times 5 \times 10^6} \qquad (3.10)$$

where

j_1 is a constant from regression analysis (j_1=15,890),
PK is the amortization period (years),
SH is the number of working shifts, and
(5×10^6) represents the number of available seconds in one shift-year assuming a plant efficiency of 70%.

(3) Cavity Number Based on Machine Parameters

Considering the limits of injection molding machines, the cavity number depends primarily on the following parameters of the injection molding machines:
- Clamping force, which has to compensate the reactive force from the maximum internal cavity pressure
- Shot size, the amount of melt that can be conveyed into the mold with one stroke of the screw or the plunger
- Plasticizing rate, the amount of plasticized material the machine can provide per unit time.

Based on the clamping force, the maximum cavity number can be calculated as follows [21]:

$$n_{t1} = \frac{F_c}{K_s \times A_R \times P_{inj}} \qquad (3.11)$$

where

K_s is the safety factor against flashing (1.2~1.5),
F_c is the maximum clamping force (N),
A_R is the projected area of a part and its runner system (mm²), and
P_{inj} is the cavity pressure (MPa).

Based on the shot capacity, the minimum cavity number n_{t2} and the maximum cavity number n_{t3} can be calculated using Eq. (3.12) as follows:

$$n_{t2} = 0.2 \times \frac{V_s}{V_p} \ ; \ n_{t3} = 0.8 \times \frac{V_s}{V_p} \qquad (3.12)$$

where

V_s is the shot capacity of injection unit (cm³), and
V_p is the volume of part and runners (cm³).

The minimum cavity number ensures that the residence time of melt in injection unit is not excessively long, while the maximum cavity number ensures that the melt is thermally and mechanically homogeneous.

Based on the plasticizing rate, the maximum cavity number n_{t4} is:

$$n_{t4} = \frac{t_{cycle} \times R_p \times 1000}{V_p \times \rho_m \times 3600} \tag{3.13}$$

where
R_p is the plasticizing rate (kg/h), and
ρ_m is the specific weight of material (g/cm³).

Thus based on machine parameters, the minimum cavity number n_{tmin} and maximum cavity number n_{tmax} are:

$$n_{t\min} = \begin{cases} n_{t2} & \text{if } n_{t1}, n_{t4} > n_{t2} \\ 0 \end{cases} \tag{3.14}$$

$$n_{t\max} = \min(n_{t1}, n_{t3}, n_4) \tag{3.15}$$

(4) Automatic Determination of Cavity Number

In order to find the most favorable number of cavities with a justifiable effort, the following procedure is proposed. Fig. 3.31 shows the flow chart of the procedure, with which one can determine the technically and economically best combination of the injection mold and machine for the products to be molded. At first, the program calculates the n_{deliv} and n_{cost} according to the delivery date and production cost. Subsequently a selection of molding machine is made. Based on the selected machine, n_{tmin} and n_{tmax} are computed according to the equations (3.14) and (3.15). If n_{tmin} is less than n_{deliv}, and n_{tmax} is greater than n_{deliv}, the selected machine meets the production requirements. Thus a range of cavity numbers can be determined, which meets the demands on delivery date and production cost on one side and can be technically realized with the selected machine on the other side. Finally, based on this range of cavity numbers, the designers can either specify the preferred numbers such as the numbers of symmetrical layout, or reselect the other machines.

Intelligent Mold Design and Assembly

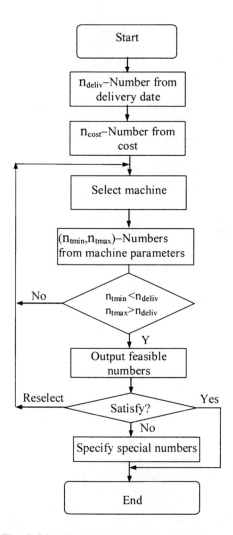

Fig. 3.31 Flow chart to determine cavity numbers.

3.5.2 Automatic Layout of Multi-Cavity

3.5.2.1 *Feature Box of Plastic Product*

To arrange cavity layout, three important factors (product shape, cavity number, and gate location) have to be considered comprehensively. For the sake of convenience, a feature box of the product is proposed, in which faces represent the geometric features and the gate information. Fig. 3.32 shows the feature box and attributes of its faces. The top face of the feature box has three attributes, namely gate type, gate location, and normal direction. Each side face of the feature box has four attributes, namely *gate type, gate location, normal direction*, and *feature information*. The attribute of feature information records all the face tags of the undercut features. If the pinpoint gate is used, the gate types in all of the side faces are set to empty; otherwise the side faces will record the gate type and position. For example, when a fan gate is located on the left side of a product, the gate information of the side face 1 is set to: *Gate_type=Fan_gate* & *Gate_pos=Matrix(9)*, where *Matrix(9)* represents the gate location and orientation. The procedure to construct a feature box is as follows:

(1) Find the bounding box of the product;
(2) Recognize undercut features of the product;

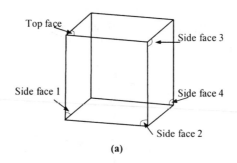

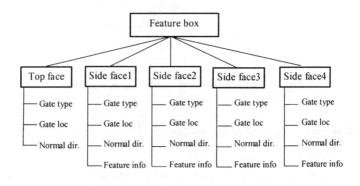

Fig. 3.32 Feature box and its face attributes.

Intelligent Mold Design and Assembly

(3) Check whether the undercut features can be freely withdrawn from any normal direction of the four side faces of the bounding box; and
(4) Save the gate and feature information to all five faces of the feature box.

To arrange the orientation of individual cavity in a mold base, the undercut feature must be taken into consideration. Detailed discussions about recognition of undercut features are presented in Section 3.6.

3.5.2.2 Design Rules for Cavity Layout

One important issue of cavity layout is arranging the orientation of an individual cavity. The design regulations for the orientation of cavities can be summarized by the following rules:

Rule 1: If a gate locates on edge 'a' in Fig. 3.33 and the gate type is "*side*" or "*submarine*", then the side face containing edge *a* must be the *entry face*.

Rule 2: If there exists an external undercut along side face 1, then side face 1 cannot be the entry face.

Rule 3: If more than two orientations are possible, then choice is the orientation with the smaller/smallest mold base.

Entry face is the one connected to the gate and runner. Three issues of cavity layout are: determining the cavity number, selecting the layout pattern, and arranging the orientation of individual cavity. Thus the procedure for automatic layout involves resolving these three issues in sequence. Fig. 3.34 shows a flow chart of the procedure for automatic layout. First, the program determines the cavity number based on the various criteria, and builds a feature box of the product. Subsequently, the layout pattern is chosen according to the cavity number. For an odd

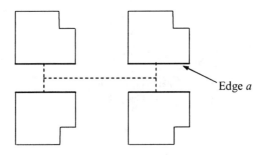

Fig. 3.33 A four-cavity layout.

number of cavities, circular layout is selected. For even numbers that are power series of 2, such as 2, 4, 8, etc., symmetrical layout is selected; while for other even numbers, the series layout is used. Then the procedure detects all of the side faces that meet the orientation rules. If a pin gate is used, only Rule 2 is set as the orientation rule. For side, submarine and fan gates, both Rule 1 and Rule 2 are set as the orientation rules. Finally the procedure finds the layout with the minimum size.

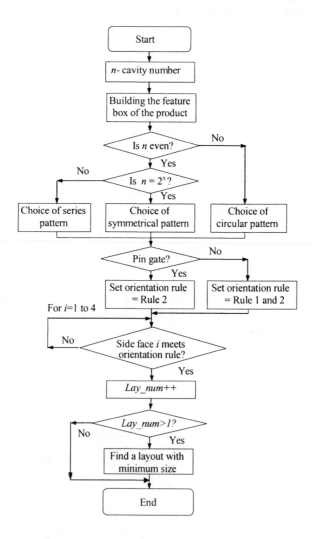

Fig. 3.34 Flow chart of automatic layout of cavities.

Intelligent Mold Design and Assembly

3.6 RECOGNITION AND EXTRACTION OF UNDERCUT FEATURES

In injection molds, the presence of undercut features would affect molding cost and the entire structure of a mold since the parting directions, parting lines and surfaces, the number of local tools and their actuating mechanisms, the core and cavity and their inserts depend very much on them. Recess or protrusion regions on a plastic product that prevent its removal from a mold along the ejection direction are called the *undercuts*. The existence of undercuts substantially influences the design and manufacturing of the mold, so the undercut region is a type of feature in the context of mold design and manufacturing. Automatic recognition of these undercut features not only aids the cost analysis of an injection mold, but also provides useful information for the layout of multiple cavities. In this section we will describe a hybrid method [4], which combines the strength of graph-based methods and hint-based methods, to recognize undercut features from plastic products.

3.6.1 Definitions and Classifications of Undercut Features

A depression or protrusion region in a plastic product that blocks the removal of the plastic product from its injection mold is referred to as an *undercut feature*. The undercut features can be further classified as *depression undercut features* or *protrusion undercut features*. The depression undercut feature is a region that consists of concave face sets and is bounded by a convex edge loop, while the protrusion undercut feature is a region that consists of convex face sets and is bounded by a concave edge loop. To form the undercut features, side-cores are required, which incur a group of auxiliary parts to fulfill the side movement. It is obvious that for every undercut feature in a plastic product, there exists a corresponding region in the injection mold, which is also referred to as undercut feature on the injection mold. Fig. 3.35 shows the undercut features in a plastic product and its injection molds. Corresponding to the depression undercut feature in the plastic product, there exists a protrusion undercut feature in the injection mold.

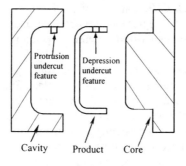

Fig. 3.35 Undercut feature.

Fu [22,23] and Nee et al. [13] classified the undercut features into two types, namely *external* and *internal undercuts*. The external undercuts were further divided into outside external undercuts and inside external undercuts, while internal undercuts were divided into outside internal undercuts, and inside internal undercuts. A group of recognition criteria for undercut features were proposed. After all the potential undercuts were extracted, the optimal ejection direction was chosen by considering the number of possible undercuts and their corresponding undercut volumes comprehensively.

Rosen et al. [25] classified undercut features into external undercut features, accessible laterally with respect to the mold closure direction, and internal undercut features, accessible only through the mold core. For a given parting direction, the first step is to ignore all the surfaces with a positive or zero dot product between the surface's normal vector and the parting direction, and all other surfaces are then projected onto the plane perpendicular to the parting direction. If the projected surfaces overlap, then the surfaces form an undercut feature. This method can be used to design side-cores and side-cavities.

3.6.2 Undercut Features Recognition

Automatic recognition and extraction of undercut features is a difficult task in CAIMDS. There are relatively few published works in this area. Chen et al. [7,8] and Mochizuki and Yuhara [8] presented the Boolean Regularized Difference Method (BRDM) to recognize and extract the potential undercut features in a molding by obtaining the regularized difference between the part to be molded and its convex hull. In Mochizuki's work, the given molding Q is a polyhedron. The convex polyhedron is generated by filling something such as clay into the concavities of Q to make it the smallest convex polyhedron, which can just fully contain the original polyhedron. After subtracting the convex polyhedron from the given polyhedron, the remaining parts are the potential undercut features, which are called the *sealed pockets* in Chen's work [7]. From the viewpoint of solid modeling, the above process is quite simple. It is the process of generating the convex hull $CH(Q)$ for a given 3D polyhedron Q and seeking the regularized difference between the convex hull $CH(Q)$ and its original 3D polyhedron Q. The regularized difference $CH(Q)-*Q$ represents the potential undercut features. Since the final step is the Boolean regularized difference operation, it is therefore called the BRDM.

From the statements above, it can be found that using the BRDM to extract the potential undercut features consists of two steps. Since the difference operation is carried out between $CH(Q)$ and Q, the convex hull $CH(Q)$ must be represented by the solid model with the same data structure as the original polyhedron Q. A suitable algorithm for determining the convex hull from a given polyhedron is needed. The second process is to perform the Boolean regularized difference between the convex hull and the polyhedron. As most commercial solid modelers use the Boolean operation for geometrical modeling, the Boolean difference op-

Intelligent Mold Design and Assembly

eration between the convex hull $CH(Q)$ and polyhedron Q can be carried out easily. Fig. 3.36 is an example of using the BRDM to extract undercut features in a molding. Fig. 3.36(a) is the 3D molding Q, (b) is the 3D convex hull $CH(Q)$, and (c) shows the undercut features generated by taking the Boolean regularized difference operation between the convex hull $CH(Q)$ and the 3D molding Q. Extracting the potential undercut features and undercut surfaces from $CH(Q)-*Q$ is also a non-trivial problem. Since $CH(Q)-*Q$ represents a 3D solid and all the undercut features in $CH(Q)-*Q$ are linked together as shown in Fig. 3.36(c), its decomposition into many single undercut features is equivalent to recognizing and extracting the features from a 3D solid object.

Besides the above methodologies, Ganter and Skoglund [26] also reported an approach in recognizing and extracting three classes of concave undercut features, *viz.*, internal voids, single and multi-surface holes, and boundary perturbations from a B-rep model for the development of casting core in casting mold design. The features recognized by the approach are limited to concave features. The core can be used to mold both convex and concave features in metal casting and plastic molding. The recognition of convex features is not mentioned in their work. The concave and convex portions of the molding can also be possible undercut features, and hence the method is not suitable for recognizing and extracting all the undercut features in a molding.

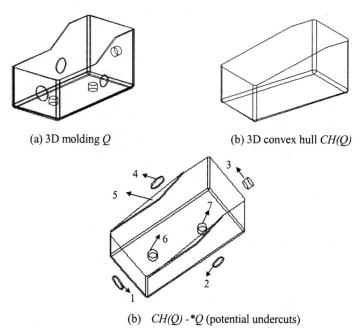

(a) 3D molding Q (b) 3D convex hull $CH(Q)$

(b) $CH(Q)-*Q$ (potential undercuts)

Fig. 3.36 Generation of convex hull and potential undercut features.

In-depth review of feature recognition shows that it is a challenge for all the existing feature recognition techniques to recognize undercut features from a plastic product with sculptured surfaces. Since the recognition and extraction of undercut features in a molding is a relatively new area, the methodologies outlined above have many limitations and cannot recognize all types of undercut features.

3.6.3 Draw Range and Direction of Undercut Features

When the core, cavity, and their inserts cannot mold the undercut features, local tools are needed. In the design of local tools, it is necessary to determine their withdrawal directions. In this section, the Gaussian map (G-map) and V-map are introduced [7,27,28]. The V-map and the draw range of undercut feature are defined. The V-map is described by the G-map and the draw range of undercut feature is determined based on the V-map of undercut features.

3.6.3.1 G-map and V-map of Surface

The G-map is a representation of the surface normal. Gauss introduced the concept of mapping the surface normal onto the surface of a unit sphere to define the local curvature of a given point [27]. There are many potential applications of G-map in the fields such as NC machining [28], geometric modeling, computer vision, etc., and applications in sheet-metal forming, and die and mold design [7]. To generate a G-map, the surface normal of any point on a given surface F_i is first transferred to the unit sphere such that the direction is the same as the original normal vector. The transferred vector passes through the centre of the unit sphere and the intersection point of the transferred normal vector with the surface of the sphere. When all the intersection points on the unit sphere are produced, all these intersection points form the G-map of surface F_i. The V-map of a surface is formed by the points on the unit sphere from which the surface is completely visible from infinity. Since every point in the V-map differs from its corresponding point in the G-map by at most 90 degrees, therefore, the V-map of a surface can be constructed by computing the intersection of hemispheres, each having its pole as a point on the G-map [29]. Fig. 3.37(a) shows how to generate the G-map and V-map for a given surface. For surface A, its surface normal is L_i. The first step is to transfer L_i in Fig. 3.37(a) to the unit sphere in Fig. 3.37(a1) and then determine the intersection point A_0. The intersection point A_0 in Fig. 3.37(a1) is the G-map of surface A. Since every point in the G-map has a hemisphere V-map, the V-map of a surface is the intersection of hemispheres with its pole as a point on the G-map. The V-map of surface A is the hemisphere in as shown in Fig. 3.37(a2). Fig. 3.37 also shows the G-maps and V-maps of surfaces B, C and D. Only the V-map of surface D is empty.

The G-map and V-map have been used to determine the optimal parting direction in Chen, Chou, and Woo's pioneering work [7]. The G-map and V-map

Intelligent Mold Design and Assembly

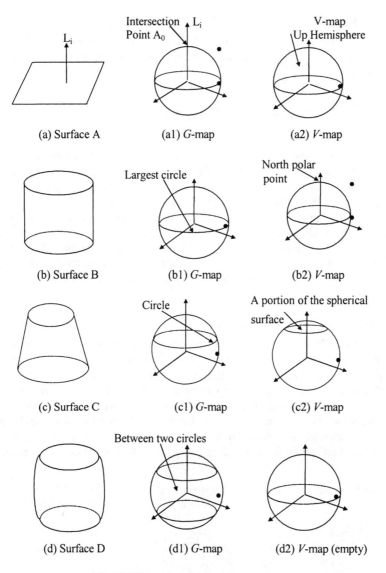

Fig. 3.37 *G*-map and *V*-map of surfaces.

can be used to determine the undercut directions in which side-cores or side-cavities or other local tools are withdrawn. In order to determine the undercut direction, the *V*-map and the draw range of undercut features are introduced in the next section.

3.6.3.2 V-map and Draw Range of Undercut Features

The V-map of undercut features V-map(U) is the intersection of all V-maps of the surfaces of which the undercut features are composed. V-map(U) can be defined as

$$V\text{-}map(U) = \bigcap_{i}^{n} V\text{-}map(F_i) \qquad (3.16)$$

where n is the number of surfaces belonging to the undercut feature. For the undercut feature shown in Fig. 3.38(a), the corresponding V-map(U) is illustrated in Fig. 3.38(b). In the figure, F_{j1} and F_{j2} are one group of the first adjacent surfaces, and H_{j1} and H_{j2} are another group of the first adjacent surfaces.

The draw range of undercut features refers to the range in which the local tools can be withdrawn from the molding. In V-map(U), the boundary of V-map constitutes the draw range of the undercut features. In practice, not all withdrawal directions in the draw range is a preferred withdrawal direction of local tools. A practical withdrawal direction should be the direction in which the local tools can be effectively withdrawn.

3.6.3.3 Direction of Undercut Features

The direction of the undercut feature can be defined as the direction in which the local tool can be withdrawn most effectively from the undercut features to avoid warpage of the molding. Based on the V-maps of undercut features, it is found that any point in the V-map can be a feasible withdrawal direction of the local tool. For any given undercut feature, there exists a most preferred withdrawal direction which is different from case to case. The preferred withdrawal direction will be the undercut direction. For external or internal undercut features, the undercut direction can be determined uniquely. For the outside undercut features,

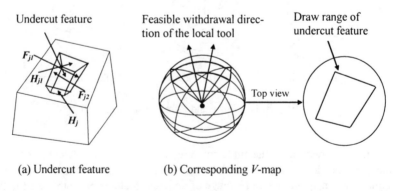

(a) Undercut feature (b) Corresponding V-map

Fig. 3.38 V-map of undercut features.

Intelligent Mold Design and Assembly

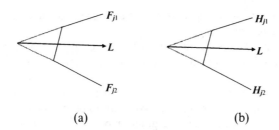

Fig. 3.39 Undercut direction.

there is more than one preferred withdrawal direction and thus more than one undercut direction. For the case shown in Fig. 3.38, the preferred direction L is in the bisection line of the angles between the two groups of the first adjacent surfaces F_{j1} and F_{j2}, and H_{j1} and H_{j2} as shown in Fig. 3.39.

3.6.4 Graph Representation of Solid Models

A boundary representation of a solid object can be regarded as a graph structure, where, for instance, faces are considered nodes of the graph and face-face relationships form the arcs of the graph. The advantage of the graph view is that the well-established techniques of graph algorithms can be readily adapted for feature recognition. On the other hand, in addition to the low-level information about faces, edges, and vertices, the graph representation of a solid object can explicitly record information regarding types of face adjacencies to assist in feature recognition.

3.6.4.1 Terminology of Graph Theory

Graph A *graph* $G=(N, A)$ consists of a set of entities $N=\{n_1, n_2, ...\}$ called *nodes*, and another set $A=\{a_1, a_2, ...\}$, whose elements are called *arcs*, such that each arc $a_{i,j}$ is identified with an unordered pair (n_i, n_j) of nodes. The nodes n_i, n_j associated with arc $a_{i,j}$ are called the *end nodes* of $a_{i,j}$.

Subgraph A graph G' is said to be a *subgraph* of a graph G if all the nodes and all the arcs of G' are in G, and each arc of G' has the same end nodes in G' as in G.

Isomorphic Two graphs G and G' are said to be *isomorphic* (to each other) if there is a one-to-one correspondence between their nodes and between their arcs such that the incidence relationship is reserved. In other words, suppose that arc a is incident on nodes n_1 and n_2 in G; then the corresponding arc a' in G' must be incident on the nodes n_1' and n_2' that correspond to n_1 and n_2, respectively.

Walk A *walk* is defined as a finite alternating sequence of nodes and arcs, beginning and ending with arcs, such that each arc is incident with the nodes preceding and following it. No arc appears more than once in a walk. A node, however, may appear more than once.

Path and **circuit** An open walk in which no node appears more than once is called a *path*. A closed walk in which no node appears more than once is called a *circuit*.

Cut-sets In a connected graph G, a *cut-set* is a set of arcs whose removal from G leaves G disconnected, provided removal of no proper subset of these arcs disconnects G. A *cut-set* always "cuts" a graph into two. In other words, suppose all the nodes of a connected graph G are partitioned into two mutually exclusive subsets, a cut-set is a minimal number of arcs whose removal from G destroys all paths between these two sets of nodes. In this section, A_c is used to represent a cut-set of a graph.

3.6.4.2   AAG of Plastic Products

Joshi and Chang [30] proposed an attribute adjacency graph (AAG) to support the feature recognition for polyhedral features and parts. The AAG is defined as a graph $G=(N, A, T)$, where N is the set of nodes, A is the set of arcs, and T is the set of attributes to arcs in A, such that:

(1) for every face f in F, there exists a unique node n in N;
(2) for every edge e in E, there exists a unique arc a in A, connecting the node n_i and n_j, corresponding to face f_i and face f_j, which share the common edge e;
(3) every arc a in A is assigned an attribute t, where $t=0$, if the faces sharing the edge form a concave angle; and $t=1$, if the faces sharing the edge form a convex angle.

For a solid part shown in Fig. 3.40(a), its AAG is shown in Fig. 3.40(b). Each node represents a face of the part, while each arc represents an edge of the part. Label '1' or '0' shows the convexity attribute of an edge. For a convex edge, its convexity attribute is '1'; while for a concave edge, its convexity attribute is '0'. Joshi and Chang have shown that AAG representation is effective for recognition of machined features from polyhedral parts. However, AAG is not suitable to represent a solid object with sculptured surfaces. For a solid object shown in Fig. 3.41, both edge e_1 and edge e_2 are incident to faces 2 and 4, and have the same convexity attribute, thus their corresponding arcs in the AAG of the solid object will result in ambiguity during feature recognition. One way to avoid this ambiguity is to add more useful attributes to the AAG representation of a part. Ye [4] proposed an extended attributed face-edge graph (EAFEG), which extends the AAG by adding several edge and face attributes, to represent a solid object. Each node of an EAFEG is corresponding to a unique face in the solid and has two attributes, face property and geometric type. Each arc of an EAFEG is corresponding to a unique edge in the solid and has the following five attributes: edge convexity, parting line information, loop type, vertex 1, and vertex 2.

Intelligent Mold Design and Assembly

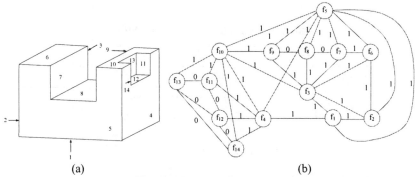

Fig. 3.40 A part and its AAG.

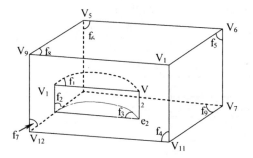

Fig. 3.41 A part with a free-form surface.

3.6.4.3 EAFEG of Plastic Parts

For a solid object, its extended attributed face-edge graph (EAFEG) is defined as a quadruple

$$G=(N, A, T_n(N), T_a(A))$$

where

$N=\{n \mid n \text{ is a node in the graph}\}$,
$A=\{a \mid a \text{ is an arc in the graph}\}$,
$T_n(N)=\{t_n \mid t_n = (C(n), C_g(n))\}$, and
$T_a(A)= \{t_a \mid t_a = (Q_c(a), Q_p(a), Q_L(a), V_1, V_2)\}$.

Node and arc attributes are summarized in Table 3.4. With the introduction of vertex connection in $T_a(A)$, the ambiguity caused by edges e_1 and e_2 in Fig. 3.41 can be avoided, because the two edges have different associated vertices. If a_1 and a_2 are two arcs corresponding to edges e_1 and e_2 respectively, one can find $t_a(a_1)=(1, 0, 4, V_1, V_4)$ and $t_a(a_2)=(1, 0, 4, V_2, V_3)$, and thus a_1 and a_2 are different.

Table 3.4 Various attributes in EAFEG

Attributes	Values
$C(n)$: face property	$C(n)=\{-1,0,1,2\}$
$C_g(n)$: geometric type	$C_g(n)=1$: the face is planar $C_g(n)=2$: the face is quadric $C_g(n)=3$: the face is free-form surface
$Q_c(a)$: edge convexity	$Q_c(a)=-1$: edge is concave $Q_c(a)=1$: edge is convex $Q_c(a)=0$: edge is shared by two coplanar faces
$Q_p(a)$: parting line info.	$Q_p(a)=1$: edge is a parting line $Q_p(a)=0$: edge is not a parting line
$Q_L(a)$: loop type	$Q_L(a)=n$: edge e is on an inner loop of node n $Q_L(a)=0$: edge e is on an outer loop
$V_1(a)$: vertex 1	$V_1(a)=(X_1, Y_1, Z_1)$
$V_2(a)$: vertex 2	$V_2(a)=(X_2, Y_2, Z_2)$

Among the arc attributes, the *edge convexity* serves a vital role for recognition of undercut features. Typically a concave edge is defined as one in which the angle between the sharing faces is less than 180 degrees; hence a convex edge is defined as one in which the angle between the sharing faces is greater than 180 degrees. However, using only the angle between the normal vectors of the faces to determine the convexity would result in an ambiguity. Fig. 3.42 shows a concave edge and a convex edge; the angles between the face-normals in both cases are between 90° and 180°. To avoid the ambiguity of edge convexity, Dong and Vijayan [32] proposed an algorithm to detect concavity and convexity of edges. Their algorithm is based on the following fact:

"For an edge shared by two planar faces, if the edge is a convex edge, the angle between the vector joining the middle points of the faces and the face normal of the second face is between 0° and 90° (hence the dot product is positive). Similarly, if the edge is a concave edge, the angle between the vector joining the middle points and the face normal of the second face is between 90° and 180° (hence the dot product is negative)."

Intelligent Mold Design and Assembly

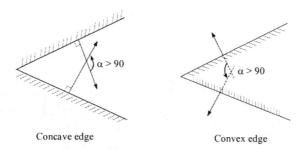

Fig. 3.42 Angles between normals of concave and convex edges.

3.6.4.4 EAFEG of Undercut Features

The existence of depressions or protrusions prevents the removal of the plastic products from a mold, it is thus natural to classify the undercut as *depression undercut* or *protrusion undercut*. The depression undercut is a region that consists of concave face sets and is bounded by a convex edge loop, while the protrusion undercut is a region that consists of convex face sets and is bounded by a concave edge loop. Furthermore, the depression undercuts can be classified into *blind depression undercut* and *through depression undercut*. If the depression region is only connected to the core or the cavity, this depression region is called a blind depression undercut (Fig. 3.43a); if the depression region is connected to both the core and the cavity, it is called a through depression undercut (Fig. 3.44a). For the undercut features to be recognized, they need to be precisely defined. In this approach, each type of undercut features is defined by a subgraph that has a set of necessary properties to classify the undercut feature uniquely.

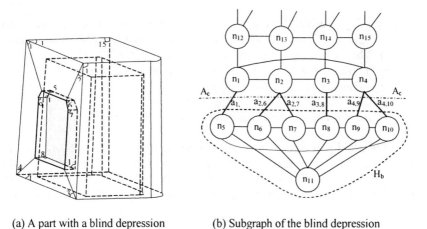

(a) A part with a blind depression (b) Subgraph of the blind depression

Fig. 3.43 A blind depression and its subgraph.

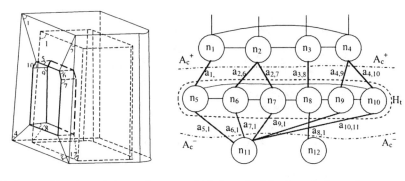

(a) A part with a through depression (b) Subgraph of the through depression

Fig. 3.44 A through depression and its subgraph.

(1) Isolated Undercut Features

Definition 3.1: For a subgraph G_1 of a graph G, A_c is the cut-set of G_1. All nodes, which are incident to A_c and belong to $G-G_1$, are called *separate nodes*, and N_s is the set of separate nodes; while all nodes, which are incident to A_c and belong to G_1, are called *construction nodes*, and N_c is the set of construction nodes. Correspondingly, there are *construction faces* and *separate faces* in plastic products. Now the undercut graph, which is the subgraph of an isolated undercut feature, can be defined as follows.

Definition 3.2: Given a global EAFEG G of a plastic product, an *undercut graph* H is a subgraph of G, i.e., $H=(N_h, A_h, T_n(N_h), T_d(A_h))$, such that H is connected and
  (1) There exists at least one positive node n_i in N_c, i.e., $C(n_i)=1$.
  (2) There exists at least one negative node n_j in N_c, i.e., $C(n_j)=-1$.
  (3) All arcs in H have a *convexity property*.
  (4) Cut-set A_c of H has a *cut-set property*.
  (5) All separate nodes of H have a *separate node property*.

For blind depressions, through depression, and protrusions, the convexity properties, cut-set properties, and the separate node properties are summarized in Table 3.5. In Fig. 3.43 a blind depression on the plastic product is shown. The blind depression that consists of faces $f_5, f_6, f_7, f_8, f_9, f_{10}$, and f_{11} is separated from

Intelligent Mold Design and Assembly

the main body of the plastic product by a set of faces, namely f_1, f_2, f_3, and f_4. Thus, the set of separate nodes is $N_s=\{n_1, n_2, n_3, n_4\}$, the cut-set is $A_c=\{a_{1,5}, a_{2,6}, a_{2,7}, a_{3,8}, a_{4,9}, a_{4,10}\}$, and the set of construction nodes is $N_c=\{n_5, n_6, n_7, n_8, n_9, n_{10}\}$.

For the through depression shown in Fig. 3.44, the set of construction nodes is $N_c=\{n_5, n_6, n_7, n_8, n_9, n_{10}\}$; the cut-set is $A_c=\{A_c^+, A_c^-\}$, where $A_c^+=\{a_{1,5}, a_{2,6}, a_{2,7}, a_{3,8}, a_{4,9}, a_{4,10}\}$, and $A_c^-=\{a_{5,11}, a_{6,11}, a_{7,11}, a_{9,11}, a_{8,12}, a_{10,11}\}$; the set of separate nodes is $N_s=\{N_s^+, N_s^-\}$, where $N_s^+=\{n_1, n_2, n_3, n_4\}$, and $N_s^-=\{n_{11}, n_{12}\}$.

For the protrusion undercut feature illustrated in Fig. 3.45, the set of construction nodes is $N_c=\{n_5, n_6, n_7, n_8, n_9, n_{10}\}$, the cut-set is $A_c=\{a_{1,5}, a_{2,6}, a_{2,7}, a_{3,8}, a_{4,9}, a_{4,10}\}$, and the set of separate nodes is $N_s=\{n_1, n_2, n_3, n_4\}$.

The above definitions of undercut features have the following two advantages; first, it can define the undercut features with free-form surfaces unambiguously; second, the definitions are flexible enough to accommodate the feature interactions, because they do not describe all constraints among nodes and arcs.

Table 3.5 Properties of undercut graph

Properties	Blind Depression	Through depression	Protrusion
Cut-set property	(1). $Q_c(a_{ij})=1$ or 0, $a_{ij} \in A_c$; (2). vertices in $T_a(A_c)$ form a closed loop.	(1). $Q_c(a_{ij})=1$ or 0, $a_{ij} \in A_c$; (2). $A_c=\{A_c^+, A_c^-\}$; (3). vertices in $T_a(A_c^+)$ form a closed loop; (4). vertices in $T_a(A_c^-)$ form a closed loop.	(1). $Q_c(a_{ij})=-1$ or 0, $a_{ij} \in A_c$; (2). vertices in $T_a(A_c)$ form a closed loop.
Convexity property	$Q_c(a_{ij})=-1$ or 0, $a_{ij} \in A_h$	$Q_c(a_{ij})=-1$ or 0, $a_{ij} \in A_h$	$Q_c(a_{ij})=1$ or 0, $a_{ij} \in A_h$
Separate node property	$C(n_i) \times C(n_j) \neq -1$, n_i, $n_j \in N_s$.	(1). $N_s=\{N_s^+, N_s^-\}$, $N_s^+=\{n \in N_s$ and n is incident to an arc of $A_c^+\}$, $N_s^-=\{n \in N_s$ and n is incident to an arc of $A_c^-\}$; (2). $C(n_i) \times C(n_j) \neq -1$, n_i, $n_j \in N_s^+$, or n_i, $n_j \in N_s^-$.	$C(n_i) \times C(n_j) \neq -1$, n_i, $n_j \in N_s$.

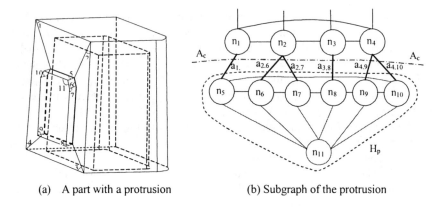

(a) A part with a protrusion (b) Subgraph of the protrusion

Fig. 3.45 A protrusion and its subgraph.

(2) Interacting Features

"Feature interactions" are intersections of feature boundaries with those of other features such that either the shapes or the semantics of a feature are altered from the standard or generic definitions. Undercut features that are involved in feature interactions are called interacting undercut features. According to the variation of cut-sets, interacting undercut features can be classified into two categories: *Type I* interacting undercut features and *Type II* interacting undercut features.

Definition 3.3: Type I interacting undercut feature
  Given an undercut feature H_1 and a feature H_G (H_G can be either an isolated undercut feature or a generic depression/protrusion feature). The result of the interaction between H_1 and H_G, namely H_I, is *Type I interacting undercut feature*, if H_I has all properties of an isolated undercut feature except convexity properties between the internal faces of an undercut feature.

  Because H_1 can be any isolated undercut features, Type I interacting undercut features can be further classified as: *Type I blind depression*, *Type I through depression*, and *Type I protrusion*. Fig. 3.46 illustrates various Type I interacting undercut features, where the bold lines show the cut-sets of the interacting undercut features. It can be seen from this figure, the cut-sets of the interacting features remain unchanged after feature interaction. The '*P*' means a local protrusion interacting with the undercut feature, while '*D*' means a local depression interacting with the undercut feature.

Intelligent Mold Design and Assembly

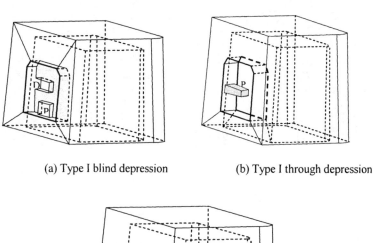

(a) Type I blind depression (b) Type I through depression

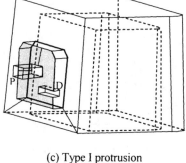

(c) Type I protrusion

Fig. 3.46 Type I interacting features.

Definition 3.4: Type II interacting undercut features

Type II interacting undercut feature H_{II} is the result of interaction between an isolated undercut feature H_1 and a feature H_G, such that H_{II} has a different cut-set from H_1; moreover, the convexity property between any two construction faces is also altered. H_G is either an isolated undercut feature or a generic depression/protrusion feature.

Similar to Type I interacting undercut features, the Type II interacting undercut feature can be further classified as: *Type II blind depression, Type II through depression* and *Type II protrusion*. Fig. 3.47 illustrates various Type II interacting undercut features, where the bold lines show the cut-sets of the interacting undercut features. It can be seen from this figure; the interactions between features alter the cut-sets of the interacting features.

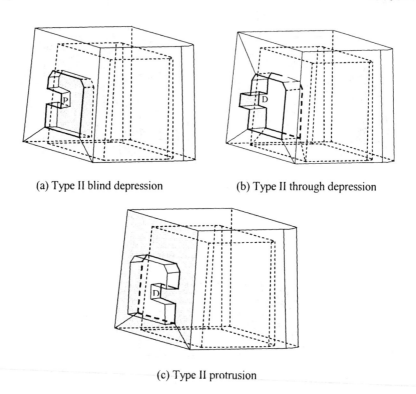

(a) Type II blind depression   (b) Type II through depression

(c) Type II protrusion

Fig. 3.47 Type II interacting features.

3.6.5 Recognition Algorithms

3.6.5.1 Hybrid Method

Based on the above definitions of undercut features, a hybrid method which combines the strength of graph-based approaches with that of hint-based recognition approaches is presented here. The new method recognizes undercut features by searching the cut-sets of undercut subgraphs from the global graph of a plastic product. The search of cut-sets is guided by various hints. Conceptually, the method consists of the following five phases:

(1) Pre-processing: determining the face property, inputting the parting lines, and constructing EAFEG of a plastic product;
(2) Identifying candidate undercut faces;
(3) Recognizing undercut features with inner cut-sets;
(4) Recognizing isolated and Type I interacting features; and
(5) Recognizing Type II interacting undercut features.

Intelligent Mold Design and Assembly

In phase (1), the B-rep model of a plastic product and its parting lines are first input, and then the program determines the face property for each surface of the B-rep model. Based on this information, EAFEG of the plastic product is established. The purpose of phase (2) is to identify all candidates of undercut faces. Since negative face f_i and positive face f_j are a pair of candidates of undercut faces, the shared edge e_{ij} is not a parting line. Correspondingly, a negative node n_i and positive node n_j are a pair of candidates if their incident arc a_{ij} has $Q_p(a_{ij})=0$.

In phases (3) to (5), various undercut features are recognized. The cut-set search plays a very important role in these phases. It is well known that even in a simple graph there are a large number of cut-sets, so it is not practical to search all cut-sets in a complex graph [33]. In this work, hints are used to guide the search of cut-sets. Based on hints, the routine first searches a specific set of arcs that satisfy the cut-set property given in the feature definitions. Then verification is conducted to make sure that this arc set is a cut-set. The verification is based on the following theorem and corollary:

Theorem 3.1: The ring sum of any two cut-sets in a graph is either a third cut-set or an arc-disjoint union of cut-sets [33]. The ring sum $(A_1 \oplus A_2)$ of two cut-sets A_1 and A_2 is the set consisting of all the elements that are either in A_1 or A_2 but not in both.

Corollary 3.1: Given a solid object and its EAFEG G, H is a connected subgraph of G, all nodes of H are $\{n_1, n_2, \ldots n_i, \ldots\}$ and A_i is the set of all arcs incident to n_i. If arc set A_c satisfies the following two conditions, A_c is the cut-set of H:
  (1) $A_c = A_1 \oplus A_2 \oplus A_3 \oplus \cdots \oplus A_i \oplus \cdots$, where $A_1, A_2, A_3,$ and $A_i \in H$
  (2) The corresponding edges of A_c in the solid object form a closed loop.

Proof: A_i is the set of all arcs incident to node n_i, so A_i is the cut-set of node n_i (A_i separates n_i from G). According to Theorem 3.1, the ring sum of A_i is either a cut-set or an arc disjoint union of cut-sets. Because of the second condition, A_c cannot be an arc-disjoint union of cut-sets.

3.6.5.2 Hints and Cut-Sets

It is the topological and geometric characteristics that differentiate an undercut feature from other regions of a part, so it is necessary to make the most use of both topological and geometric characteristics of an undercut during feature recognition. In addition to the important characteristics of undercut features given in Definition 3.2 and Table 3.5, some other geometric and topological characteristics are very useful for recognition of undercut features. These geometric and topological characteristics are called *hints* of undercut features. Some examples of hints are as follows.

Hint 3.1 - For a pair of candidate undercut faces f_i and f_j, $e_{i,j}$ is the shared edge between f_i and f_j, and $e_{i,j}$ is a convex edge. If the following three conditions are satisfied, negative face f_i is the construction face of a depression undercut feature, and the depression undercut feature is separated from the part by a set of cavity faces, i.e.,
  (1) there exists another convex edge $e_{k,l}$ that is connected to $e_{i,j}$,
  (2) faces f_k and f_l are adjacent to f_i and f_j, respectively, and
  (3) $e_{k,l}$ is lower than $e_{i,j}$.

Hint 3.2 - For a pair of candidate undercut faces f_i and f_j, $e_{i,j}$ is the shared edge between f_i and f_j, and $e_{i,j}$ is a concave edge. If the following three conditions are satisfied, negative face f_i is the construction face of a protrusion undercut feature, and the protrusion undercut feature is separated from the part by a set of cavity faces, i.e.,
  (1) there exists another concave edge $e_{k,l}$ that is connected to $e_{i,j}$,
  (2) faces f_k and f_l are adjacent to f_i and f_j respectively, and
  (3) $e_{k,l}$ is higher than $e_{i,j}$.

Hint 3.3 - For a Type II interacting undercut feature, f_i is one of its construction faces, f_j is the separate face, and $e_{i,j}$ is shared by f_i and f_j. If at vertex V_p (one vertex of $e_{i,j}$), there exist two optional edges for the cut-set, choose the edge that does not belong to f_i as the next edge of the cut-set.

Fig. 3.48 illustrates this hint. The centered lines show the separate faces, while the bold lines show the edges in the cut-sets. $e_{i,j}$ is the current edge in the cut-set, and at the point V_p, there exist two edges $e_{i,k}$ and $e_{k,j}$ that are convex. According Hint 3.3, edge $e_{k,j}$ is selected as the next edge in the cut-sets.

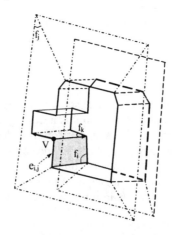

Fig. 3.48 Hint 3.3.

Intelligent Mold Design and Assembly

Hint 3.4 - For a pair of candidate undercut faces f_i and f_j, $e_{i,j}$ is the shared edge between f_i and f_j, and $e_{i,j}$ is a convex edge. If the following three conditions are satisfied, positive face f_i is the construction face of a depression undercut feature, and the depression undercut feature is separated from the part by a set of core faces, i.e.,

(1) there exists another convex edge $e_{k,l}$ that is connected to $e_{i,j}$,
(2) faces f_k and f_l are adjacent to f_i and f_j, respectively, and
(3) $e_{k,l}$ is higher than $e_{i,j}$.

Hint 3.5 - For a pair of candidate undercut faces f_i and f_j, $e_{i,j}$ is the shared edge between f_i and f_j, and $e_{i,j}$ is a concave edge. If the following three conditions are satisfied, positive face f_i is the construction face of a protrusion undercut feature, and the depression undercut feature is separated from the part by a set of core faces, i.e.,

(4) there exists another concave edge $e_{k,l}$ that is connected to $e_{i,j}$,
(5) faces f_k and f_l are adjacent to f_i and f_j respectively, and
(6) $e_{k,l}$ is lower than $e_{i,j}$.

3.6.5.3 Recognition of Isolated and Type I Interacting Undercut Features

Both isolated undercut features and Type I interacting undercut features have the convexity property between two construction faces. An algorithm consisting of the following steps is used to search their cut-sets and subgraphs:

(1) Get a pair of undercut nodes (n_i, n_j, a_{ij}).
(2) For a plastic product shown in Fig. 3.43, three pairs of candidates can be found, i.e., $(n_4, n_{10}, a_{4,10})$, $(n_1, n_5, a_{1,5})$, and $(n_2, n_6, a_{2,6})$.
(3) Generate hints for (n_i, n_j, a_{ij}). If no hint generated, go to Step 1, otherwise go to next step.
(4) Search the cut-set A_c according to the specific cut-set property. For example, if Hint 3.1 is generated, search A_c with $Q_c(a) = 1$ or 0; if Hint 3.2 is generated, search the cut-set A_c with $Q_c(a) = -1$ or 0.
(5) When a closed loop is found, the search stops. The sets N_c (construction nodes) and N_s (separate nodes) are then found. In the example of Fig. 3.43, $N_s = \{n_1, n_2, n_3, n_4\}$ and $N_c = \{n_5, n_6, n_7, n_8, n_9, n_{10},\}$. If the search reaches a parting line, it also stops. In this case, the search is unsuccessful.
(6) Find all nodes, which are adjacent to N_h and do not belong to N_s; and add these nodes to N_h. For the first time to run this step, $N_h = N_c$.
(7) Iteratively execute Step (4) until all adjacent nodes of N_h are found, or N_h reaches an arc of A_{pool}. If A_c is a cut-set of N_h, N_h and its arcs form undercut subgraph H, which is either a blind depression or a protrusion. If A_c and another set A_c' in A_{pool} compose a cut-set of N_h, N_h and its arcs form a through depression feature. Otherwise, save A_c to candidate pool A_{pool} of cut-sets. In Fig. 3.43, all nodes of H are $N_h = \{n_5, n_6, n_7, n_8, n_9, n_{10}, n_{11}\}$.

(8) Further classification. According to the convexity property between any two internal faces of undercuts, the recognized undercuts can be further classified as isolated undercut features or Type I-interacting undercut features.

3.6.5.4 Recognition of Type II-interacting Undercut Features

As mentioned in Definition 3.4, the Type II-interacting undercut features do not necessarily maintain the convexity property between two construction nodes. This results in more than one candidate arc for the cut-set at some nodes. To successfully recognize Type II-interacting undercut features, the recognition procedure should make decision to choose the correct element arc of the cut-set. Hint 3.3 can be used to choose the next correct arc of the cut-set. With the help the hint, the recognition procedure is similar to that of recognition of the isolated undercut features. But sometimes, hints may result in an unsuccessful search or wrong result of candidate cut-set. So the procedure to recognize Type II-interacting undercut features is a trial-and-error process and can be time-consuming.

3.7　GENERATION OF SIDE-CORES FOR SLIDERS AND LIFTERS

As discussed in last section, the existence of an undercut feature in a plastic product prevents the removal of the plastic product from its injection mold. For any undercut feature of a plastic product, there exists a corresponding undercut feature in the injection mold. To form the undercut region, a *side-core* is required, which can be withdrawn along the side direction. For an undercut region attached to the cavity, the side-core is driven by the *slider* mechanism. For an undercut region in the core, a *lifter* system is required to produce side movement for the side-core. This section first introduces the slider and lifter mechanisms, and then discusses the automatic generation of side-cores for the undercut features of cores and cavities. In this section, the undercut feature is referred to as the one in cores or cavities, unless specified specially.

3.7.1 Slider and Lifter Mechanism

3.7.1.1　Slider Mechanism

Any undercut region of a cavity will block the separation of the plastic product from the cavity. Thus, one has to split the undercut region from the cavity. The split portion from the cavity is called a side-core. During the opening of a mold, a side-core not only moves along the ejection direction, but also slides along the side direction, which may be perpendicular to the ejection direction. The most popular way to implement this compound movement is the slider mechanism, as illustrated in Fig. 3.49. In Fig. 3.49(a), the heel block on the fixed-half locks up the side-core operated by the cam pin. The cam pin is mounted at not more than 25 degrees to keep its bending stress low. When the injection mold is open, the side-

Intelligent Mold Design and Assembly

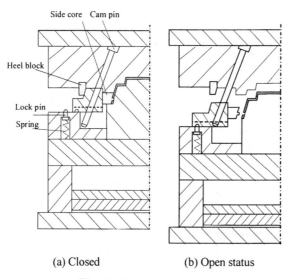

Fig. 3.49 Slider mechanism.

core slides along a "T"-slot on the moving half of the mold. At the end of the opening stroke, the side-core is withdrawn completely from the plastic product, and is captured by a spring operated lock pin. Fig. 3.49(b) shows the open status of the slider system.

3.7.1.2 Lifter Mechanism

The undercut portion in a core can be freed by introducing a separate core block, which can be removed to allow ejection of the plastic product in the usual way. Such a core block is called side-core for an internal undercut. It is the *lifter* mechanism that produces the driven force for the removal of the side-core from the main core. An illustration of one such lifter mechanism is indicated in Fig. 3.50. The side way movement of the side-core is controlled by the pillar, which slides in and out of the plate. One end of the pillar is secured to the side-core and the other end terminates at the roller, which engages a hardened block secured to the ejector retainer plate. On the opening of the mold, the fixed-half of the mold separates from the product, leaving the product in the moving half and with the side-core still engaged. Towards the end of the opening stroke, the ejector plate begins to move forward, ejecting the product in the ejection direction. At the same time, the forward action of the ejector plate forces the pillar to move inwards. This inward movement also forces the side-core to move inwards, thus the undercut portion is withdrawn from the plastic product. Fig. 3.50(b) illustrates the open status of the lifter system.

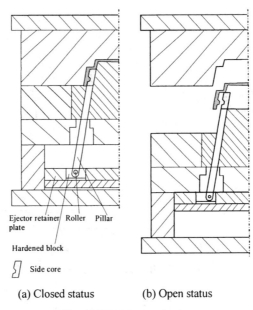

Ejector retainer / Roller / Pillar
plate

Hardened block

Side core

(a) Closed status (b) Open status

Fig. 3.50 Lifter mechanism.

3.7.2 Designing Side-Cores

For every undercut feature of a core or a cavity, there exists a portion that blocks the removal of the plastic product from the injection mold. This portion is called the blockage portion. To design a side-core, one needs to determine the blockage portion of the undercut feature. Depending on the types of undercut features and the attachment modes (to core or cavity), the blockage portions are different. Fig. 3.51 illustrates blockage portions of undercut features attached to cavities. In Fig. 3.51(a), there exists a protrusion region in a cavity. The blockage portion is the whole protrusion undercut. But in Fig. 3.51(b), the region that blocks removal of the product is the one from the positive face of the undercut feature to the parting surfaces. In Fig. 3.51, regions labeled 'b' are the blockage portions. To move blockage portions in side directions, one must connect them to the slider bodies that are driven by the cam pins. Regions labeled 'l' are between the slider bodies and the blockage portions, so they must be split from the cavities. Thus, the minimum side-core for cavities is region b plus region l.

Intelligent Mold Design and Assembly

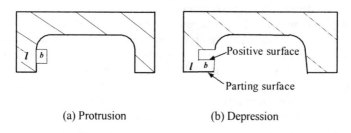

Fig. 3.51 Side-cores for undercuts of the cavities.

Fig. 3.52 shows side-cores for undercut features of the cores. For the protrusion undercut shown in Fig. 3.52(a), the blockage portion is the whole protrusion region. In Fig. 3.52(b), the blockage portion is the one from the negative face of the undercut feature to the top face of the core. To accommodate the connection between blockage portions and pillars, regions labeled '*l*' are required. Thus, the side-cores consist of regions *b* and *l*. It can be seen from the above discussion, designing a side-core can be carried out in four steps, i.e.,

(1) Identifying the undercut feature;
(2) Finding the region that blocks removal the product;
(3) Specifying the releasing direction of the side-core; and
(4) Cutting out a portion that contains the blocking region from the cavity.

3.7.3 Recognition of Undercuts from Core and Cavity

To design side-cores, one needs to recognize the undercut features first. The hybrid method proposed in Section 3.6.5 can be adapted to recognize undercut features from cores and cavities.

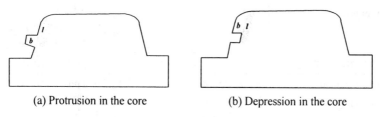

Fig. 3.52 Side-cores for undercuts of the cores.

3.7.3.1 EAFEG Definitions of Undercuts

It is the depressions or protrusions in cores and cavities that prevent the removal of plastic products from injection molds. Thus, undercut features can be classified into two categories, namely, *depression undercut* features and *protrusion undercut* features. The depression undercut feature is a region that consists of a set of concave faces and is bounded by a convex edge loop. The protrusion undercut feature is a region consisting of convex face sets and is bounded by a concave edge loop. To recognize undercut features from cores and cavities, it is necessary to precisely define various undercut features. Similar to the definitions of undercut features in plastic products, each kind of undercut features of cores and cavities can be defined uniquely by a subgraph of the global EAFEG. The EAFEG proposed in the last chapter for plastic products can also be used to represent a core or a cavity.

3.7.3.2 Isolated Undercuts

Definition 3.5: Given a global EAFEG of the core or cavity, an *undercut graph* H is a subgraph of G, i.e., $H=(N_h, A_h, T_n(N_h), T_a(A_h))$, such that H is connected and
 (1) There exists at least one positive node n_i in N_c, i.e., $C(n_i)=1$.
 (2) There exists at least one negative node n_j in N_c, i.e., $C(n_j)=-1$.
 (3) All arcs in H have *convexity property*.
 (4) Cut-set A_c of H has *cut-set property*.
 (5) All separate nodes of H have *separate node property*.

For depressions and protrusions, the convexity properties, cut-set properties, and the separate node properties are summarized in Table 3.6. Fig. 3.53 illustrates an undercut graph for the protrusion of the cavity. The protrusion that consists of faces $f_5, f_6, f_7, f_8, f_9, f_{10}$, and f_{11} is separated from the main body of the cavity by a

Table 3.6 Properties of undercut graph

Properties	Depression	Protrusion
Cut-set property	(1) $Q_c(a_{ij})=1$ or 0, $a_{ij} \in A_c$ (2) Vertices in $T_a(A_c)$ form a closed loop	(1) $Q_c(a_{ij})=-1$ or 0, $a_{ij} \in A_c$ (2) Vertices in $T_a(A_c)$ form a closed loop
Convexity property	$Q_c(a_{ij})=-1$ or 0, $a_{ij} \in A_h$	$Q_c(a_{ij})=1$ or 0, $a_{ij} \in A_h$
Separate node property	For a depression of a cavity, $C(n_i)=-1$, $n_i \in N_s$ For a depression of a core, $C(n_i)=1$, $n_i \in N_s$	For a protrusion of a cavity, $C(n_i)=-1$, $n_i \in N_s$ For a protrusion of a core, $C(n_i)=1$, $n_i \in N_s$

Intelligent Mold Design and Assembly

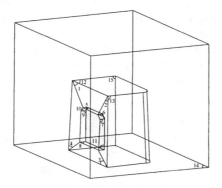

(a) A protrusion undercut of the cavity

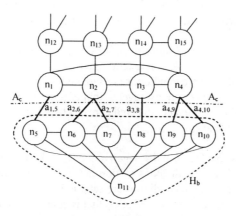

(b) Subgraph of the protrusion undercut

Fig. 3.53 Protrusion undercut and its subgraph.

set of faces, namely f_1, f_2, f_3, and f_4. Thus, the set of separate nodes is $N_s=\{n_1, n_2, n_3, n_4\}$, the cut-set is $A_c=\{a_{1,5}, a_{2,6}, a_{2,7}, a_{3,8}, a_{4,9}, a_{4,10}\}$, and the set of construction nodes is $N_c=\{n_5, n_6, n_7, n_8, n_9, n_{10}\}$.

3.7.3.3 Interacting Undercuts

In Section 3.6.4, interacting undercut features of plastic products are classified into two categories, namely *Type I* and *Type II* interacting undercut features. Similarly, an interacting undercut feature of a core or a cavity can be classified as a *Type I* interacting undercut feature or a *Type II* interacting undercut feature. The interacting undercut is the result of interaction between an isolated feature and another feature. If the cut-set of the resulting interacting feature maintains the original cut-set of the isolated undercut feature, the interacting undercut feature is *Type I interaction*, otherwise, it is *Type II interaction*. Corresponding to the isolated undercut features, an interacting undercut feature can also be classified as an interacting depression or an interacting protrusion. Fig. 3.54 illustrates the classification of interacting undercut features.

3.7.3.4 Recognition of Undercut Features

Based on the above definitions of undercut features, the hybrid method proposed previously can be used to recognize various undercut features from cores and cavities. Compared to the recognition of undercut features from plastic products, the hints to guide the recognition processes from cores and cavities are more direct. When recognizing undercut features from cavities, the hints for cavities are used. To recognize undercut features from cores, the program selects the hints for cores. Below are two hints for undercut features of cavities.

Hint 3.6: For a positive face f_i and a negative face f_j, e_{ij} is shared by f_i and f_j, and e_{ij} is a convex edge. Face f_i is the construction face of a depression undercut feature attached to the cavity if three conditions are satisfied, i.e.,
   (1) there exists another convex edge $e_{k,l}$ that is connected to $e_{i,j}$,
   (2) faces f_k and f_l are adjacent to f_i and f_j, respectively, and
   (3) $e_{k,l}$ is higher than $e_{i,j}$.

Hint 3.7: For a positive face f_i and a negative face f_j, e_{ij} is shared by f_i and f_j, and e_{ij} is a concave edge. Face f_i is the construction face of a protrusion undercut feature attached to the cavity if three conditions are satisfied, i.e.,
   (1) there exist another concave edge $e_{k,l}$ that is connected to $e_{i,j}$,
   (2) faces f_k and f_l are adjacent to f_i and f_j, respectively, and
   (3) $e_{k,l}$ is lower than $e_{i,j}$.

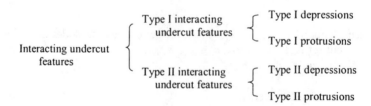

Fig. 3.54 Classification of interacting undercut features.

Intelligent Mold Design and Assembly

3.7.4 Automatic Generation of Side-Cores

3.7.4.1 Overview of the Method

Based on the above discussion, a procedure [4] with the following steps can be used for automatically generating side-cores:
(1) Recognizing the undercut features from cores and cavities;
(2) Specifying and verifying the release direction for every undercut feature;
(3) Creating a bounding box for each undercut feature;
(4) Trimming the bounding boxes with core surfaces or cavity surfaces; and
(5) Resolving side-core intersections.

After recognizing all the undercut features from cores or cavities, one needs to specify a releasing direction for each undercut feature, along which all surfaces of the undercut feature can be accessed freely. The releasing direction actually determines the movement of the side-core. To make sure that the specified releasing direction is accepted for the undercut feature, verification is required. The purposes of Steps 3 and 4 are to create side-cores for the undercut features. Sometimes, two or more than two side-cores intersect each other. This scenario might cause side-core duplication. In Step 5, the program resolves the side-core intersection.

3.7.4.2 Releasing Direction

The releasing direction is a side direction along which the side-core has movement. To freely withdraw the side-core from the undercut feature, all surfaces of the undercut feature must be accessible along the side direction. According to the theory of visibility map [28], if a surface of the undercut feature is oriented such that the releasing direction lies in the same direction as one in the visibility map (V-map), the corresponding surface is accessible freely. The visibility map for a given surface is an image on the unit sphere whereby every point in the V-map deviates from a corresponding point in a Gaussian map by an angle less than 90 degrees. Thus the condition for freely accessing a surface along the releasing direction is that the angle between the releasing direction and the normal direction at any point of the surface is less than 90 degrees.

For planar, cylindrical, and conic surfaces, it is not difficult to check if the surface can be freely accessed along the releasing direction. But for a free form surface, the task is not straightforward, because the normal directions of the surface can change from point to point. It is not practical to calculate the angle between the releasing direction and the normal directions at all points of the freeform surface. Therefore, it is necessary to explore practical algorithms to check if the surface can be freely accessed along the releasing direction. The convex-hull algorithm proposed is extended to check if a Bezier surface can be accessible freely along the releasing direction. Based on Theorem 3.1, one can immediately have the following corollary:

Corollary 3.2: For a $p \times q$ Bezier surface that is specified by a $(p+1) \times (q+1)$ mesh of control points $\{b_{i,j}=(X_{i,j}, Y_{i,j}, Z_{i,j}), i=0,1, \ldots p; j=0,1, \ldots q\}$, $R_{i,j}^{k,l}$ is a cross product vector between any two forward difference vectors $\Delta^{1,0}b_{i,j}$ and $\Delta^{0,1}b_{k,l}$ ($i=0, 1, \ldots, p-1; j=0, 1, \ldots q; k=0, 1, \ldots, p; l=0, 1, \ldots, q-1$). If cosine of any angle θ between $R_{i,j}^{k,l}$ and the releasing direction r is positive, the Bezier surface is freely accessible along the releasing direction.

The proof of the corollary is simple. Replacing the ejection direction of Theorem 3.1 with releasing direction r, one can immediately conclude that if any $\cos\theta$ is positive, the angle between releasing direction r and the normal vector at any point of the Bezier surface is less than 90 degrees. Thus the Bezier surface is accessible along releasing direction r.

3.7.4.3 Bounding Box of the Undercut Feature

The bounding box of an undercut feature is the smallest cubic block that contains the blockage portion and the additional portion of the undercut feature. To create a bounding box, one needs to determine the local coordinate system of the cubic block first. Upon specifying the releasing direction, one axis of the local coordinate system is determined. So it is reasonable to assume the releasing direction as the x-axis of the local coordinate system of the cubic block. Let $r = (r_x, r_y, r_z)$ represents the releasing direction, (i, j, k) and (i_l, j_l, k_l) represent the global and local coordinate systems respectively. The positive x-axis can be calculated by the following equation:

$$\vec{i}_l = \frac{\vec{r}}{|\vec{r}|} \tag{3.17}$$

Considering the convenience of manufacturing, the x-z plane of the local coordinate system is constrained to be perpendicular to the x-y plane of the global coordinate system. With the help of this constraint, the y-axis of the local coordinate system can be determined by the following equation:

$$\vec{j}_l = \frac{\vec{i}_l \times \vec{k}}{|\vec{i}_l \times \vec{k}|} \tag{3.18}$$

Thus, the z-axis of the local coordinate system can be determined by the equation below:

$$\vec{k}_l = \vec{i}_l \times \vec{j}_l \tag{3.19}$$

Once the local coordinate system is determined, the bounding box of the undercut feature is defined by the extreme coordinates of the bounding boxes for all of the related surfaces. For a protrusion undercut feature of the cavity, the related surfaces are the constituent surfaces of the undercut, while for a depression under-

Intelligent Mold Design and Assembly

cut of the cavity, the related surfaces are the positive surfaces of the undercut feature and the related parting surfaces. Fig. 3.55(a) shows a bounding box (the box with dashed lines) for protrusion undercut features.

3.7.4.4 Trimming Bounding Boxes to Create Side-Cores

The bounding boxes are cubic blocks which do not reflect the geometric shape of the undercut features. To create side-cores, one needs to modify the boundaries of the bounding boxes by using the surfaces of the undercut features. Such a modification can be carried out by the following Boolean operation:

$$SC = BB \cap P \qquad (3.20)$$

where

SC is the side-core
BB is the bounding box
P is the core or cavity

As mentioned in the above discussion, a bounding box is the smallest cubic block that contains blockage and additional portions of an undercut feature. In practice, mold designers need to cut some portions from cores or cavities to create a bigger side-core. Because the bounding box is actually represented by a box feature, and there exists associativity between the side-core and the main core/cavity, changing the side-core is not a difficult task. What the designers need to do is simply to adjust the size of the bounding box, and the side-core will be changed to the desired size. Fig. 3.55(b) illustrates a side-core of the undercut feature shown in Fig. 3.55(a).

3.7.4.5 Resolving Side-Core Intersection

When two or more than two side-cores overlap in space, the side-core intersection occurs. To resolve side-core intersection, Zhang et al. [34] gave the following two rules:

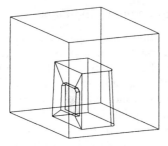 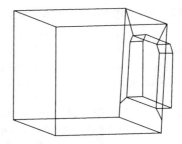

(a) A bounding box for the protrusion undercut (b) A side-core for the protrusion undercut

Fig. 3.55 A bounding box and a side-core.

Rule 3.1:   IF $SC_1 \cup SC_2 = SC_1$
THEN   delete side-core SC_2

Rule 3.2:   IF $SC_1 \cap SC_2 \neq 0$
THEN $SC_3 = SC_1 \cup SC_2$
Delete side-cores SC_1 and SC_2.

In above rules, SC_1 and SC_2 are the two side-cores, SC_3 is the new side-core. Symbol $\cup$ represents '*union*' operation, and $\cap$ represents '*and*' operation. Rule 3.1 is used to eliminate redundant heads, i.e., heads that are completely enclosed by other heads. Rule 3.2 is used to unite the heads which are intersected with each other.

3.8 SYSTEM IMPLEMENTATIONS AND CASE STUDIES

The concepts and methodology presented in the earlier sections have been used to develop a feature and associativity-based computer-aided injection mold design system called FA-MOLD [4]. This section first describes the development platform and system architecture of FA-MOLD, and then introduces various functional modules of FA-MOLD. Case studies for illustration are also presented in the end.

3.8.1 System Architecture

Fig. 3.56 illustrates the modularized architecture of FA-MOLD. Each module of FA-MOLD consists of methods and procedures to carry out a group of specific functions in computer-aided injection mold design. These modules share and use the same object-oriented representation of an injection mold.

3.8.2 Development Platforms and Programming Languages

Choosing an appropriate development platform and support software is very important for developing an effective computer-aided injection mold design system. The FA-MOLD system was developed on HP-C100 workstations that use the HP-UX operating system and the HP Visual User Environment (VUE). HP-UX is a versatile operating system that one can run application programs and perform a variety of tasks. HP VUE also provides a graphical interface to HP-UX that simplifies many of the involved tasks. Convenient application tools of HP VUE, such as the text editor, the icon editor, the compilers of high-level languages, etc., provide a favorable development environment for CAD/CAM software. On the other hand, computer-aided injection mold design requires many geometric modeling functions, and thus the FA-MOLD system used a commercially available CAD/

Intelligent Mold Design and Assembly

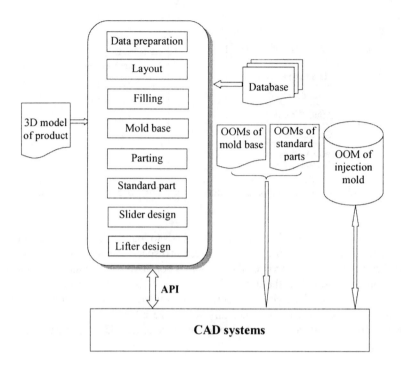

Fig. 3.56 System architecture of FA-MOLD (OOM: object-oriented model).

CAM system, namely *Unigraphics* [35] to develop the mold design application.

Since Unigraphics supports programming with C and FORTRAN languages, C language was then chosen to implement FA-MOLD system based on the algorithms presented in the previous sections. UG/Open API [36] is an easy interface between Unigraphics and third party developers. It consists of a large set of user callable subroutines that access the graphic terminals and the database of Unigraphics. With the help of UG/Open API, the developers can easily enhance the capabilities of Unigraphics. Interfaces of FA-MOLD were implemented using C language to enhance the capabilities of Unigraphics. The classes discussed in Section 3.3 are also defined using the structure of C language.

3.8.3 Functional Modules and Graphical User Interfaces

(1) Data Preparation

'Data preparation' is a module to initialize the injection mold design. This module provides functions to specify the ejection direction and input the shrinkage rate and other design information for a plastic product. It also recognizes the undercut features from a plastic product. Fig. 3.57 illustrates the interface of *Data Preparation* module.

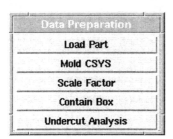 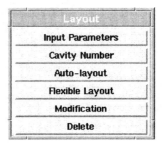

Fig. 3.57 Interface of data preparation.   **Fig. 3.58** Layout interface.

To start a new mold design, the user needs to load in the 3D solid model of a plastic product by using the *load part* button. To specify the ejection direction, the user has to align the +Z axis of the working coordinate system along the ejection direction, and then click the *Mold CSYS* button. This is because FA-MOLD uses the +Z direction as the ejection direction. The *undercut analysis* function in *Data Preparation* module can automatically recognize all undercut features from the plastic product, and save them to the feature box of the plastic product for layout arrangement.

(2) Layout Module

'Layout module' provides the functions to calculate the cavity number, and automatically arrange the cavity layout. The interface of layout module is shown in Fig. 3.58. To calculate the cavity number for a plastic product, the user has to input production parameters, cost parameters and machine parameters. Fig. 3.59 illustrates the interface of *Input Parameter* function in details. Based on these parameters, FA-MOLD calculates the preferable cavity numbers and recommends an optimal cavity number to the user. *Auto-layout* function in *Layout* module can automatically arrange the cavity layout.

In practical design, the user would like to have more flexibility to control the layout arrangement, such as layout patterns, layout references, and sizes of layout. To meet these requirements, Flexible Layout function is added to Layout module. Fig. 3.60 illustrates the detailed interface of Flexible Layout function. With the help of this function, the user can select balanced layout, series layout, or single layout, and then specifies the cavity number and layout reference.

(3) Filling Module

With the help of the 'Filling module', mold designers can design runners and gates efficiently. When designing a runner, what the mold designers need to

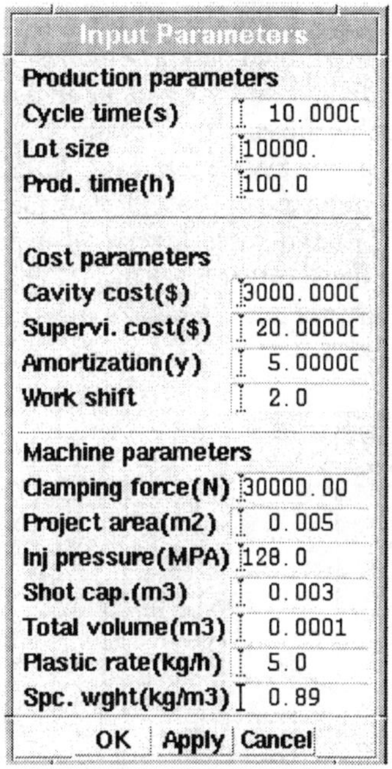

Intelligent Mold Design and Assembly

Fig. 3.59 Input interface. 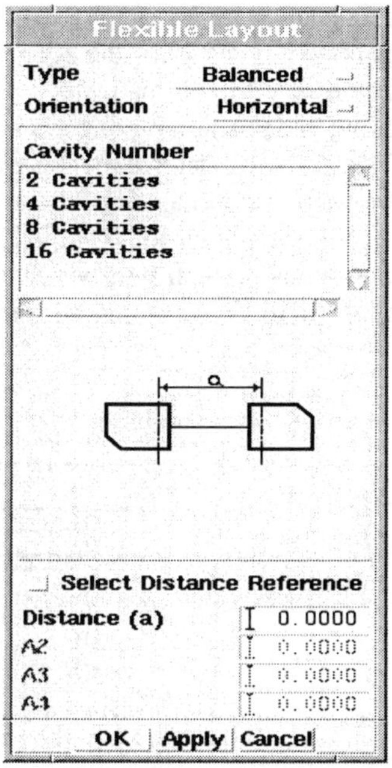   **Fig. 3.60** Interface of flexible layout.

do is selecting the section type of the runner and specifying the path of the runner, then FA-MOLD automatically creates a 3D solid object for the runner. The associativity between gates and the mold layout propagates the changes of the cavity layout to the gate locations. Fig. 3.61 shows the interface of *Filling* module.

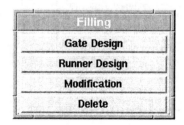

Fig. 3.61 Filling interface.

(4) Mold Base Module

'Mold Base' module provides a favorable design environment for assembly modeling of mold bases. The top-down browser of *mold base* module supports compatible paradigm for the conceptual design of mold base assemblies. Because of the facility to automatically infer the configuration of a part in an assembly, orienting and positioning of parts in the assembly are easy to carry out.

Fig. 3.62 illustrates the top-down browser for DME mold base. At the top of the browser is a bitmap that reflects the structure of a mold base. Below the bitmap is the option buttons for type and sub-type of mold bases. Every sub-type has a unique bitmap, which will be updated when the sub-type is changed. The values highlighted in *Dimension List* and displayed in *Plate Thickness* are set to the

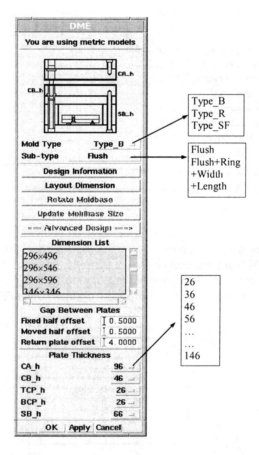

Fig. 3.62 Top-down browser for DME mold base.

Intelligent Mold Design and Assembly

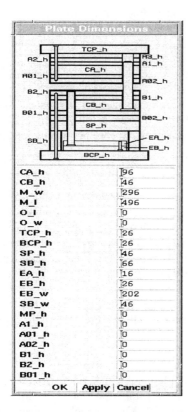

Fig. 3.63 Plate dimensions.

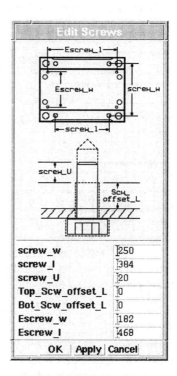

Fig. 3.64 Screw positioning.

smallest possible sizes according the layout dimensions. Options displayed in *Dimension List* represent the widths and the lengths of the mold base, respectively. For example, option 296×496 means 296 millimeters in width and 496 millimeters in length. Once the conceptual design of mold base is completed with the help of the top-down browser, mold designers can further adjust the dimensions of plates and bushes, and change the positioning dimensions between screws and pins. Figs. 3.63 and 3.64 illustrate the interfaces for changing plate dimensions and editing screw positions.

(5) Parting Module

'Parting module' is a powerful tool to help mold designers generate cores and cavities according to the geometric shape of plastic products. The geometric linkages between plastic products and the cavity system make sure that the cores and cavities are changed with the plastic products. Fig. 3.65 shows the interface of *Parting* module.

'Patching' function of Parting module is developed to block all through holes in plastic products. 'Freeform' function helps mold designers to create cores and cavities for a plastic product with free-form parting lines. Mold designers have to select parting lines and specify the extending directions interactively, then the program automatically creates the core and cavity for the plastic product. At the same time, the associativity between the core/cavity and the plastic product is established. Whenever the plastic product is modified, its core and cavity are automatically updated.

(6) Standard Part Module

'Standard Part' module establishes a component library, in which standard parts are represented in object-oriented models. It is this library that makes high-level objects of injection molds ready to use when designing an injection mold. All function buttons of *Standard Part* module are illustrated in Fig. 3.66. The interface of "Add Standard Part" is shown in Fig. 3.67. While adding a standard part, what a designer needs to do is selecting the dimensions of the standard part from the option lists. Then the program instantiates the predefined object-oriented model of the standard part, and automatically determines the orientation and position of the standard part according to the high-level mating conditions specified in the object-oriented model.

(7) Slider and Lifter Design

In the 'Slider Design' and 'Lifter Design' modules, the undercut features are automatically recognized from cores and cavities, then the side-cores and form pins to shape these undercut features are generated. What the mold designers need

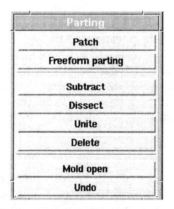

Fig. 3.65 Parting interface.

Fig. 3.66 Interface for standard part.

Intelligent Mold Design and Assembly 127

to do is select the structure types of sliders and lifters, and specify the profile dimensions of the sliders and lifters. The geometric linkages between the side-cores and the undercut features of the plastic products keep the side-cores updated, when the plastic products are changed. Fig. 3.68 illustrates the functions of *Slider design* module.

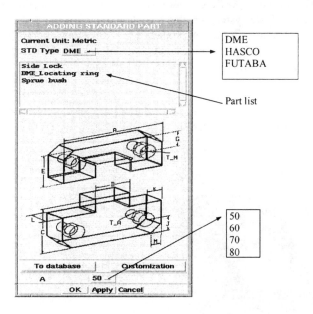

Fig. 3.67 Adding standard part.

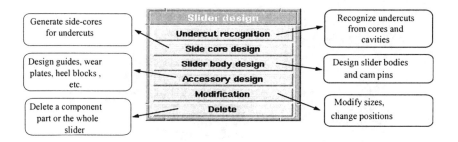

Fig. 3.68 Slider design module.

3.8.4 Case Study

(1) Preparation of the Mold Design

To design an injection mold for the test product, the mold designer has to load the 3D model of the product first, then adjusts the coordinate system of the product, inputs some basic information of the plastic product, such as material type and property. FA-MOLD automatically aligns the ejection direction to the working coordinate system; creates a containing box for later use, and recognizes all undercut features in the test product. Fig. 3.69 shows the 3D geometric model of the test product. Fig. 3.70 illustrates the new coordinate system, the containing box, and the undercut features of the plastic product.

(2) Arranging Cavity Layout

Subsequent to the initialization of mold design, the mold designer proceeds to the layout design. To determine the optimum cavity number, production and

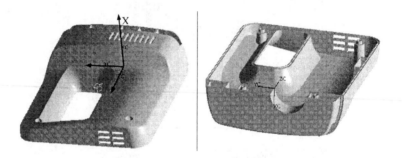

Fig. 3.69 Two views of the test product.

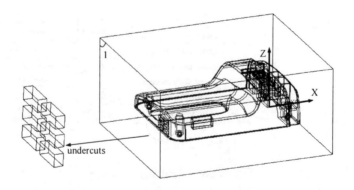

Fig. 3.70 New orientation of the test product.

Intelligent Mold Design and Assembly

machine parameters are required. In this design, the injection machine is Engel ES 500 [37]. For the parameters shown in Table 3.7, it is found by FA-MOLD that the optimum cavity number is 2. FA-MOLD also finds six depression undercuts, which can be formed by an external side-core. Because of the existence of undercut features along side face 1 of the feature box, the layout must be arranged such that side face 1 can be accessed freely from outside of the cavity. FA-MOLD thus recommends the layout arrangement as illustrated in Fig. 3.71.

Table 3.7 Production and machine parameters

Parameters	Values
Lot size	200,000
Cycle time (s)	20
Specific weight (g/cm^3)	1.03
Projected area (cm^2)	245.64
Volume of the product (cm^3)	139.652
Production time (h)	40×15
Cost of one cavity ($)	3000
Amortization period (y)	6
Number of working shifts	2
Supervision cost ($/y)	20,000
Clamping force (KN)*	4444.3
Injection pressure (MPA)*	213.9
Plasticating rate (kg/h)*	330.2
Shot capacity (cm^3)*	1192.5

(*Transferred from English unit.)

(3) Designing Mold Base

After the layout design, a mold base can be added to the top assembly of the injection mold. Using the top-down browser of mold base, Type A mold base is selected. The sizes of the cavity layout is 300×500, thus FA-MOLD recommends a mold base with standard sizes 396×496. Taking into consideration of the sliders, a mold base with sizes 496×596 is selected (Fig. 3.72). Because the mold base consists of parts that are represented by feature models, it is convenient to change

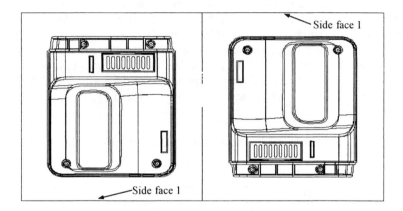

Fig. 3.71 Cavity layout for the test product.

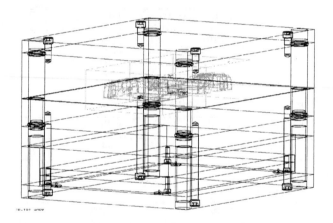

Fig. 3.72 Mold base for the test product.

the sizes of the plates. Associativity between the component parts of the mold base makes it easy to propagate changes in one part to the whole mold base.

(4) Designing Core, Cavity, and Slider

With the help of *parting* module, the core and cavity for the test product are easily created. Firstly, all through holes are patched; then parting lines and extending directions are specified; FA-MOLD automatically extracts surfaces from the test product to generate a core and a cavity (Fig. 3.73). Associativity between

Intelligent Mold Design and Assembly

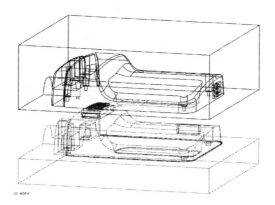

Fig. 3.73 The core and cavity.

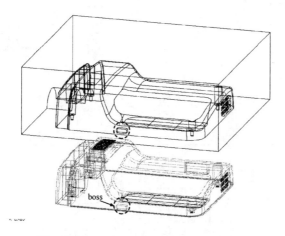

Fig. 3.74 New cavity for the modified product.

the test product and its cavity is also established. If a boss is added to the test product, the geometric shape of the cavity is changed accordingly (Fig. 3.74).

To design sliders, FA-MOLD first recognizes all undercut features from the core and the cavity, then creates side-cores for the undercut features. In this design, there exist six undercut features in the cavity. If one creates a side-core for each undercut feature, the side-cores would intersect each other. FA-MOLD can automatically combine six side-cores into one single side-core by using *Rule* 3.2. Fig. 3.75 illustrates a single side-core with six undercut features.

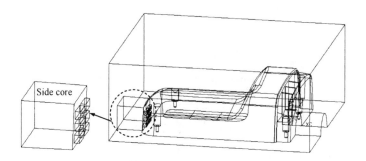

Fig. 3.75 A side-core for the undercuts.

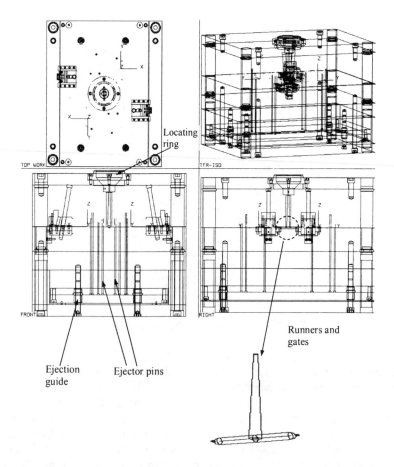

Fig. 3.76 A completed injection mold for the test product.

(5) Designing Runners, Gates, and Standard Parts

Using 'Filling' module, runners and gates can be easily created. In this design, side gates and runners with a circular section are used. With the help of Standard part module, various standard parts can be added to the mold assembly. Fig. 3.76 shows the injection mold with runners, gates and side locks, support pillars, etc.

3.9 SUMMARY

The existing commercial CAD/CAM systems are usually generic mechanical CAD/CAM software and do not provide the specific functions for injection mold design. Therefore, there is a need to develop a computer-aided injection mold design system, which enables mold designers to design injection molds effectively and efficiently. This chapter presents the main algorithms on automating CAD-based injection mold design applicable to complex industrial components. The key methodologies on optimal parting deign that includes parting direction, parting surface, and generation of core and cavity are presented in details. The introduction of undercut feature definitions and extraction greatly assist the effective design of slider and lifer mechanisms in a mold. Various undercut features are defined uniquely using a hybrid EAFEG. Each undercut feature corresponds to a unique undercut graph, which is a subgraph of the global EAFEG and has the convexity property, cut-set property and separate node property. The new method of undercut feature recognition is very suitable for handling the free-form surface features of dies and molds. Besides parting design, the optimal layout design for multiple cavities is also presented. An object-oriented representation scheme for 3D injection molds, which can encapsulate both feature and associativity information into an assembly object, is also described. Implementation and case studies based on the algorithms are described to facilitate possible commercialization of certain functions.

REFERENCES

1. J.J. Shah and M. Mantyla, Parametric and feature-based CAD/CAM, John Wiley & Sons, Inc., New York, 1995.
2. J.J. Shah, "Assessment of features technology", Computer-Aided Design, Vol.23, No.5, pp. 331-343, 1991.
3. R.A. Malloy, Plastic part design for injection molding, Hanser Publishers, Munich, Vienna, New York, 1994.
4. X.G. Ye, Feature and Associativity-based Computer-Aided Design for Plastic Injection Molds, PhD dissertation, National University of Singapore, 2000.
5. S.R. Gorti, A. Gupta, G.J. Kim, R.D. Sriram, and A. Wong, "An objection-oriented representation for product and design process", Computer-Aided Design, Vol. 30, No.7, pp. 489-501, 1998.

6. J. Rumbaugh, M. Blaha, et al., Object-oriented modeling and design, Prentice-Hall, Englewood Cliffs, NJ, pp. 21-56, 1991.
7. L.L. Chen, S. Y. Chou, and T.C. Woo, "Parting directions for mold and die design", Computer-Aided Design, 25, pp.762-768, 1993.
8. L.L. Chen, S.Y. Chou, and T.C. Woo, "Partial visibility for selecting a parting direction in mold and die design", J of Manufacturing Systems, 14, pp.319-330, 1995.
9. T. Mochizuki, and N. Yuhara, "Methods of extracting potential undercut and determining optimum withdrawal direction for mold designing", International J Japan Soc Prec Eng, 26, pp.68-73, 1992.
10. K.C. Hui and S.T. Tan, "Mold design with sweep operations-a heuristic search approach", Computer-Aided Design, Vol.24, No.2, pp.81-91, 1992.
11. M. Weinstein and S. Manoochehri, "Geometric influence of a molded part on the draw direction range and parting line locations", ASME, J of Mechanical Design, 118, pp.29-39, 1996.
12. M. Weinstein, and S. Manoochehri, "Optimum parting line design of molded and cast parts for manufacturability", J of Manufacturing Systems, 16, pp.1-12, 1997.
13. A.Y.C Nee, M. W. Fu, J.Y.H Fuh, K.S. Lee and Y.F. Zhang, Determination of optimal parting directions in plastic injection mold design. CIRP Annals: Manufacturing Technology, 46, 1997, No. 1: 429-432.
14. A.Y.C. Nee, M.W. Fu, J.Y.H Fuh, K.S. Lee and Y.F. Zhang, Automatic determination of 3-D parting lines and surfaces in plastic injection mold design. CIRP Annals: Manufacturing Technology, 47, 1998, No. 1: 95-98.
15. M.W. Fu, Determination of optimal 3-D parting in plastic injection mold design, Ph.D. dissertation, National University of Singapore, 1998.
16. S.T. Tan, M.F. Yuen, W.S. Sze and W.K. Kwong, "A method for generation of parting surface for injection molds", Proc. of 4th Int Conf on Computer Aided Production Engineering, Edinburgh, UK, 1988, pp.401-408.
17. S.T. Tan, M.F. Yuen, W.S. Sze, and W.K. Kwong, "Parting lines and parting surfaces of injection molded parts", Proc Instn Mech Engrs, 204, pp.201-219, 1990.
18. L. Kong, "Development of a Windows-based Computer-Aided Plastic Injection Mold Design System", MEng thesis, National University of Singapore, 2001.
19. G. Menges, and P. Mohren, How to make injection molds, 2nd Edition, Munich: Hanser Publishers, 1993.
20. P. Dewhurst, and D. Kuppurajan, "Determination of optimum processing conditions for injection molding", International J of Production Research, Vol. 27, No. 1, pp.21-29, 1989.
21. G. Menges, and P. Mohren, How to make injection molds, 1st Edition, Macmillan: Hanser Publishers, 1986.
22. M. W. Fu, J.Y.H. Fuh and A.Y.C. Nee, "Undercut feature recognition in an injection mold design system", Computer-Aided Design, Vol. 31, No. 12, pp.777-790, 1999.
23 M.W. Fu, J.Y.H. Fuh and A.Y.C. Nee, Generation of optimal parting direction based on undercut features. IIE Transactions, 31, No. 10:947-955, 1999.
24. M.W. Fu, J.Y.H Fuh and A.Y.C. Nee, Generation of optimal parting direction based on undercut features. IIE Transactions, 31, No. 10: 947-955, 1999.
25. D. W. Rosen, "Towards automated construction of molds and dies", ASME Computers in Engineering, Vol. 1, pp. 317-323, 1994.
26. M.A. Ganter and P.A. Skoglund, "Feature extraction for casting core development", ASME J of Mechanical Design, Vol. 115, pp. 744-750, 1993.

27. D. Hilbert and S. Cohn-Vossen, "Geometry and the imagination", Translated by P. Nementi, Chelsea, NY, pp. 172-271, 1983.
28. L.L. Chen, and T.C. Woo, "Computational geometry on the sphere for automated machining", ASME J Mech Des, 114, pp.288-295, 1992.
29. J.G. Gan, T.C. Woo and K. Tang, "Spherical Maps: their construction, properties, and approximation", ASME J of Mechanical Design, 116, pp.357-363, 1994.
30. S. Joshi and T.C. Chang, "Graph-based heuristics for recognition of machined features from a 3D solid model", Computer-Aided Design, Vol.20, No. 2, pp.58-66, 1988.
31. X.G. Ye, J.Y.H. Fuh and K.S. Lee, A hybrid method for recognition of undercut features from molded parts, Computer-Aided Design, 33:1023-1034, 2001.
32. J. Dong and S. Vijayan, "Manufacturing feature determination and extraction Part II: a heuristic approach", Computer-Aided Design, Vol. 29, No. 7, pp. 475-484, 1997.
33. N. Deo, Graph theory with applications to engineering and computer science, Prentice-Hall, Inc., Englewood Cliffs, NJ, 1974, pp. 68-87.
34. Y.F. Zhang, K.S. Lee, Y. Wang, J.Y.H. Fuh, A.Y.C. Nee, "Automatic side core creation for designing slider/lifter of injection molds", Proc. of CIRP International Conference and Exhibition on Design and Production of Dies and Molds, Istanbul, Turkey, 1997, pp. 33-38.
35. Unigraphics essentials user manuals, Vol. 1, Unigraphics Solution Co., Maryland Heights, MO, 1997.
36. UG/open API reference, Vol. 1 to Vol. 4, Unigraphics Solution Co., Maryland Heights, MO, 1997.
37. Engel, http://www.engelmachinery.com, 1999.

4

Semi-Automated Die Casting Die Design

4.1 INTRODUCTION

After introducing injection molding and mold design for plastic components, a very similar process and mechanism to make metal injection parts is discussed. *Die casting* is a manufacturing process for producing precisely dimensioned simple or complex-shaped parts from non-ferrous metals such as aluminum, magnesium, zinc, copper, etc. [1]. In this process, molten metal is injected into the die cavity under high pressure, and the die-casting is ejected from the die halves after it has solidified. In this chapter, the term *die-casting* is used to describe the part produced, while *die casting* refers to the process. Focus will be given to gating, runner, filling, die base design, etc. and semi-automated die design methodology that can lead to the commercialization on CAD-based die design systems. The parting, ejection, and cooling designs are basically quite similar to the plastic injection mold design as described in the previous chapters.

Die casting is widely used in the automobile, aerospace, electronics, and household appliance industries due to its high strength and good dimensional accuracy. It is usually categorized into two main types in accordance with the machines used: (1) hot chamber die casting and (2) cold chamber die casting. In the hot chamber machine, the injection mechanism is immersed in molten metal in a furnace attached to the machine. The molten metal fills the chamber from the port and the plunger moves downward to force molten metal through the gooseneck and nozzle into the die. This machine deploys a sprue to deliver the metal from the nozzle into the die, and the shot hole is positioned in the center of the cover platen. A hot chamber machine is normally used for casting alloys such as magnesium, zinc, lead, and tin.

In the cold chamber machine, the plunger and cylinder are not submerged in molten metal, and the molten metal is poured into a "cold chamber" through a port or pouring slot by a hand or automatic ladle. The machine uses a shot sleeve to deliver the metal into the die, and the shot hole usually offsets for a clearance down from the centre of the cover platen. A cold chamber machine is used for alloys such as aluminum, magnesium, and copper.

Compared to the injection tooling for plastic, the die casting tool for injecting metal is often called *die casting die* instead of die casting mold. The quality of die-casting is essentially determined by the die casting die, in which the die-casting is formed. A die consists of two sections called cover die half and ejector die half. The former is attached to the injection system of the machine, and the latter is connected to the ejection mechanism. Other auxiliary systems include a gating system, ejection system, cooling system, etc. Fig. 4.1 shows a typical die casting die assembly. Moving cores and sliders are usually employed if a die-casting has undercuts. Traditionally, die designers use 2D-based CAD systems for designing the die casting die due to the lack of an application software developed specially for die design. The traditional method depends largely on the designers' experience and expertise, and takes a relatively longer time to design. The die casting industry has been imperatively demanding a specialized software application, which can automate or semi-automate the design of die casting die so as to reduce the lead-time and the cost.

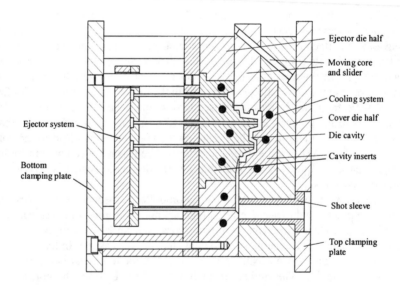

Fig. 4.1 A die casting die assembly.

4.2 PRINCIPLES OF DIE CASTING DIE DESIGN

4.2.1 Die Casting Die Design

A die casting die does not only consist of the main components such as the die base, but also has other features that control metal flow, heat flow, and the forces applied by the machine. Die design is a complex process that contains several critical tasks, which are described briefly in the following:

(1) Setting shrinkage: As the molten metal contracts during solidification, the original die-casting part geometry must be scaled by a certain factor to reflect the material shrinkage.
(2) Determining cavity number and layout: They are determined based on the part shape and dimension, machine type, machine size limitation, machine clamping force, and machine pumping capacity, etc.
(3) Designing gating system: Flow paths and filling conditions are analyzed at this stage. The type, size, and location of the gate, runner, and overflow are determined in accordance with the part geometry and cavity layout to achieve proper filling in the die cavity.
(4) Designing die-base: With the cavity number and cavity layout decided, a suitable die-base is selected to accomplish the proposed layout. The criterion generally used in establishing the overall size of the die-base is that the ejector plate must completely cover or contain all of the cavity area within its bounds.
(5) Parting design: It is to create parting surfaces along the selected parting lines and eventually split the containing box into two halves – a core block and a cavity block, in which the negative impression of the die-casting part is formed.
(6) Designing moving core mechanism: If the die-casting part has any undercut, which will block the die halves from opening, moving cores and angled pins should be designed to facilitate the die opening and the removal of die-casting from the die.
(7) Designing cooling system: Cooling system is an essential feature of the die casting die, which is composed of a set of waterlines drilled within the dies and inserts that conduct the heat away from the die cavity. The cooling system should be positioned and sized properly so as to achieve rapid and uniform cooling without interfering with the ejection system and moving core mechanism.

Fig. 4.2 indicates the flow chart of CAD-based die design process from inputting the die-casting part model to output the designed die assembly for die making. The main design tasks include: layout design, gating design, die-base design, parting design, ejection design, cooling design, moving core design, standard component design, etc.

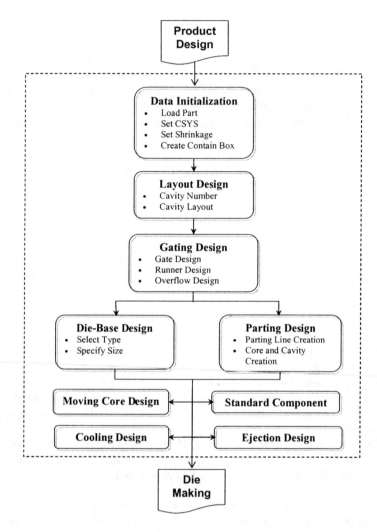

Fig. 4.2 The CAD-based die casting die design process [2].

4.2.2 Gating and Runner System Design

The gating system is a series of passages through which the molten metal enters and fills the die cavity (Fig. 4.3). It is composed of several gating elements as follows:

Gate: as the runner approaches the cavity, it blends into a slit-like opening into the cavity. The slit-like opening is a gate, and the blended portion that links the gate and

Semi-Automated Die Casting Die Design

runner is a gate-runner. The gate and gate-runner have an important function in directing the metal flow so that the cavity is correctly filled. Here, the term gate refers to the combination of the gate and gate-runner.

Runner: a runner is a channel located at the parting line to route liquid metal from the sprue or shot sleeve to the gate, and it may split into two or more as required to direct the liquid to various places.

Overflow: it is a chamber usually located at the opposite side of the gate to absorb undesirable non-metallic inclusions.

Shot sleeve (Sprue): the cavity inside the shot sleeve or the sprue forms a part of the gating system.

Due to the flow and thermal effects in the die, the size, shape, and placement of gating elements are very critical to the performance of the die. Not only should the gating system be designed to successfully deliver metal to fill the cavity, it must also be compatible with the die casting machine because a die with a specific gating system performs differently on different machines. Usually, the design of gating system is a case-by-case process, but researchers have gathered and summarized many guidelines for designing the gating system; therefore, a solution could be found to automate and semi-automate the design of these repeatedly used gating elements.

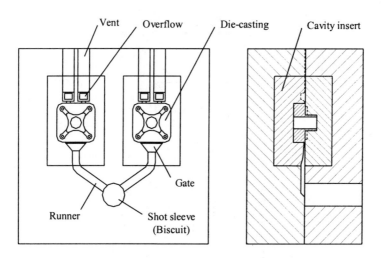

Fig. 4.3 The gating system of a die casting die [3].

4.2.3 Die-Base Design

A typical die-base (also called die set) of a die casting die is an assembly consisting of up to fifty component parts. Therefore, it is very tedious and time consuming to model all the parts one by one, so it is necessary to find a method to automatically generate the die-base in the design system. In order to facilitate the automatic generation of standard die-base assembly, various parts and die-base assemblies are pre-modeled and stored in a die-base library. All the parts in the assemblies are parametric models, and their dimensions will vary according to the type and dimension of the die-base specified. The automatic generation of die-base from the catalogue-based dimensions requires a database support. Since standard die-bases are commercially available, the establishment of such a database becomes very convenient. A user interface is also necessary for the user to select the desired type and dimension series of die-base. The user just needs to click on the interface, and the system will automatically import the desired die-base. The flowchart of this method is shown in Fig. 4.4.

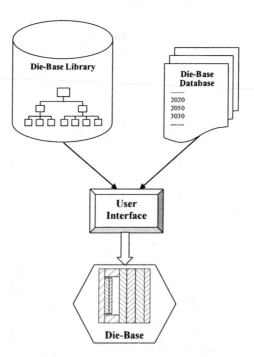

Fig. 4.4 The flow chart of automatic generation of die-base.

4.3 COMPUTER-AIDED DIE CASTING DIE DESIGN

Since computer-aided design (CAD) software is employed in injection mold and die casting die design, the design efficiency has been increased significantly because it is convenient to use the knowledge-based data library and also easy to modify the design structure and details. With further development of CAD technology, i.e., solid-based and feature-based modeling technology, 3D mold and die design becomes feasible in practice. At present, 3D CAD software such as Unigraphics [4], Pro/Engineer [5], and SolidWorks [6] have broad applications in the area of mold and die design.

However, most of the CAD systems provide only the geometric modeling functions that facilitate the modeling and drafting operations, but they do not offer designers with the necessary knowledge and modules for mold and die design. Therefore some "add-on" software applications have been researched and developed on these high-level 3D modeling platforms to facilitate the injection mold design process, for example, IMOLD^{TM}, MoldWizard [4] and IMOLDWorks^{TM} [7], etc. The development of the add-on software applications is made easier and faster as these platforms provide the developers with a library of functions, an established user interface, and style of programming. These add-ons are more advantageous than the CAD platforms to an extent from the viewpoint of time and cost-saving. However the development of a similar add-ons software application for die casting die design has not been reported yet.

4.3.1 Automated Design of Die Casting Die

Traditionally, die casting die design and configuration is performed manually depending on the die engineers' experience. There is no report on the automated or semi-automated design of the whole or a portion of die casting die design process. Though the die casting industry has a desperate need for an automated or semi-automated approach for die design and configuration, the discussion of this issue was not aroused until 1997. Mehalawi and Miller [8] concluded that there is almost no approach available for computer-aided die design that extracts the casting information from its CAD model, and little attention is received in this area. In their research, they proposed an approach for automating die configuration using case-based reasoning (CBR). This approach requires a smart part database in which die casting parts are classified based on their shapes and represented by faces. Die configuration of each part is attached to the record in the database describing the part's shape. When designing die configuration for a new part, the system looks up the database and picks the part most similar to the new one through similarity assessment and selects its configuration as a starting point for the new die. The selected configuration will help both in selecting the set of components required to build a die and determining their spatial arrangement. In other words, a new die design will be based on the historical cases stored in the database. This research focused on the automatic configuration of the whole die, but no details are given on the design of individual die components. No further development based on this approach is presented, nor is there any software application reported.

4.3.2 Computer-Aided Design of Gating System

The design of gating system is an important step in the whole die design, and computer-aided design of gating system is receiving interests. Some work has been done on the design of gating system, and software packages like CASTFLOW^{TM} [9] and MAGMA^{TM} [10] are available in the market. CASTFLOW^{TM} can quickly develop a wire-frame 3D model of the gating system that matches the capacity of a machine with the metal pressure and speed requirements of the die, and the wire-frame can be provided to the toolmaker to use for cutter path generation. However this software application is focused on the flow analysis and limited to the design of runner and gating system. Furthermore, the wire-frame 3D model only defines the edges of an object, and its appearance lacks definition and may produce an optical illusion. MAGMA^{TM} is a pure simulation software application that can help avoid gating and feeding problems, but it lacks the solution for a complete die design.

To improve the process of die design, an approach must be reached for the automated or semi-automated design of the gating system. However, the result of further literature review shows that very little has been done in this area. On the contrary, researchers had carried out some work on the automated design of feed and runner system of injection molds. Irani et al. [11] presented an approach towards automating the design of the feed system of injection molds. Their research integrated CAE simulation with an iterative redesign and feature-based database in one system (AMDS) and aimed the design goal at filling all the cavities simultaneously based on the flow simulation output. Based on their system, the gating design involved the determination of number, location, and type of gates. Runner layout was decided during the configuration stage, and the size of gating elements was specified in the parametric stage. The automated design was accomplished by initiating a trial gating design, evaluating the new design using some performance parameters, and redesigning it repeatedly until an acceptable design was obtained. Lee and Kim [12] modified the previous approach by taking into consideration the transformation of the melt as it fills and packs the cavities based on packing simulation and making sure that all the cavities fill under a uniform pressure. The improved approach could achieve a more uniform cavity quality in a multiple-cavity mold and is also applicable to a family mold with different cavities.

The restrictions are that the approaches mentioned above were proposed for the design of the feed and runner system of injection molds, and they all depend on the use of a simulation package. However, these approaches deploy the parametric design and feature-based design paradigms so that it could provide some general guidelines on the automated design of the gating system of die casting die.

Semi-Automated Die Casting Die Design 145

Tu et al. [13] deployed parametric design for the gating system of investment-casting mold. In this research, parameterized solid models (primitives) of gating geometries are pre-constructed and stored in a directory (gating library). When the CAD model of the component is retrieved from the original equipment manufacturers (OEMs), the gating primitives are selected and retrieved from the gating library, specifying desired dimensions and locations. These gating pieces are then joined to the solid model of the component using a Boolean operation. This procedure can efficiently reduce the solid-model construction time required to modify the OEM model into a casting model when standard gating features are used. In their work, the approach to accelerate the model construction is proposed and hence addresses the mesh generation for the simulation process.

4.4 DESIGN OF CAVITY LAYOUT AND GATING SYSTEM

When designing a die casting die, it is important that the molten metal fills the die cavity properly. Therefore, the design of gating system is very crucial. To achieve a good gating system design, the following aspects should be taken into consideration. The first is the total volume of molten metal to be injected into the die during a shot, which is generally determined by the cavity number of the die. Secondly, the size of the gating system directly influences the performance of the die, so it should be calculated accurately. Thirdly, it is important to match the shape of the gate to the filling patterns of the die-casting part. This section shows how these three aspects are analyzed and determined.

4.4.1 Determination of Cavity Number

Using a multi-cavity die will save labor cost and improve the production efficiency. However, increasing the number of cavities requires a larger die, hence a larger machine, which will result in higher investment in the die and machine.

The cavity number is affected by several factors, such as the delivery requirement, the allowable die cost, and the machine's capacity. The first two are dependent on the dynamic market condition, and the designer must review the delivery and cost issues based on the latest information available. The capacity of a certain machine is predictable when the machine is selected for use. Therefore, the cavity number is selected which must be economically acceptable and technically permissible. This section focuses on the impact of the machine's capacity on the cavity number. The parameters related to the machine capacity, including the clamping (locking) force, the maximum flow rate, and the machine size, will primarily determine the cavity number.

4.4.1.1 Cavity Number Based on the Clamping Force

The force applied by the machine is required to hold the die halves together and must therefore exceed the force generated within the die [14]. Thus, the maximum cavity number (N_{cf}) can be calculated as in the following:

$$N_{cf} = \frac{F_c}{I \times A_p \times P_m} \qquad (4.1)$$

where
  F_c is the maximum clamping force (N),
  I is the dimensionless impact/freeze factor (1.5~3.0).
  A_p is the projected area of each cavity, its overflow and runner (mm^2),
  P_m is the maximum metal pressure (MPa).

4.4.1.2 Cavity Number Based on the Maximum Flow Rate

Flow rate represents the power of the shot system of the die casting machine. It describes the volume of molten metal that is pushed into the cavity per second by the shot system. The flow rate required by a die should not exceed the power that the machine can provide [14]. Based on this principle, the maximum cavity number (N_{fr}) can be determined as follows:

$$N_{fr} = \frac{Q_{max} \times T_{max}}{V} \qquad (4.2)$$

where
  Q_{max} is the maximum flow rate that the machine can supply (mm^3/sec),
  T_{max} is the allowed maximum filling time (sec),
  V is the volume of each cavity, its overflow and runner (mm^3).

4.4.1.3 Cavity Number Based on the Machine Size

The distance between the tie bars of the machine decides the maximum size of the die that can be used. When choosing the cavity number, one should assure that all cavity inserts be contained within the die and that adequate margins be provided. Assuming that the cavities are arranged in a rectangle array in the die, the maximum cavity number will be:

$$N_{ms} = (\text{int})\left(\frac{L_{die} - L_{mar}}{L_{ins} + 0.5 L_{dis}}\right) \times (\text{int})\left(\frac{W_{die} - L_{mar}}{W_{ins} + 0.5 L_{dis}}\right) \qquad (4.3)$$

where
  L_{ins} and W_{ins} are the length and width of the cavity inserts (mm),

Semi-Automated Die Casting Die Design

L_{die} and W_{die} are the allowable maximum length and width of the die used (mm),
L_{mar} is the minimum margin between the edges of the die and insert (mm),
L_{dis} is the minimum distance between the cavity inserts (mm), and
int is a mathematic function to round off the given numeral to an integer.

4.4.1.4 Selection of Cavity Number

Given the necessary information, the system can calculate N_{cf}, N_{fr}, and N_{ms} using the above equations. In practice, a single-cavity die is chosen frequently in the die casting process. If a multi-cavity die has to be used, the designed cavity number should not exceed the smallest value of N_{cf}, N_{fr}, and N_{ms}. The designer should consider the economic and technical issues comprehensively in order to select an acceptable cavity number that meets the overall requirements.

4.4.2 Automatic Creation of Cavity Layout

When the cavity number has been determined, the next issue of cavity layout is to arrange the cavities in the die. Some factors that may affect the cavity arrangement include product shape, cavity number, gate position, machine type, etc.

The shape of a die-casting can be very complex, so it can cause much difficulty in arranging the cavity layout. To tackle this problem, a containing box is created for the product so that the product will be treated as a single block regardless of its shape complexity. A containing box is created through these steps:
(1) Search the boundary of the part and determine the maximum 3D dimensions X_p, Y_p, and Z_p.
(2) Input the clearances between the containing box and the product's outmost boundaries, i.e., $dx1$, $dx2$, $dy1$, $dy2$, $dz1$, and $dz2$.
(3) Define the side face, which is connected to the gate and made to face to the runner inlet.
(4) Generate the containing box with derived dimensions X_c, Y_c, and Z_c.
(5) Record the information in the containing box in the forms of attributes

A containing box with a die-casting part is illustrated in Fig. 4.5. The containing box will be split into two halves along the parting surface, and impressions of the part are created on both halves. In die casting die design, die cavities can be planned in circular, balanced, or unbalanced patterns according to the spatial relationship between cavities. The patterns used in the die casting die are quite similar to those of the injection mold. However, an injection mold can have up to 96 cavities, while the cavity number of a die casting die will not exceed six normally. A die casting die with three or four cavities are most commonly used. To simplify the design of cavity layout, some layout patterns are pre-defined in the system. Table 4.1 summarized some commonly used layout patterns and their applications. The designer may select a suitable pattern in line with the cavity number and machine type, specify related

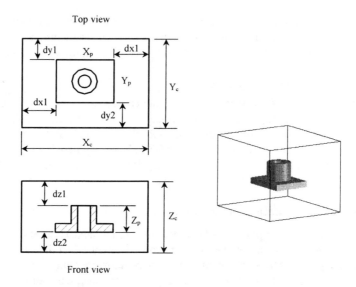

Fig. 4.5 A die-casting part and the containing box.

clearances, and then the system will arrange the cavities in the desired array corresponding to the layout pattern automatically.

4.4.3 Determining the Parameters of Gating System

For a long time, the gating system of a die-casting die was designed with the aid of past experience or a few simple empirically derived formulae. As the researchers and die-casters get better understandings of the laws of fluid flow and the machine pumping characteristics, correct design of the gating system can now be achieved. Allsop and Kennedy [15] proposed a method for calculating the cavity filling time, gate velocity, pumping rate (flow rate), gate area, etc., so as to be in line with the performances of both the die and the machine. This method is known as the P-Q^2 technique or P-Q^2 diagram. This technique combines the characteristic curves of both machine and die in one diagram and is able to estimate how a die will perform on a die-casting machine.

The P-Q^2 technique is now the most commonly adopted technique for determining how a die will perform on a particular machine. Traditionally, die designers either utilize the specially designed scales or pre-drawn diagrams. These scales or diagrams are based on the P-Q^2 relationship and reflect the characteristics of a specific material and die casting machine.

Table 4.1 Patterns of cavity layout and their applications

	Patterns	Applications
Unbalanced Patterns		The patterns in this group are only suitable for a cold chamber machine whose shot hole is not centered on the platen. The number of cavities may be odd or even.
Balanced Patterns		The series of patterns may be used in a hot chamber machine, as well as a cold chamber machine whose shot hole could be centered on the platen. The number of cavities is an even.
Circular Pattern		These patterns may be used in a hot chamber machine as well as a cold chamber machine. All cavities have an equal flow distance. Extra cost is needed for manufacturing the die.

Manual work is involved in using these scales or diagrams. In the following sections, algorithms based on the P-Q^2 technique are described to computerize the gate design application. With the user inputs, the computer will calculate and compare the data, and output the optimal result automatically. By this means, users need not draw the diagrams and calculate the results manually. A P-Q^2 diagram determines how much pressure the machine can apply to the metal to push it through the gating system for any given flow rate (plunger displacement rate).

4.4.3.1 P-Q² Diagram of Die Casting Machine

Experiments have shown that metal pressure drops as the flow rate increases in the machine's shot system during die filling. If the vertical axis represents the metal pressure, and the horizontal axis is the flow rate on squared scale, the machine's characteristic curve is a straight line (Fig. 4.6(a)). The point at which the line crosses the vertical axis is the metal pressure immediately at the end of plunger stroke, and the point at which the line crosses the horizontal is the dry shot flow rate. The relationship can be expressed mathematically by the following equation [14]:

$$P_m = P_{max} - \frac{P_{max}}{Q_{max}^2} \times Q^2 \qquad (4.4)$$

where P_m is the metal pressure (MPa) at a given flow rate Q, P_{max} is the maximum metal pressure (MPa) when the plunger reaches the end of stroke, Q is the flow rate (mm³/s), and Q_{max} is the maximum flow rate (mm³/s) when the plunger advances at dry shot speed.

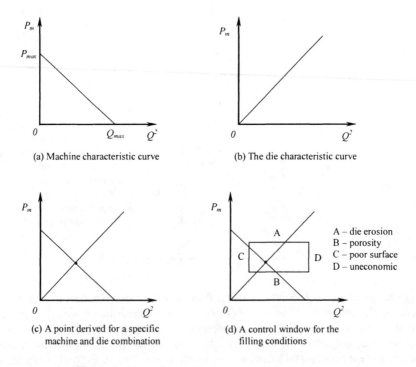

(a) Machine characteristic curve

(b) The die characteristic curve

(c) A point derived for a specific machine and die combination

(d) A control window for the filling conditions

A – die erosion
B – porosity
C – poor surface
D – uneconomic

Fig. 4.6 The P-Q^2 diagram and the control "window."

4.4.3.2 P-Q² Diagram of Die

Allsop and Kennedy [15] concluded that the metal velocity at the gate, or the gate velocity, is proportional to the square root of the metal pressure as expressed by the equation:

$$v_g = C_d \times \sqrt{\frac{0.002 P_m}{\rho}} \qquad (4.5)$$

where v_g is the gate velocity (m/s), C_d is the discharge coefficient of the die, ρ is the metal density (g/cm³), and P_m is the metal pressure (MPa). Since the flow rate (Q) is the product of gate area and gate velocity, Eq. 4.4 becomes:

$$P_m = \frac{\rho}{2 \times 10^9 \times A_g^2 \times C_d^2} \times Q^2 \qquad (4.6)$$

where A_g is the given gate area (mm²). Therefore, another straight line through the origin can be plotted on the same scale to represent the die's characteristic curve (Fig. 4.6(b)). Contrary to the shot system, metal pressure rises as flow rate increases in the die.

4.4.3.3 Determination of the Filling Time and Gate Velocity

The machine characteristic diagram or the die characteristic diagram may not make any sense to the die designer if the two are separated from each other. By adding the die characteristic diagram to the machine characteristic diagram, just like putting the die on the machine, a flow rate (Q') for a specific die and machine combination can be derived from the intersection point of the two lines (Fig. 4.6(c)). Then the filling time (T) and gate velocity (V_g) can be calculated using Eqs. (4.7) and (4.8):

$$T = \frac{V_t}{Q'} \qquad (4.7)$$

where T is the filling time (sec), V_t is the total volume (mm³) of all the parts and overflows in the die cavity, and Q' is the derived flow rate (mm³/sec).

$$v_g = \frac{Q'}{1000 \times A_g} \qquad (4.8)$$

where v_g is the gate velocity (m/sec), Q' is the derived flow rate (mm^3/sec), and A_g is the gate area (mm^2).

Gate velocity and filling time are two important factors that reflect the filling condition of the die and affect the die-casting quality significantly. A low metal pressure, i.e., a low gate velocity will lead to high air entrapment and porosity in the casting; a high metal pressure, hence a high gate velocity, will cause aggressive die erosion. On the other hand, a low flow rate, i.e., a long filling time, will result in poor casting surfaces; a short filling time gives good surface finishes, however, it will not be economical for the die caster if the filling time is too short, so its minimum value depends on the surface finish required. Table 4.2 summarizes the rules for determining boundary values of T and v_g for the filling conditions of various alloys [16]. It will be more explicable to use a "window" in the P-Q^2 diagram based on these rules (Fig. 4.6(d)). If the intersection point of the machine curve and the die curve falls within the control window, the filling conditions are acceptable, but may not be optimal.

4.4.3.4 Controlling the Optimal Filling Conditions

There are several factors that can change the machine and die curves, i.e., the interaction point, and then change the filling time and gate velocity subsequently.

(1) **Gate area**: according to Eq. 4.6, increasing the gate area will result in a die characteristic curve with reduced slope, so the derived filling time and gate velocity vary accordingly. A larger gate area will decrease both the filling time and gate velocity. Therefore, trying with a series of values of the gate area will result in various combinations of the filling time and the gate velocity (Fig. 4.7(a)).

(2) **Hydraulic pressure**: higher machine hydraulic pressure will produce higher P_{max} and Q_{max}, and the result is that the machine curve shifts away from the origin (Fig. 4.7(b)).

Table 4.2 Rules for determining boundary values of T and v_g

Alloys	Gate velocity v_g (m/s)	Filling time T (millisecond) vs. surface quality		
		Good	Medium	Common
Aluminum Alloys	$20 \leq v_g \leq 60$	$10 \leq T < 25$	$25 \leq T < 40$	$40 \leq T < 60$
Magnesium Alloys	$40 \leq v_g \leq 60$	$10 \leq T < 20$	$20 \leq T < 30$	$30 \leq T < 40$
Zinc Alloys	$30 \leq v_g \leq 50$	$10 \leq T < 20$	$20 \leq T < 30$	$30 \leq T < 40$

(3) **Shot speed setting**: the plunger speed can be adjusted through the fast shot speed valve. Changing the shot speed influences Q_{max}, but does not affect the final metal pressure P_{max}. Q_{max} increases as the shot speed goes up, but the effect is usually not linear (Fig. 4.7 (c)). The relationship between the flow rate and shot speed setting (Ψ) can be established through calibrating the machine's shot system.

(4) **Plunger size**: plunger size affects both maximum metal pressure P_{max} and maximum flow rate Q_{max} of the machine curve. Q_{max} is proportional to the plunger size while P_{max} is inversely proportional to it. With other machine's settings fixed, a larger plunger will give a lower metal pressure, and, however, a greater flow rate (Fig. 4.7(d)). Every die-casting machine has a different P-Q^2 diagram for every size of plunger that might be used. There exists an optimum plunger size for each machine and die combination such that machine will give a high flow rate and short filling time and maintain sufficient gate velocity while metal pressure will not be too high to produce excessive flash.

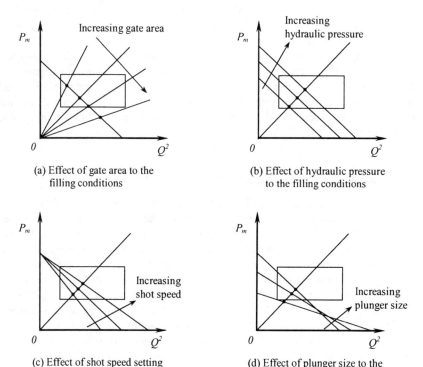

(a) Effect of gate area to the filling conditions

(b) Effect of hydraulic pressure to the filling conditions

(c) Effect of shot speed setting to the filling conditions

(d) Effect of plunger size to the filling conditions

Fig. 4.7 The factors that influence the filling conditions.

Considering the influences of the above factors on the P-Q^2 diagram, an optimum filling condition can be achieved. Under each machine setting, one may try with different values of gate area, calculate the flow rate, filling time, and gate velocity, and then compare the values of the filling time and gate velocity with the boundary values. If no satisfactory result is obtained, one may adjust the machine settings, i.e., change the values of plunger size, shot speed and hydraulic pressure, and then repeat the calculation again until a suitable combination of filling time and gate velocity is obtained. The machine settings and the gate area, which achieve the filling conditions, are then finally determined. This process is analogous to carrying out a series of die trials on the machine.

4.4.3.5 Automatic Determination of the Filling Conditions

Based on the above principles and guidelines, an algorithm is proposed for automatically determining the gate area and machine settings. In this study, the hydraulic pressure of the machine is regarded as fixed. Assuming that a die casting machine has been tested using a certain plunger and the resultant P_{max} and Q_{max} are obtained, if given the total part volume (V) in the die cavity, the optimal flow conditions can be estimated, and the related parameters can be determined using the algorithm as shown in Fig. 4.8, in which the symbol Com (v_g, T) represents a combination of gate v_g and filling time T.

The derived result is a set of data including plunger diameter, shot speed setting, gate area, filling time, flow rate, and gate velocity. The value of the gate area could then be utilized to determine the geometric parameters of gating system, and some other data are used for setting up the machine. These data are recorded in the gating part as attributes, and they can be extracted for calculation through some automatic routines.

4.4.4 Design of Gating Features

To automate the generation of the gating system, a *gating feature library* must first be built, and adequate user-defined features for the gating system must be constructed and stored in the library before the gating system design proceeds. A user-defined gating feature is a parametric model, which contains a principal shape, a group of driving parameters of dimension and a set of origin and axis for positioning. The gating feature library stores the parametric models of gate, overflow, runner and shot sleeve bush/sprue, and the application-specific features can be added to the library whenever necessary. The geometries suitable for parametric and feature-based design are those that can be standardized. Generally, the gating elements, especially the gate-runner, have relatively complex geometry. However, it is possible to classify the gating elements that are commonly used into different types by their geometric structures and functionalities.

Semi-Automated Die Casting Die Design

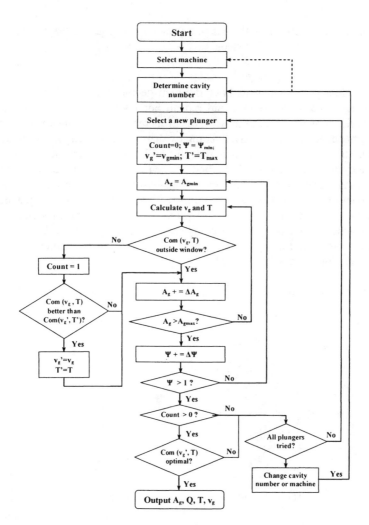

Fig. 4.8 The flow chart for determining the filling conditions and machine settings.

The gating features are defined based on the considerations of both product functionality and design intent in order to make the design and redesign easy. First, it must be assured that the gating element is able to fulfill its functions when the molten metal fills the die. Secondly, it must be certain that the parametric models will change as expected when the parameters are modified. The approaches presented by Hoffmann and Arinyo [17] and Chen and Wei [18] were implemented on Pro/Engineer CAD system. In contrast to Pro/Engineer, Unigraphics allows more than one solid object to exist separately in one part, and their shape and placement do not necessarily depend on the existing geometry. The gating features can be created through the following five steps.

(1) *Creation of solid models*

Different gating elements have different geometrical complexity, so various methods are used to create the solid model. For instance, the gate-runner, which connects the runner and gate to achieve an increase of pressure by reducing the cross-sectional area, has a relatively complex shape, and is often jointly created with the gate. Street [19] used five cross section profiles to represent the different zones of the gate-runner, among which the first profile (S_0) and the fifth (S_4) are gate opening and runner cross section respectively. Fig. 4.9(a) shows the process of creating a solid model of a gate with a fan gate-runner. First, three basic datum planes DP_{xy}, DP_{yz} and DP_{zx} are set up, and S_0 is sketched in DP_{zx}. Then the profiles S_1, S_2, S_3, and S_4 are sketched respectively in four equidistant planes DP_{zx1}, DP_{zx2}, DP_{zx3}, and DP_{zx4} offsetting from DP_{zx}. Finally, all the profiles are formed into the solid model through lofting operations (Fig. 4.9(c)).

(2) *Definition of constraints*

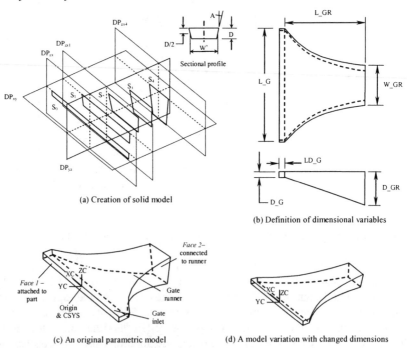

Fig. 4.9 Creation process of gating feature [2].

Semi-Automated Die Casting Die Design

Spatial constraints, e.g., *parallel, point-on-line, offset*, etc., are applied while creating the datum planes and sketching the cross-sectional profiles. Now the dimensional constraints can be defined to capture the mathematical relationships between the geometric dimensions and engineering variables. In this gate-runner model, there is a linear reduction of cross-sectional area as well as a linear decrease of depth from S_4 to S_0 along the length, so the equation constraints between the five cross-sectional profiles are defined as follows:

$$D_g = A_g/W'_g; \quad D_r = A_r/W'_r;$$
$$D_1 = D_g + \tfrac{1}{4}(D_r - D_g); \quad W'_1 = (A_g + \tfrac{1}{4}(A_r - A_g))/D_1;$$
$$D_2 = D_g + \tfrac{1}{2}(D_r - D_g); \quad W'_2 = (A_g + \tfrac{1}{2}(A_r - A_g))/D_2; \quad (4.9)$$
$$D_3 = D_g + \tfrac{3}{4}(D_r - D_g); \quad W'_3 = (A_g + \tfrac{3}{4}(A_r - A_g))/D_3;$$
$$D_r > D_g; \quad A_r > A_g$$

where A, W' and, D represent cross-sectional area, average width, and depth, and subscripts g and r stand for the gate and its connected runner.

(3) *Definition of dimensional variables*

The driving dimensions of a user-defined gating feature are specified as dimensional variables. Usually these dimensions are of functional significance to a gating element, e.g., for the gate shown in Fig. 4.9(b), the gate length (*L_G*), gate depth (*D_G*), gate land (*LD_G*), gate-runner length (*L_GR*), gate-runner depth (*D_GR*), and gate-runner width (*W_GR*), require well-evaluated values to match the filling conditions. The user can derive a gating element from the user-defined feature with a desired size by specifying the values of dimensional variables through the interface. The values are recorded in the generated gating element in the form of expressions. When the user modifies the variables externally, the new values are passed to the expressions, and the shape of the gating element will be changed accordingly upon updating (Fig. 4.9(d)).

(4) *Specification of origin and CSYS*

Each gating feature has its own origin and CSYS (coordinate system) to facilitate the attachment of the feature to the part. When the feature is retrieved from the library, it is attached by matching both its origin and CSYS with a designation point and CSYS derived from the part. The origin and CSYS of the gating feature also function as references when the geometry is rotated or translated.

(5) *Definition of attributes*

The feature information can be recorded as attributes. The gating features are categorized into various families according to their functions: gate, runner, overflow, sprue, etc. Each family is also composed of features of different types and subtypes. Therefore, textual attributes, including name, type, subtype, and series number, can be assigned to establish a unique identity for the feature. Topological attributes are assigned to the boundary faces to determine which object they must be connected to, e.g., *face* 1 and *face* 2 must be linked to a part and runner respectively (Fig. 4.9(c)). In addition, the information of position and orientation of the feature must be recorded due to the loss of such information after the feature is imported into from a CAD system.

Fig. 4.10 illustrates some of the gating features that are stored in the library of the design system. Among them, (a), (b), and (c) are fan gate features, (d) is tangential gate feature, (e) is a compound gate feature composed of two tangential gates, (f) and (g) are overflow features. (h) and (i) are the two runner template features suitable for a two-cavity and a four-cavity die respectively in a cold chamber machine.

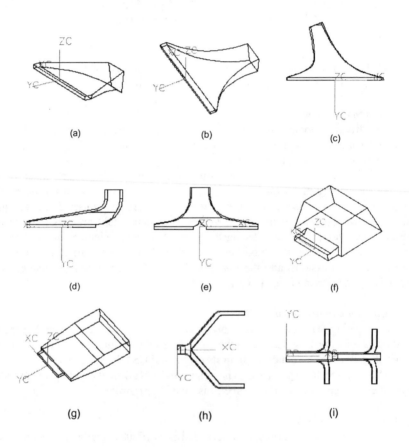

Fig. 4.10 The gating features [2].

Semi-Automated Die Casting Die Design

4.4.5 Conforming the Gate Geometry to the Die-Casting Part

After the value of the gate area is determined using the preceding algorithms, one has to decide how the molten metal would flow into the cavity through the gate. The shape of the die casting part varies a great deal. Herman [14] listed some common shapes to which a die-casting part may be similar. They are: (a) general flat plate, (b) open frame, (c) nearly flat disc, (d) shallow box, (e) deep box or doom, and (f) tubular, each of which represents a unique metal flow pattern and requires a certain feeding style. There are also some guidelines for planning the flow of the molten metal, such as:

- To make the metal flow in the direction of the shortest dimension of the cavity.
- To make all the flow paths parallel or diverge from each other.
- To make all parts of all the cavities in the die be filled at the same distance.

The flow pattern depends on the geometric complexity of the die-casting part. The desired flow pattern can be achieved by directing the molten metal through the gate using different gate-runners. The gate-runner forms a transition from the main runner to the gate opening. There are two most commonly used gate-runners, namely *fan* gate-runner and *tangential* gate-runner. Some rules may be summarized for selecting the gate-runner type to match the shape of the die-casting part as follows:

Rule 1: If some portion of the die-casting is like a trapezoid or square, then the fan gate-runner is used.
Rule 2: If the region of a die-casting approximates a parallelogram or rectangle, then the tangential gate-runner is used.
Rule 3: If the die-casting has a complex shape, then a hybrid gate-runner is used.

For example, the shape of the die-casting shown in Fig. 4.11 approximates a square, so it is properly filled using a fan gate-runner. The telephone case shown in Fig. 4.12 is die cast using magnesium. Its shape is like a rectangle, and it is better to use a tangential gate-runner.

4.4.6 Design of Shot Sleeve, Sprue, and Spreader

The cold chamber machine requires a shot sleeve inside which the plunger pushes the metal into the die. The chamber shaped by the shot sleeve and the plunger forms a part of the gating system. A biscuit is formed at the tapered end of the shot sleeve when the metal solidifies completely. Standard shot sleeves are available commercially in the tooling industry.

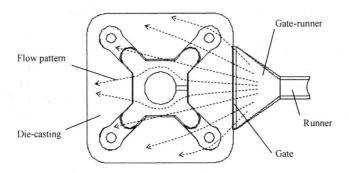

Fig. 4.11 A simple fan gate-runner and the flow pattern in the die-casting.

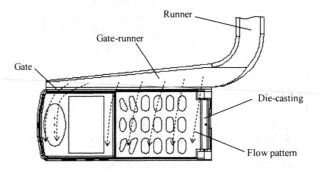

Fig. 4.12 A single tangential gate-runner and the flow pattern in the die-casting.

The molten metal delivery system on hot chamber machine allows for the use of a sprue rather than an inserted shot sleeve. Usually a sprue spreader is used together with the sprue to allow smooth delivery of the metal into the die. The cavity between the sprue and spreader forms a part of the gating system. The die casting industry has also made the sprues and spreaders in standard sizes.

Semi-Automated Die Casting Die Design

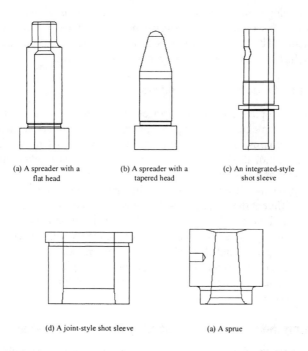

(a) A spreader with a flat head (b) A spreader with a tapered head (c) An integrated-style shot sleeve

(d) A joint-style shot sleeve (a) A sprue

Fig. 4.13 The parametric models of standard components.

In this study, the shot sleeve, sprue, and spreader are treated as standard parts. To facilitate the design of these standard parts, parametric models of different series are constructed according to the data in [20]. Some of these standard parts are shown in Fig. 4.13. The standard parts are also instantiated and inserted to the die assembly through a user interface. Various types and dimension series are ready for selection. The shot sleeve must be selected so that the inner diameter is equal to the outside diameter of the plunger. The length series must be in line with the thickness of the die plates. The sprue and spreader must be selected so that the most restrictive cross-sectional area in the metal flow is equal to or greater than the total cross section of the runners leading away from the sprue.

4.5 DIE-BASE DESIGN

4.5.1 Design of Die-Base

Die-base is an essential part of a complete die casting die. It is the structural frame that contains all the die components like the die inserts, cooling system, ejection system, moving core mechanism, etc. A two-plate die is adopted as the standard die-base because this particular die construction is the most widely used design in industrial practice. The following sections describe a systematic approach to design a die-base, which can be automated through a computer system.

4.5.2 Die-Base Structure and Variables

A typical two-plate die-base comprises two main plates—a cavity plate and a core plate, plus a top clamping, a support plate, an ejector-retainer plate, an ejector plate, a bottom clamping plate, guide pins, guide bushes, and screws. Fig. 4.14 illustrates the cross-sectional view of a typical two-plate die-base (excluding the guide pins, guide bushes and screws).

Basically, the die-base varies in three dimensions, i.e., length, width, and overall height. For different die casting dies that have various cavity, there exists a structural similarity in their ejection system, the size of its components, i.e., the ejector plate, ejector-retaining plate, and ejectors. They are relatively independent on the shape of die-casting part, but dependent on the overall size of the die-base. With reference to Fig. 4.14, the dimensional variables of a typical two-plate die may be summarized in Table 4.3, among which the indirect variables are dependent upon the primary direct variables.

4.5.3 Creating the Parametric Assembly Models of Die-Base

To automate the die-base design, a parametric die-base assembly, which includes all the plates, guide pins, guide bushes, and screws, is pre-modeled and stored in the library. The die-base assembly is divided into two subassemblies, namely Cavity Half and Core Half. The former consists of the Top Clamping Plate and Cavity Plate A, and the latter includes the Core Plate B, Support Plate, Ejector Plate, etc. The assembly structure of the die-base and the whole assembly are illustrated in Fig. 4.15 and Fig. 4.16, respectively.

In the die-base assembly, all the parts have a relationship with one another. These relationships are represented in terms of mating constraint and dimensional constraints. *Mating Constraint* is a mating relationship between two geometric objects on two different components in an assembly. There are three types of mating constraints for mating a component to another [21]:

- *Mate* (one face to another). *Mate* positions two objects of the same type so that they are coincident, and an offset distance between planar objects is allowed.
- *Align* (faces or axes). For planar objects, this constraint places two objects so that they are coplanar and adjacent, and an offset is allowed between the two planes. For axi-symmetric objects, *Align* makes their axes concentric or parallel.

Semi-Automated Die Casting Die Design

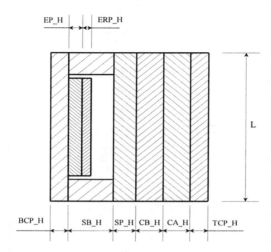

Fig. 4.14 A typical two-plate die and its variables.

Table 4.3 Dimensional variables of a typical two-plate die-base

Variable Symbol	Meaning of the variable	Remarks
L	Length of die-base	Primary direct variable
W	Width of die-base	Primary direct variable
CA_H	Height of Cavity Plate A	Primary direct variable
CB_H	Height of Core Plate B	Primary direct variable
TCP_H	Height of top clamping plate	Indirect variable
SP_H	Height of support plate	Indirect variable
BCP_H	Height of bottom clamping plate	Indirect variable
ERP_H	Height of ejector-retaining plate	Indirect variable
EP_H	Height of ejector plate	Indirect variable
SB_H	Height of space blocks	Indirect variable

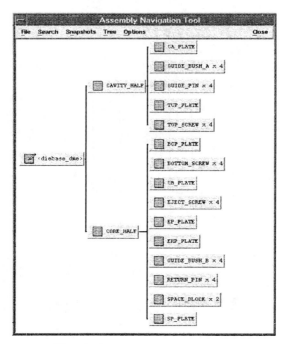

Fig. 4.15 The assembly structure of a die-base.

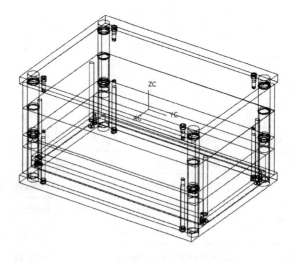

Fig. 4.16 The parametric assembly of a die-base.

Semi-Automated Die Casting Die Design 165

- *Orient* (lock down the angle). *Orient* positions two objects using a rotational angle between them, and allows the mating between mixed geometry types such as faces and edges.

To position a component in an assembly, a combination of these constraints type is used. Usually, two or three mating constraints are enough to determine the position and orientation of a component. For example, to constrain a rectangle block, usually three mating constraints are applied in succession: *Mate* (face) =>*Align* (face) =>*Align* (face); usually a cylinder is sufficiently orientated by two constraints: *Mate* (face) =>*Align* (Axis).

Dimensional Constraint - To maintain the dimensional associativity among all parts in the assembly, all the dimensional variables of the die-base are defined in the parent part of the assembly in the form of expressions, and dimensional variables of individual child part are linked to those in the top-level part. Fig. 4.17 shows that the variables of child parts *ca_plate* and *top_screw* are linked (symbol '::') to the parent part *diebase_dme*. During a design process, the input values are passed to the parent part solely, and this associativity updates all the expressions linked to the parent part. When the values of expressions in the parent part are changed, those in the child parts are also changed.

4.5.4 Building the Die-Base Database

As standard die-bases are available commercially, die-base design becomes easier. The D-M-E Company is one of these companies that supply standard die-bases of various sizes [14]. The size of its B-series die-base ranges from 156mm×156mm to 596mm×896mm, and the plate thickness varies in a large range [22]. A database is established to contain the information in the die-base catalogue, including the primary dimension list, plate thickness, spatial dimensions, and sizes of the screws, pins, bushes, etc. When a certain type and size of die-base is chosen, a corresponding record in the database is retrieved and passed to the parametric assembly model upon instantiation.

4.5.5 Automatic Generation of Die-Base

To facilitate the design of die-base, a user interface needs to be designed for the user to select the type, primary dimension, and other related data. The parametric assembly model of die-base can be retrieved from the library and instantiated with the corresponding parameters accessed from the catalogue database. The die-base is to be positioned and oriented by the following methods:

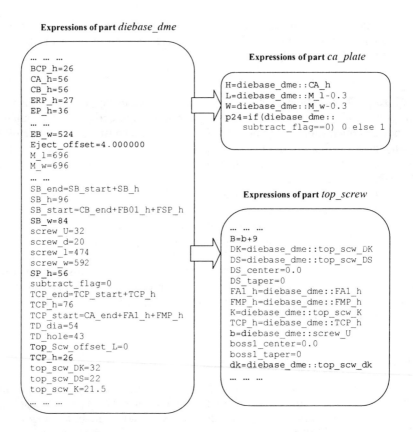

Fig. 4.17 Dimensional variables associativity among the parts.

- Mate the interface between cavity and core plates to the potential parting surface;
- Align the Z+ axis to the die-open direction; and
- Align the center of the die-base to that of the cavity layout.

The process of instantiating the parametric assembly to the whole die assembly was illustrated in Fig. 4.4.

4.6 GENERATION OF CORE AND CAVITY

Similar to the case of injection mold, generation of core and cavity is a very important step in die casting die design. The proposed die design approach can utilize the functions provided by any mold design system, e.g., IMOLD™, to create the core and cavity for a die since the design tool is independent of the product geometry. Using this tool, the core and cavity are generated as follows:

(1) Patch the through holes on the part surface that may block the subtract operation;
(2) Create the parting lines and then generate the parting surface;
(3) Subtract the containing box with the die-casting part model; and
(4) Split the containing box into two halves using the derived parting surfaces and patching surfaces.

4.7 AUTOMATIC SUBTRACTION OF DIE COMPONENT

The die plates in the parametric die-base assembly are blank solid blocks, i.e., no hole or cavity is created in the plates yet. Corresponding holes and cavities should be created on the plates after the gating system, the ejector system, the cavity inserts, etc., are designed. This is accomplished by subtracting the plates using the related components and features.

The subtract operation is to subtract the volume of one object from another. For example, to create a hole in a plate, one first models an object in the shape of the desired hole and use it as the modifier object (tool object) to perform a Boolean subtract operation on the original object (target object). Let P represent the target object and Q the tool object, the subtract operation can be represented mathematically by Eq.(4.10) [23]:

$$P - Q = P \cap \overline{Q} \tag{4.10}$$

where $\overline{Q} := \{X \mid X \text{ out of } Q\}$, which is the complement of Q. In a subtract operation, the plates, such as the cavity plate and the core plate, usually function as target objects, while the bushes, shot sleeve, sprue, screws, gating features, ejector pins, etc., function as tool objects.

However, in practice, the hole, into which a component is inserted, is not exactly the same as the outside boundary of the components. For example, there is component assembly containing a plate and a shot sleeve, the latter is to be mounted into the hole on the plate. The geometry of the shot sleeve is shown in Fig. 4.18(a). It is not acceptable to subtract the plate using the bush directly due to these reasons:

- The groove adjacent to the step produces a circular protrusion in the hole, which may block the shot sleeve from inserting in and pulling out.
- The hole in the shot sleeve may prevent the two objects from subtracting properly.

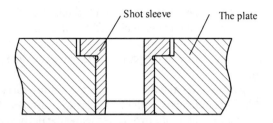

(a) The assembly of a plate and a shot sleeve

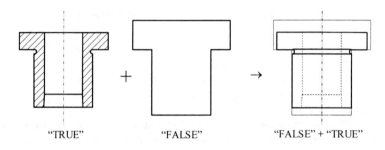

"TRUE"　　　　　"FALSE"　　　　"FALSE" + "TRUE"

(b) The two reference sets within one component

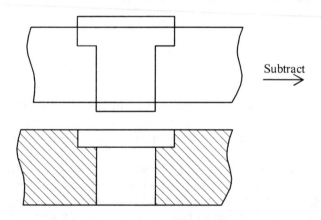

(c) Subtract with the "FALSE" reference set

Fig. 4.18 The process of subtracting a plate using a shot sleeve.

- It is reasonable to make the upper part of the hole larger than the head of the shot sleeve.

To solve the problem, another object, which reflects the shape of the hole, is created and linked with the original model of shot sleeve. This accessory object acts as a reference set for the subtract operation. To differentiate the two objects, the original and the accessory are named "TRUE" and "FALSE" reference set separately (Fig. 4.18(b)). When a component is chosen as the tool object during a subtract operation, the system will take the "FALSE" reference set automatically, and subtract the target object with it correctly. Fig. 4.18(c) shows the process of subtracting the plate using the "FALSE" reference set of the shot sleeve.

4.8 SYSTEM IMPLEMENTATION AND EXAMPLES

The concepts and methodology presented in the previous sections have been implemented into a feature-based parametric computer-aided design system for the die casting die, i.e., *DieWizard* [3]. In this section, the development platform, system architecture, and various functional modules of the system are introduced first, and then two case studies are presented to demonstrate the approach and prototype system.

4.8.1 Development Platforms and Languages

The reported die design system was developed on the Unigraphics platform in a HP-UX environment using C language. Unigraphics was chosen in this case because it allows users to build a customized system on it using UG /Open API and it also supports design with user-defined features due to its parametric nature and the ability to extract design information from the features. Unigraphics provides the User Function (UFUN) as a software tool to enable an easy interface between Unigraphics and user-developed routines. Similar to UG, the Pro/Engineer or SolidWorks systems all have the following advantages from the viewpoint of software development:

(1) API contains a large set of user callable functions/subroutines that access the graphics terminal, file manager, and database of the CAD module. It offers an easy and friendly interface between the CAD platform and the "add-on" application developer.
(2) provides a versatile 2D sketch tool and some fundamental form features, which can be used to construct more complex user-defined features according to various application-specific needs.
(3) provides a routine for the user to record and extract certain types of information in the form of attributes. Attributes can be assigned to parts and geometries,

including solids, faces, and edges, and such information can be extracted when necessary through calling certain subroutines/functions.
(4) It also has a friendly user-interface builder, with which the developers can design the menus and dialogues for the communication between the system and the end users.
(5) The User Function makes it very easy to interface to the solid kernal from high-level languages like C/C++, and the User Function programs can run in either an external or an internal environment.

The details on system development platforms for die and mold design will be further discussed in Chapter 9.

4.8.2 System Architecture—*DieWizard*

The architecture of the proposed die design system is illustrated in Fig. 4.19. The database, the gating feature library and the die-base library are embedded in the CAD platform. The sub-menu of each module contains a group of methods and procedures to carry out specific functions.

4.8.2.1 Data Initialization Module

This module provides the user functions to load a die-casting part, input process data, specify die open direction, create containing box, etc.

The *Load Part* option enables the user to start a new project of die design by loading the 3D model of a die-casting part. The user may continue with a half-done project or make modification to an existing one by loading the whole assembly using this option. The *Set Die CSYS* option lets the user specify the die-open direction. The system defaults the Z+ as the die open direction. The *Input Process Data* option lets the user input the process data including the material properties, the characteristics of the product; and the machine parameters. The shrinkage allowance is to be assigned and the part is scaled using the *Apply Shrinkage Allowance* button (Fig. 4.20). The containing box of the cavity insert is created using the *Contain Box* option.

4.8.2.2 Cavity Layout Module

This module provides routines for calculating the cavity number based on the process data input previously and automatically arranging the cavity layout upon user selection. Fig. 4.21 shows the interface for selecting the layout pattern and specifying spatial parameters.

Semi-Automated Die Casting Die Design

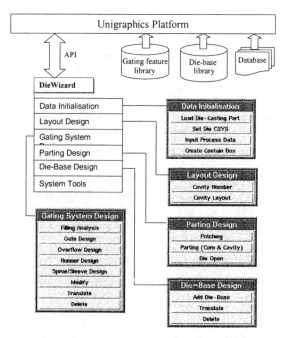

Fig. 4.19 The system architecture of *DieWizard*.

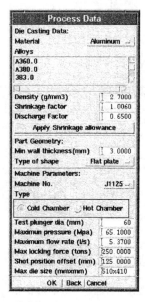

Fig. 4.20 The dialogue for data input.

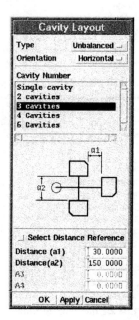

Fig. 4.21 Interface for cavity layout.

4.8.2.3 Gating System Design Module

Before gating system design proceeds, the user needs to analyze the filling conditions using the *Filling Analysis* option (Fig. 4.22). With the result, the user may design the gate, overflow, runner, and sprue/sleeve using related option. The user needs an interface to retrieve the gating features from the library and attach them to the gating part. Fig. 4.23 shows an interface designed using Unigraphics User Interface Styler (UIStyler) for designing the overflow. The user can select the type and sub-type of the feature, choose the location option, and input related data. The interface will display a template of the feature of selected type so that the user can identify the dimensional variables visually.

The interfaces of gate, runner and sprue/sleeve design are illustrated respectively in Fig. 4.24, Fig. 4.25, and Fig. 4.26. The sizes of the sprue and sleeve are dependent on the thickness of the die plates, and are usually designed after the die-base is built.

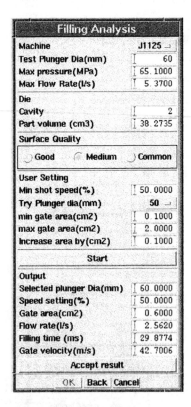
Fig. 4.22 Interface for filling analysis.

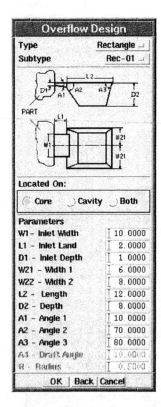
Fig. 4.23 Interface for overflow design.

Semi-Automated Die Casting Die Design

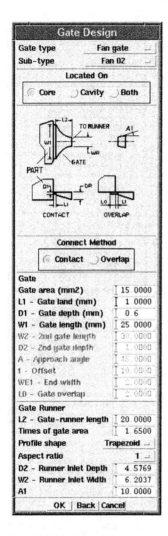

Fig. 4.24 Interface for gate design.   **Fig. 4.25** Interface for runner design.

4.8.2.4   Die-Base Design Module

The die-base may be designed before or after the parting is accomplished. Fig. 4.27 demonstrates the interface of conceptual design of the die-base. Using this dialogue, the user can specify the type and subtype of die-base used, and choose the primary dimensions of the die-base and the thickness of various die plates. Using the detailed design dialogues, the designer may also further adjust the dimensions of the plates and change the spatial dimensions of the screws, pins, and bushes.

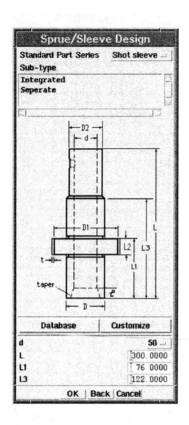 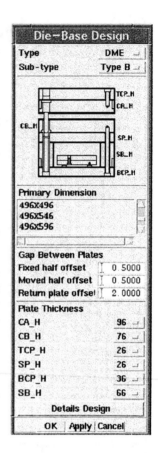

Fig. 4.26 Interface for sleeve/sprue design.   **Fig. 4.27** Interface for die-base design.

4.8.2.5   Parting Module

This module provides the tools for patching and parting. After the through holes are patched using the *Patch* option, the user may use the *Parting* option to select the parting lines, and the system will then generate the core and cavity for the die-casting product automatically. *Die Open* option can be used to simulate the process of die opening and closing.

4.8.2.6   System Tools Module

This module provides tools for subtracting the related cavities and holes in the plates and cavity inserts. When the subtraction function is selected, the user will be prompted to choose the target objects and tool objects separately, and then the system will subtract target objects using the tool objects automatically.

4.8.3 Examples

(1) *Case Study 1*

This case study uses a simple die-casting part to illustrate the whole process of gating, layout and die-base design. First, the 3D model of the sample part, a cover plate, is loaded, necessary process data are input and shrinkage allowance is applied (Fig. 4.28). In this case, the machine considered is based on a cold chamber die casting machine J1125 with clamping force of 250 tons and maximum allowable die size of 580mm x 580mm. The machine has a maximum metal pressure of 65.1MPa and a maximum flow rate of 5.37 liter/sec when using a plunger of 60mm in diameter [20].

The shape of the cover plate is similar to a symmetric trapezoid. When the die cavity is filled, the molten metal starts flowing in from the side, scatters in the middle and then ends at the other side. Hence one fan gate is used to deliver the molten metal as the whole part can be considered as one segment. An overflow is designed to absorb the air and inclusion at the point where the flow reaches last.

After considering the size limitation, injection, and clamping capacities of the machine, a four-cavity die is used. A four-cavity layout pattern is loaded and the cavities are planned in the die in conformity with the pattern after the intervals between the cavity inserts are specified (Fig. 4.29).

The filling analysis module is then activated to calculate the required gate area, filling time, and gate velocity (Fig. 4.30). The results show that the filling time and gate velocity reach 31.35m/s and 39.32m/s respectively when a $\Phi 50$ plunger is used; the shot speed is set at 65% and total gate area comes to $0.60cm^2$. This set of data is chosen for the specific machine and die combination, and they are then recorded in the part file as attributes.

The next step is to design the gating elements. Each of the four cavities has a single gate, and all four gates are created simultaneously. As the cross-sectional area of each gate is $15.0mm^2$, dimensional parameters could be determined accordingly and the gate geometries are generated using the gate feature (Fig. 4.31). A runner feature that matches the cavity layout pattern is chosen (Fig. 4.32). As there is an intended reduction in area from the runner inlet to the gate along the flow path, the sectional area of each runner segment is determined based on the total area of the next segment. The user needs only to specify each segment's area factor, cross-sectional profile type, and aspect ratio; the system will then calculate all the dimensional parameters of the main runner and sub-runners, and the runner geometry is generated and attached to the gating part automatically. A die-base with a size of 446×546 is selected for the four-cavity die (Fig. 4.33). The shot sleeve is designed last, and the whole die assembly is shown in Fig. 4.34.

176   Chapter 4

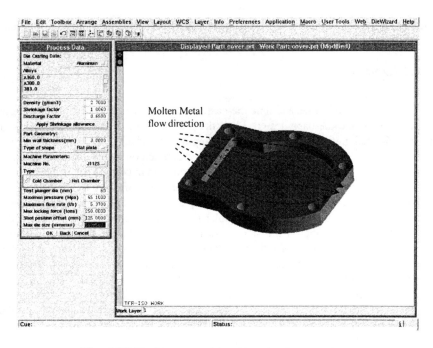

Fig. 4.28 The 3D part model and its related process data.

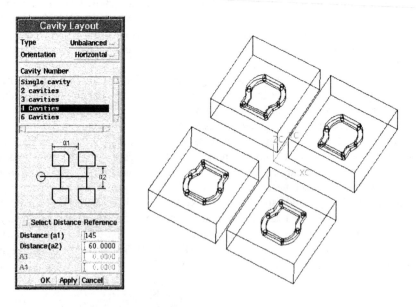

Fig. 4.29 The cavity layout created from a four-cavity pattern.

Semi-Automated Die Casting Die Design

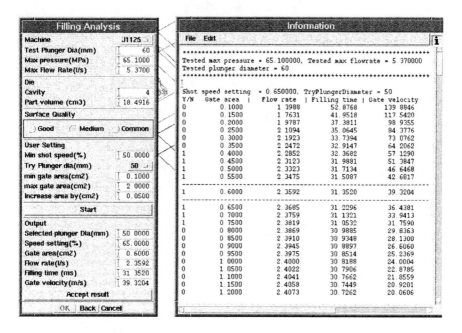

Fig. 4.30 The interface for filling analysis and the result generated.

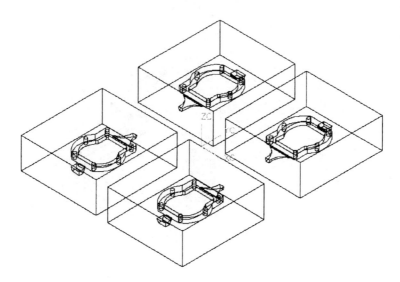

Fig. 4.31 Design of gate and overflow.

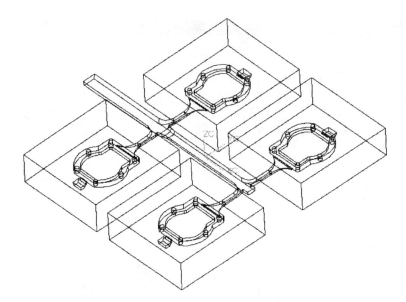

Fig. 4.32 Design of runner.

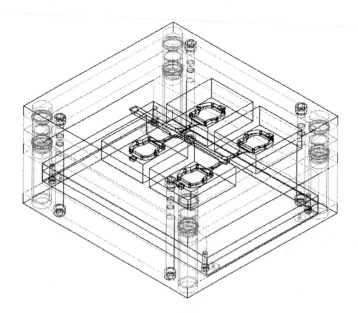

Fig. 4.33 Design of die-base.

Semi-Automated Die Casting Die Design

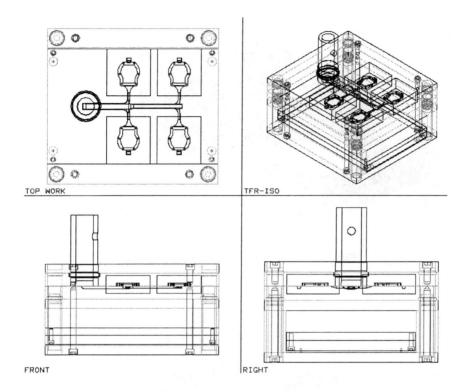

Fig. 4.34 The designed die-casting die assembly, Case 1.

(2) *Case Study 2*

This case study uses a relatively complex die-casting part, a hard disk support plate. This case emphasizes how to select proper gating elements for a complex die casting part, how to generate core and cavity, and how to subtract different die plates and components to form the related impressions, holes, and channels.

This case study is based on a cold chamber machine J1140 with a clamping capacity of 400 tons and allowable die size of 760mm x 660mm. The 3D model of the support plate is shown in Fig. 4.35. A three-cavity layout pattern is used for this part (Fig. 4.36).

As the big hole in the part will split the metal flow into two streams, two flow paths are established. The molten metal will go in from one side, divide into two streams, and meet at another side (Fig. 4.37). Hence one compound gate with two tangential gates are used in this case. Gate areas of the two sub-gates are in proportion to the volume of the two portions they fill. Two overflows are posi-

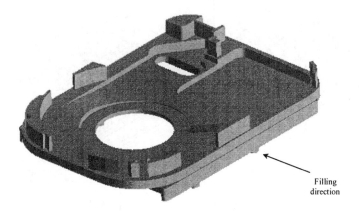

Fig. 4.35 The 3D model of a hard disk support plate.

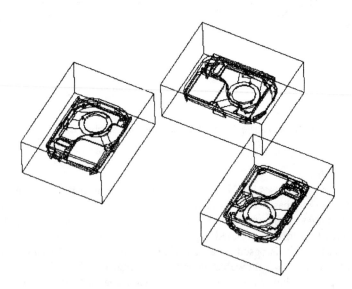

Fig. 4.36 The three-cavity layout.

tioned where the flow fronts end to absorb the air and inclusions. After the filling conditions are analyzed and related parameters are derived, the gates and overflows are created. Then the runner is designed in accordance to the layout pattern. In this case, the overflows and gates are located in the fixed die half, while the runner is positioned in the moving die half, the gates and runner are connected through over-

Semi-Automated Die Casting Die Design

lapping each other for a certain length (Fig. 4.38). The die base is then designed and the shot sleeve is added; the whole die assembly is shown in Fig. 4.39.

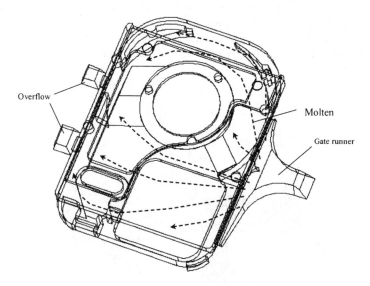

Fig. 4.37 Design of gate and overflow.

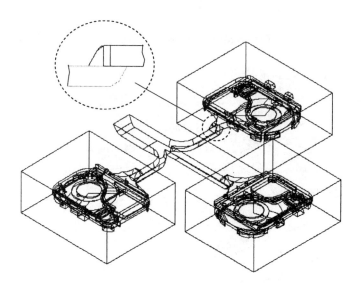

Fig. 4.38 The entire gating system.

The parting module is now used to create the core and cavity. First, the two holes in the part are patched. Then the parting lines and parting directions are specified manually or automatically to generate the parting surfaces. Lastly, the containing box is split into two halves by the parting surfaces, and impressions of the part shape are formed in the two halves automatically. The two halves must be further subtracted using the gating system, and then the cavity and core inserts are generated completely (Fig. 4.40).

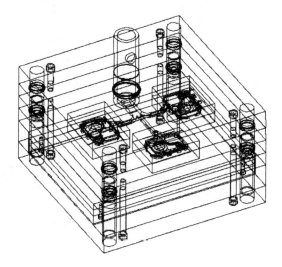

Fig. 4.39 The die assembly, Case 2.

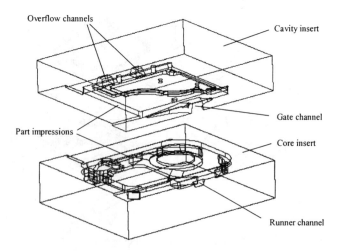

Fig. 4.40 The core and cavity inserts.

Semi-Automated Die Casting Die Design

The next step is to subtract the die plates to create the related impressions, holes and channels. Fig. 4.41 and Fig. 4.42 show the subtracted core plate and cavity plate.

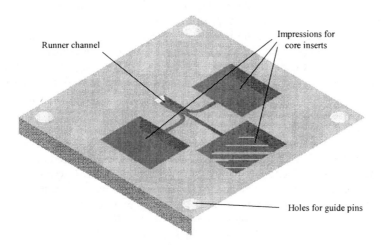

Fig. 4.41 The core plate.

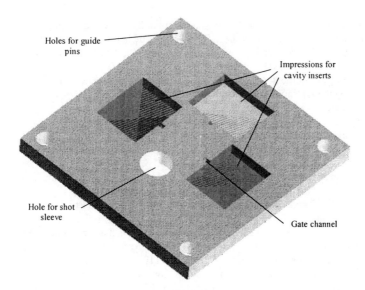

Fig. 4.42 The cavity plate.

4.9 SUMMARY

This chapter summarizes the research and development of the computer-aided design approaches for die casting die design. The die casting die design process involves the design of cavity layout, gating system, die-base, parting, ejection system, cooling system, etc. The proposed approach and its reported prototype system aim to facilitate these processes by automating the generation of the gating system, cavity layout, parting, die-base, etc. To achieve the goal, parametric and feature-based design paradigms are employed to develop the die casting die design system. Parametric models of gating elements, die-bases, and layout templates are pre-constructed and stored in the library, and they are retrieved from the library and instantiated into the die assembly automatically after related dimensional and positioning parameters are specified. Algorithms based on the P-Q^2 technique are proposed to carry out filling analysis and estimate the process parameters, and the dimensional parameters of the gating system are determined based on the estimated result. Based on these concepts and methodologies, a prototype design system, *DieWizard*, was reported and developed based on a commercial CAD platform. The system is able to semi-automate the processes of cavity layout design, gating system design, die-base design, etc., and greatly shorten the design time needed. However, the methodology can be similarly applied to and implemented with other CAD/CAM systems.

REFERENCES

1. NADCA, North American Die Casting Association, http://www.diecasting.org, 2001.
2. J.Y.H. Fuh, S.H. Wu and K.S. Lee, Development of Semi-Automated Die Casting Die Design System", Proc. of Inst. of Mech Engr, Part B.: J of Engr Mfg, Vol. 216, No. V12, pp. 1575-1588, 2003.
3. S.H. Wu, Development of Feature-based Parametric Design System for Die Casting Die, MEng thesis, National University of Singapore, 2001.
4. Unigraphics, UG Solutions, http://www.ugsolutions.com, 2002.
5. Pro/Engineer, PTC Corporation, http://www.ptc.com, 2002.
6. SolidWorks 2002, Solid Works Corporation, http://www.solidworks.com, 2002.
7. IMOLD, ManuSoft Technology Pte. Ltd., http://www.manusoftcorp.com, Singapore, 2002.
8. M.E. Mehalawi, and R.A. Miller, "Towards computer-assisted configuration of die casting dies", Die-Casting Engineer, January/February, 2000, pp.41-46.
9. CASTFLOW™ and CASTHERM™ Software, Castec Australia Pty. Ltd., 2000.
10. MAGMA Software, Germany, http://www.magmasoft.com/, 2001.
11. R.K. Irani, B.H. Kim, and J.R. Dixon, "Towards automated design of the feed system of injection molds by integrating CAE, iterative redesign and features", Journal of Engineering for Industry-Transactions of the ASME, Vol.117, No.1, pp.72-77, 1995.
12. B.H. Lee and B.H. Kim, "Automated design for the runner system of injection molds based on packing simulation", Processing Annual Technical Conference–ANTEC, Conference Proceedings, Indianapolis, Indiana, Vol.1, 1996, pp.708-713.

13. J.S. Tu, R.K. Foran, A.M. Hines, and P.R. Aimone, "Integrated procedure for modeling investment castings", JOM, Vol. 47, No.10, pp.64-68, 1995.
14. E.A. Herman, Designing Die Casting Die, North American Die Casting Association, 1996.
15. D.F. Allsop, and D. Kennedy, Pressure Die Casting Part 2: the Technology of the Casting and the Die, Oxford, New York: Pergamon Press, 1983.
16. Die Design Manual, MMI Holding Ltd, Singapore, 1999.
17. M. Hoffmann, M. and R.J. Arinyo, "On user-defined features", Computer-Aided Design, Vol.30, No.5, pp.321-332, 1998.
18. Y.M. Chen, and C.L. Wei, "Computer-aided feature-based design for net shape manufacturing", Computer Integrated Manufacturing Systems, Vol.10, No.2, pp.147-164, 1997.
19. A. Street, The Die Casting Book, Portcullis Press Ltd, 1977.
20. B.Y. Feng, T.R. Han, Z.H. Yin, and W.S. Jiang, "Die casting die design", A Concise Handbook for Mold Design and Manufacture, Shanghai Technological Press, China, 1996.
21. Unigraphics, Assemblies User Manual, Unigraphics Solution Co., Maryland Heights, MO, 1997.
22. D-M-E Mold Base Catalogue, D-M-E Company, Madison Heights, MI, 1997.
23. G. Glaeser, and H. Stachel, Open Geometry: OpenGL + Advanced Geometry, Springer-Verlag, New York, 1998.

5
CAE Applications in Mold Design

5.1 INTRODUCTION

In most polymer molding processes, the final product quality is determined by the melting, flowing, and mixing of the polymer during the process in which the design of molded parts, molds, and molding processes (referred to as the molding activities) play an important role. The traditional trial-and-error approach on the optimal design of the molding activities is not sufficient to support contemporary polymer product development due to the increasing demand for the high product quality and shorter tooling and production lead-times.

Injection molding, one of the widely used polymer-processing operations, is integral to many of today's mainstream manufacturing processes. As more and more plastic molded parts are increasingly being used in industries such as telecommunications, consumer electronics, medical devices, computers, and automotive, plastic mold-making and molding industries are quickly expanding. The production runs of plastic injection molds, however, are typically of small lot-sizes with great varieties. It is imperative that (a) the tooling lead-time be reduced, (b) overall product quality be increased, and (c) the design of the molding activities be optimal [1].

To achieve the above-mentioned goals in the current product development paradigm, the CAD/CAM approach is a preferred method to provide an essential part of the solution. It provides the technologies for the representation of design intent, design schemes and solutions and helps the realization of the design physically. CAD/CAM technologies greatly enhance product design quality and shorten design and manufacturing lead-times. However, it is difficult to address some of the critical issues in optimal design of the molding activities and product quality assurance using CAD/CAM technologies alone.

CAE technology, however, fills this gap in plastic injection molding production. It uses software tools to help designers generate, verify, validate, test, and optimize design solutions before they are practically implemented and physically realized. CAE software bridges the gap between the part designers, mold makers, and molding engineers. In plastic injection production, the production design requires many issues to be considered in the early design stage. In addition to the 3D geometrical and functional design of a plastic molded part, other factors must be accommodated simultaneously. Some key influences such as part structure, wall thickness, plastic materials and material properties, mold structure and layout, the number and position of gates, molding processes, shrinkage and warpage allowances, etc., are all interrelated. These factors would further influence product quality. To develop an efficient plastic injection production system, CAE technology can be used in these four aspects as depicted in the following.

- *Molded part design:* In part design, there are two critical issues to be addressed. One is the part geometry and structure optimization and the other is the part moldability analysis. CAE applications, in this aspect, address both issues. They verify and validate whether the designed part meets its functional requirements and design specifications in terms of structure layout and geometry dimensions. Besides, they also reveal whether the designed part has good moldability on the basis of filing analysis. CAE applications in part design provide practical advices for part structure and geometry modification, as well as moldability improvements.
- *Injection mold design:* CAE applications in mold design help mold designers and mold makers analyze part design, identify critical manufacturing and quality issues, and recommend appropriate solutions. In detail, they help to decide mold layout, gate location, and runner systems in single-cavity, multi-cavity and family molds. They also help to design the cooling system for optimal cycle time.
- *Molding process determination:* This activity decides processes and process parameters. CAD applications in process design ensure good quality of the molded part. They provide quantitative analysis for production parameters determination such as clamping tonnage, shot size, and cycle time and reveal the temperature distribution, injection velocity, and polymer flowing phenomena.
- *Product quality assurance:* Plastic moldings may contain defects due to the poor design of the molding activities. In practical production, the efficient control of molding processes can help prevent some common production defects from occurrence, but the defects caused by poor design can only be avoided or eliminated with better design at the design stage. CAE applications in product quality assurance reveal the defect and suggest appropriate measures for its avoidance.

CAE Applications in Mold Design 189

CAE technology for plastic mold design deploys numerical techniques to provide solutions. Numerical techniques use physical, mathematical, and numerical models generated on the basis of the practical engineering problems as well as software tools for solving the engineering problems. The physical model idealizes the real engineering problems and abstracts them to comply with certain physical theory with assumptions. For example, the physical model of a beam complies with the continuum mechanics and assumptions describing the physical characteristics of the beam, such as static, 3D, elastic type of material, etc. The mathematical model specifies the mathematical equations such as the differential equations in FEM analysis the physical model should follow. Besides, the mathematical model details the boundary and initial conditions and constraints. In numerical techniques, the numerical model for FEM or BEM describes the element types, mesh density, and solution parameters. The solution parameters provide detailed calculation tolerances, error bounds, iteration, convergence criteria, etc.

CAE solutions provide engineers with visual and numerical feedback in injection molding production for the molding activities design. In the early design stage, it is critical to ensure "design right the first time" since the first 10-15% of the design activities commit to more than 80% of product development cost. CAE applications in the injection molding activities have a significant leverage on the design and production costs as well as design and production lead-times. It avoids late design changes. Since the design process is iterative and the boundary of CAE applications in the above four different areas are not clear-cut, the above classification of the CAE applications is proposed to facilitate the delineation and presentation of CAE applications in mold design in this chapter.

In this chapter, the various levels of the CAE applications in the molding activities and product quality assurance are presented. In each application scenario, the role of CAE technology in seeking optimal solutions is presented. The chapter provides an overview of CAE technology for the mold design professionals. In addition, the challenges and limitations of this technology in the current molding product development paradigm are also presented.

5.2 CAE ANALYSIS PROCEDURES AND FUNCTIONALITIES

CAE analysis tools are a suite of software toolkits which utilize numerical analysis techniques such as FEM and BEM. CAE analysis tools provide engineers with visual and quantitative analysis results of the polymer behavior inside the mold core and cavity blocks during molding processes. This allows the designer to have better understanding of the molding behavior and the flow phenomenon of the material inside the designed part such that errors in the design stage can be avoided or eliminated. In an integrated mold CAD/CAM/CAE development paradigm, CAE analysis tools can directly import CAD models into the CAE analysis package for molding and structural analysis and simulation. With the development of software and computing technologies, it has become increasingly simpler and more convenient to conduct part and mold design analysis. Previously, the CAE tools users need to perform a lot of pre-processing work for modeling and data preparation, and post-processing tasks for obtaining results and performing output analysis. Currently, most of these tasks can be done automatically and the final results and outputs can be exported visually, but the interpretation of the analysis results presents a major challenge since it requires expert knowledge and experience in mold design and molding production. In this section, the CAE analysis procedures and the functionalities for mold design are presented.

5.2.1 Analysis Procedures

Fig. 5.1 shows the CAE analysis procedures in detail. To conduct a CAE analysis, firstly, a part CAD model needs to be created, and it should ideally represent all the design intent of the designers and the specifications. In the current product development scenario, a model can be directly imported into the CAE analysis systems or via the data exchange technologies to convert the part CAD model into a neutral data format such as *IGES* and *STEP*. After that, the designers need to decide on the physical, mathematical, and numerical models of the specific mold design problems. For the physical model, the CAE molding analysis tools have built-in physical entities with which the plastics melt should comply in the molding process, the users are only required to provide assumptions such as whether the problem is 2D, 3D, or axisymmetric, selection of material type, specifying the physical boundary conditions and constraints, etc. In terms of the mathematical model, the users need to deal with the choice of the coordinate system, computing units, geometric boundary conditions and constraints. Under the context of the numerical model, the users need to consider the element types, mesh density, calculation parameters such as iteration number and convergence criteria, etc. All such information would support the pre-processing of CAD design data and further generate a CAE analysis model. Before conducting analysis and simulation, the users need to specify the detailed molding information, process parameters and molding conditions. After simulation and analysis, the calculated results need to be analyzed and evaluated. If the results and solutions are not satisfactory, the suggested changes regarding part geometry and structure, mold structure, and molding conditions will be made for the next round of computation. The process is iterative until all the design requirements for moldability, productivity, and product quality are met and an optimal solution is obtained.

5.2.2 CAE Functionalities

In employing CAE technology for injection mold design, the most common functionalities can be grouped into four types of simulation and analysis activities, viz., flow simulation, cooling simulation, structural analysis, and fiber orientation prediction. Currently, most CAE mold design tools are able to provide these functions. These functionalities can be summarized in the following sections.

CAE Applications in Mold Design

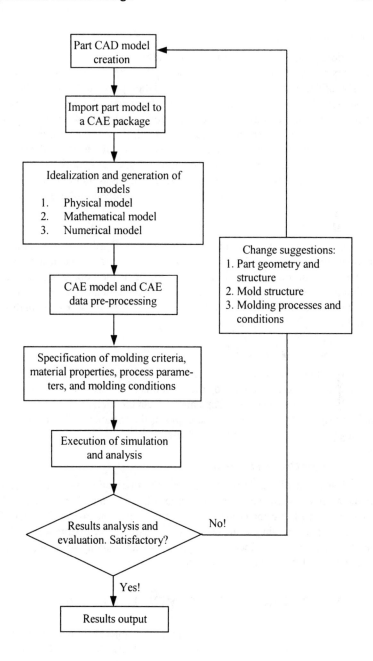

Fig. 5.1 CAE analysis procedures.

5.2.2.1 Flow Simulation

Flow simulation is used to dynamically analyze the plastic melt flowing in the runner systems, filling of the cavities and packing in the mold. From the point of view of part and mold design, flow simulation provides the moldability analysis to predict whether the part can be completely molded. It also helps to provide optimal design solutions of gates, runners, and any combinations thereof. From the process optimization aspect, on the other hand, flow simulation helps to estimate the production cycle time, clamping tonnage, etc., and determine the optimal process conditions and operation configuration. From the quality assurance point of view, it helps to identify the molding defects such as weld lines, melt lines, and air traps and reveal the causes of the defects and recommend solutions to avoid them.

5.2.2.2 Cooling Simulation

Cooling simulation reveals the temperature distribution of a mold, the plastic melt, and the solidified melt during the molding process via heat transfer analysis between the plastic, mold, and the coolant in the cooling circuit. It helps to design the cooling system for achieving optimal cooling cycle times. In addition, it also helps the designers to conduct the shrinkage and warpage analysis of the molding.

5.2.2.3 Structural Analysis

Molding structural analysis validates whether the part structure and dimensions meet the requirements and specifications in the context of strength and rigidity. It estimates the final part shape and reveals the underlying causes of shrinkage and warpage in the molding. It also explores the associativity of the mold and part cooling process to the final product shape.

5.2.2.4 Fiber Orientation Prediction

The effect of plastic melt flow on fiber orientation has a significant impact on the mechanical and structural properties of injection molded products. The fiber orientation in plastic materials also plays a significant role in the degree of final warpage that occurs in the injection molded parts. Fiber orientation prediction can be used to estimate the mechanical properties and quality of an injection molded part and reveal the root causes of warpage.

5.3 CAE IN THE MOLD DEVELOPMENT LIFE CYCLE

One of the greatest benefits of the plastic injection molding process is that parts of extremely complex 3D shapes can be fabricated in "net-shape" or "near-net-shape". Once a part is molded, cooled, and ejected from a mold, it is usually in the final form required for the next manufacturing process. Net-shape production can be achieved in high volumes with shorter manufacturing lead-time. However,

CAE Applications in Mold Design 193

it requires the entire consideration of part geometry design, part material selection, mold design, and processing conditions and their combined effects on part moldability and product quality. Using CAE analysis tools to analyze the part and mold design and simulate the molding process, product engineers can achieve optimal design in the early stages. CAE analysis tools play this critical role well in the mold development life cycle.

Fig. 5.2 shows an overview of the entire mold design and development life cycle and the role played by CAE. The process begins with the input of the plastic molded part created in a CAD system. The part model can be a solid, surface or wire-frame model. The model can be input into a mold design CAD systems in native CAD formats or in standard data exchange formats such as *IGES* or *STEP*. After inputting the part model into the mold design system, conversion of surface and wire-frame models into solid models is also needed since most design activities in a computer-aided mold system are 3D-based. Before the conceptualization of the mold structure can be materialized, part design analysis and moldability evaluation are needed. It is an iterative process from part design analysis and moldability evaluation to mold conceptualization. CAE analysis is critical here since these design activities need part structural analysis, plastic flow simulation, cooling simulation, and fiber orientation prediction. Once the preliminary mold structure is determined, the process goes to the detailed design of core and cavity blocks, inserts development, and assembly design. In the development of core and cavity blocks, the cavity number and layout, gate, runner, and cooling channels need to be designed. Some mold components including side-cores, side-cavities and form pins are also considered. In these detailed design stages, CAE analysis helps to determine the detailed design solutions via the above-mentioned four types of simulation and analysis procedures. After the design of the entire mold assembly and detailed components are completed, the process goes to CNC programming, machining, assembly, testing, molding inspection, and fine-tuning of critical components, which could significantly affect the molding dimensions and quality.

5.4 CAE DETAILS IN MOLD DEVELOPMENT

The development of injection molded part production involves the issues in the following four areas:
- Plastic materials selection;
- Injection molded part design;
- Injection mold design; and
- Molding process determination.

The issues in each area are equally important, and changes in any one will impact the remaining areas. In the past, decision-making in the above activities was usually made using either rules-of-thumb, trial-and-error, and experience based on

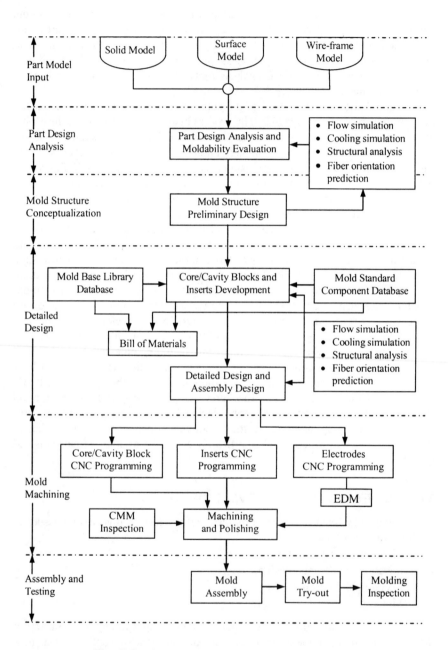

Fig. 5.2 CAE in the mold development life cycle [1].

previous designs or any combination thereof with some theoretical guidelines [2]. For this kind of mold development paradigm, it is difficult if not impossible to generate optimal and practical design solutions since the cross-impact of the variables in the above-mentioned four areas is difficult to evaluate, analyze, and predict. With effective computer technologies and CAE tools, it is possible to conduct hybrid analysis of these interplaying issues and the time and cost of the analysis have been reduced significantly. CAE technologies help to address the issues in those four areas with the analysis of the various physical variables and phenomena in the molding activities. The following sections will focus on discussing what CAE can reveal.

5.4.1 What CAE Reveals

CAE handles the objective variables of engineering problems using the laws of physics and mathematics to help designers seek scientific solutions. In molding activities, CAE analyses can be classified into the following categories of variables, revealing the physical phenomena in the molding process and providing the basis for realistic solution generation.

5.4.1.1 Physical Variables

In the injection molding process, the physical variables are important since they relate the molding phenomena inside the cavity and help the designers to have an epistemological and ontological understanding of these phenomena. In the following sections, these physical variables are articulated and explained [3].

(1) Pressure

In CAE analysis, pressure in a plastic melt during molding is a field variable which is defined as the normal force acting on a unit area. Pressure is a function of time and location since it changes during the molding process and varies from location to location. At the onset of filling, the pressure is 1 atm. The pressure at a specific location starts to increase only after the plastic melt flow-front reaches that location, and continues to increase as the melt-front moves past the specific location. In CAE analysis, pressure is a nodal variable, but the pressure inside an element can be determined using the interpolation approach based on the values at the nodes of the element.

Pressure variable relates to some flow concepts in molding. Pressure gradient is one of them. Like water or other liquids, plastic melt always flows from the place with higher to lower pressure. In other words, it flows in the direction with negative pressure gradient. In molding, the constant pressure gradient principle indicates that the more constant is the pressure gradient along the flow path, the more efficient will be the filling pattern during the molding process. The pressure gradient has a close relationship with some flow phenomena including flow hesitation, over-packing and under-packing, etc. The non-uniform pressure gradient would cause these problems and affect product quality.

(2) Temperature

Temperature in a plastic melt is a nodal variable, and the value inside an element can be calculated based on the nodal values of the element via interpolation. It is the direct output of a cooling analysis. In the molding process, the temperature of the plastic melt varies with time and location. The concepts such as bulk temperature and average temperature derived from temperature analysis are very critical in the molding process and quality control.

Uniform temperature distribution during filling is important to product quality as non-uniform temperature can lead to hot spots. The temperature of the hot spots can be excessive and cause pre-gel. In addition, it could cause differential shrinkage and warpage after the molding has cooled down.

(3) Shear Rate

The shear rate in a plastic melt is also a nodal variable, and its value inside the element can also be determined using the interpolation approach based on the nodal value of the shear rate. Shear rate is defined as the derivative of the velocity in the thickness. It is equal to zero at the neutral layer of the plastic melt in the flow channel and reaches a peak value at the interface between the plastic melt and the flow channel.

Shear rate represents the flow or deformation rate of the plastic melt in the cavity. From the rheological point of view, shear rate closely relates to the viscosity of the plastic melt. Therefore, the shear rate distribution in filling affects the fluidity of the plastic melt in the molding process. For designers, it is therefore necessary to be aware of the shear rate distribution.

(4) Shear Stress

Shear stress in a plastic melt is the shear force acting on a unit area. It is also a nodal variable, and the value inside the element can be determined based on the nodal value. Similar to shear rate, the shear stress is zero at the center of a runner or at the neutral layer of the plastic melt. It usually has the maximum value at the interface between the plastic melt and the wall of the runner.

Shear stress affects fiber orientation. The higher the shear stress, the higher the polymer orientation is. Since the shear stress has the maximum value at the interface of the plastic melt and the wall of mold cavities, the highest fiber orientation would be near the surface of the molding. The shear stress could also cause melt-fracture, and thus its value should be efficiently controlled under a certain level in the molding process to avoid molding cracks, to reduce fiber orientation, and to improve the overall quality of the molding surface.

(5) Velocity

The flow velocity of a plastic melt is again a nodal variable, and it represents the instant moving direction and speed of a specific point in a molding. The velocity inside an element can be determined based on the value of the nodal velocity in the element. In CAE analysis, the average velocity is adopted to describe the plastic melt flow. The average velocity is usually defined as the average velocity of the element.

Velocity has a very close relationship with the molding flow pattern in the filling process and is directly related to the flow phenomena such as weld and meld lines, hesitation, underflow, etc. In the molding process, the uni-directional flow principle requires that the polymer melt should flow in one direction with a straight flow front throughout the filling stage. The polymer melt velocity determines the flow pattern and further helps the designers to efficiently design the feed system to achieve a balanced mold and runner design.

(6) Clamp Force

Clamp force can be determined according to the pressure and the projected area of runners and cavities on the parting plane. It is a function of time and varies during the filling process. The clamp force reaches the peak value at the end of the filling stage. The exact clamp force is difficult to calculate, but an estimated one is also useful. The predicted clamp force can be used for the selection of the molding machine, determination of number of cavities, and the detailed design of a mold.

5.4.1.2 Hybrid Variables

In injection molding, some molding phenomena cannot be described by the above physical variables individually. These phenomena can only be interpreted by the combination of the above variables. These combined variables are the so-called hybrid variables in this book. Some of the main hybrid variables are presented in the following sections [3].

(1) Melt-Front Advancement

Melt-front advancement describes the movement of the plastic melt-front and the arrival-time distribution of the melt-front at each node. Melt-front advancement is a nodal variable, and the value inside the element can be determined based on the node value. It reveals the flow phenomena taking place inside the mold. One of the important phenomena is flow balancing, which requires that all the flow paths within a mold be balanced, viz., filling in equal time with uniform pressure. If the molding process is unbalanced, the area filled earlier could be over-packing, and the area filled later could be short-shot. Melt-front advancement reveals whether the unbalanced flow could exist or not and identifies some defects such as weld-lines, meld lines, air traps, etc. In addition, the detailed design schemes can be modified based on the melt-front advancement such as re-position of gate locations and the adjustment of the injection molded part dimensions to achieve uniform and balanced flow.

(2) Core and Skin Orientations

The core and skin orientations of the polymer are element variables. The core orientation for each element is in the transverse direction to the velocity vector at the end of filling. This is the most probable fiber orientation. The other possible orientation is in the direction of the velocity vector. As for the skin orientation of an element, they usually align with the velocity vector of the element, and this would give the most probable fiber orientation on the external surfaces of a molded part.

Core and skin orientations reveal the fiber orientation, which is critical for the prediction of the mechanical properties of a molded part. There are certain skin orientations which could lead to good mechanical properties, such as the impact and tensile strength. Based on this, mold designers can position the suitable gate location and pre-design the plastic flow pattern to ensure that the molded parts have good mechanical properties in certain directions.

(3) Quality Defects

The prediction of defects in injection molded parts is one of the main functions of CAE applications. Defect prediction is also considered to be one of the main output variables in CAE analysis. The output usually includes the prediction of some main defects such as weld lines, meld lines, air traps, etc. Weld lines are formed when the separate plastic melt fronts meet up in opposite directions. Meld lines occur if two plastic melt fronts flow parallel to each other and create a bond between them. Air traps are formed when the air in the plastic melt gathers together. Based on the CAE analysis results on velocity distribution and melt-front advancement, these defects can be predicted in terms of occurrence and location. Measures could be adopted to improve the flow uniformity to avoid or eliminate the defects, or to try to locate the defects to some less sensitive areas by changing the gate locations, part structure, and gate and runner sizes. For air traps, CAE analysis also helps to identify the venting hole locations.

5.4.2 CAE in Part Design

CAE in part design helps part designers to optimize their part designs and investigate the impact of the design solutions on the moldability and quality of a product [4]. Generally, CAE's role in part design can be summarized in the following three areas.

5.4.2.1 Part Structure Optimization

Plastic molded parts are usually used directly as components in assembled products. They have specific structural and dimensional specifications. CAE applications in part design helps to optimize the part structure and dimensions through the simulation of the part in a virtual working environment. Besides, it also helps to verify and validate whether the design meets strength and rigidity specifications.

5.4.2.2 Part Moldability

To ensure that a product has good moldability, designers need to consider some fundamental issues at the part design stage. The first issue is the gate location in the part. Current CAE tools can help designers determine the optimal gate location through plastic flow simulation. It could reveal areas which are the best locations for gating and areas totally unacceptable. After the gate location has been decided, the designers can further investigate the moldability of the part and analyze the areas which may have some potential filling problems.

5.4.2.3 Part Quality

Molding quality is an important issue for designers to consider. Even though the final geometric shape and the overall dimensions can meet all the specifications, the product may not be free from some physical and internal defects. All the defects could be caused by the irrational value of some variables such as shear stress, shear rate, cooling time, flow front advancement, temperature and pressure distribution, etc. Excessive shear stress, for instance, would drastically reduce the mechanical strength of a molded part and cause premature part failure, reducing the service life of the product. To ensure good product quality, CAE tools can help designers identify whether the molded part has any potential quality problems and explore some quality-affected variables in the early design stage. All of these efforts help to avoid and eliminate the quality problems at the product realization stage.

In general, CAE tools allow part designers to identify and address some critical moldability and quality issues at the early design stage. Besides, designers can also adopt some heuristic solutions to improve the product quality based on CAE analysis results. Such efforts are critical for ensuring product development on the right track and avoid or eliminate later design changes during the mold fabrication stage.

5.4.3 CAE in Mold Design

In mold design, designers need to build up the best design practice, use previously proven design experience and know-how, discuss the design schemes and issues with all disciplines involved in the project team, and use CAE analysis tools to help address the design issues and generate the optimal design solutions. CAE analysis software provides mold designers with the tools to verify and validate injection molded part design, help designers determine mold layout, and optimize the design of the feed and cooling systems [5].

5.4.3.1 Part Design Verification

Before mold designers proceed to design a mold, they need to verify whether the designed part meets the manufacturing and quality requirements. As mentioned in the part design section, the part moldability and quality analysis are a must for an optimal part design from the product design point of view. From the manufacturing point of view, mold designers also need to ensure part manufacturability and product quality on the basis of the specific mold manufacturing facilities and the conditions in the mold-making company.

5.4.3.2 Mold Layout

Once the part design is acceptable, the actual mold design work can begin. In mold layout design, the first step is to determine the number of cavities, the orientation of the part model, and the layout of the cavities in the cavity block. Basically, the number of molding cavities can be decided according to production requirements. Whether the requirements can be fully met further depends on the actual scenario. In detailed design, it needs to orientate a single part in the cavity block, specify the offset between the molding cavities, and replicate the single part in the cavity block, which will create the multiple molding units in the mold [5]. In these processes, CAE tools help to determine the mold layout based on the plastic flow simulation and balancing analysis. Usually, mold layout design should be done simultaneously with the feed system design.

5.4.3.3 Feed System Design

The feed system, also called the runner system or delivery system, consists of a sprue, runners, and gates. Through the feed system, the polymer melt flows into the individual cavities. The feed system is important since it has great impact on the quality of the molded parts, molding productivity, and production costs [5-10]. With the demand for good part performance and part quality, design of the feed system can no longer be done on an intuitional level alone. In the past, the design process of the runner system was based on a trial-and-error approach. The dimensions of the runner system was progressively enlarged and adjusted by re-machining and testing of the mold until good part quality was achieved. With the advent of plastic flow and cooling simulation programs, mold designers can iteratively explore the performance of the runner system until a satisfactory design is obtained.

In the design of the feed system, CAE tools only provide supporting information for the designers in their decision-making process. The entire design follows the related design rationale, formulae and user-defined design criteria of the feed system. Taking the gate design as an example, the number of gates designed should be based on the pressure to fill the cavity and the minimum number of gates to be selected. The gate location has a large effect on part quality and proper working of the mold. Some of the effects attributable to the gate location include mold venting, warpage, shrinkage, over-packing, and short-filling. Therefore, the gate location can be determined based on the flow balancing principle and user-defined design criteria which could include molded part defects (warpage), aesthetic requirements (weld and meld lines), and engineering property (impact strength) of the polymer material [9]. To support the gate design, CAE analysis tool can be used to analyze whether the gates number, locations, and dimensions are optimal to ensure the related design requirements are met. Similarly, in runner design, the ideal case is to ensure that the runners are in good dimensions and locations such that the plastic flow front reaches the end of runners at the same time and the same temperature. Besides, the runners are actually flow controllers to balance the plastic flow rate such that the each cavity is filled at the same pace, although it is quite difficult to achieve since the plastic flow in each runner may face different resistance and conditions. In addition, the good flow pattern should always be met in the design of the feed system to avoid any defects occurrence. Fortunately, CAE analysis tools are used as a supporting tool to analyze and validate whether the design schemes meet the above requirements, and thus optimizing the design to a large extent. Although the process is iterative, the final optimal feed system could be obtained via modification of the design schemes and dimensions through a few rounds of flow and cooling simulation.

5.4.3.4   Cooling System Design

Mold cooling time is the biggest contributor to the overall cycle time in almost all the plastic molding applications and takes up more than 75% of the cycle time in an injection molding process. Hence, there is a substantial potential for reducing cycle time and thus increasing the productivity by improving the cooling design. In addition, a design aimed at more uniform and balanced cooling will improve part quality by reducing differential shrinkage, internal stresses, and warpage [11-13]. Therefore, the design objective of a cooling system is to achieve a rapid, uniform, and balanced cooling. Design variables typically include the sizes of the cooling channels, their locations and connectivity, cooling fluid, and its inlet temperature and flow rate in each manifold. For an optimal design, the designer needs an analysis tool which can predict cooling time and thus cycle time, total heat released by the polymer melt per cycle, temperature and its distribution within the plastic melt, and the fraction of heat extracted by each cooling channel [11].

Traditionally, designers take the traditional path of regular cooling principles and past experiences and know-how to design a cooling system. However, using the traditional approach, it is difficult to analyze the heat exchange pattern among the plastic melt, mold and coolant and the temperature distribution in the mold. It is also difficult to search for optimal solutions for the design of cooling channel number, size, locations, and the relationship of cooling design with product quality and productivity. Currently, CAE analysis tools help designers in the design of cooling system via heat transfer analysis for obtaining the temperature distribution from the plastic part to the mold cavity surface and from the mold to the coolant.

In detailed design of the cooling system, the design objective functions, variables, and constraints need to be specified. They should be practical and the solution easily achievable. Taking the design variables as an instance, only four variables can be included, viz., cooling channel number, size, locations, and coolant flow rate. The generation of optimal design solution needs CAE tools for cooling simulation together with the optimization approach. The hybrid CAE analysis and optimization approach can produce better design solutions.

5.4.4 CAE in Process Design

The injection molding processes consist of three main phases, viz., mold filling, melt packing, and part cooling. The molding processing cycle starts with filling the mold cavities with plastic melt at the molding temperature. After all the cavities in the mold are filled, some additional polymer melt is needed for packing into the cavities at a very high pressure to compensate for the shrinkage of the molded part. The mold cooling process follows the processes where the polymer melt cools down and solidifies essentially due to the conductive heat transfer from the part to the mold surfaces and to the coolant until the part is sufficiently rigid to be ejected from the cavity block. Ejection is the last process in the injection processing cycle. The first three processes are the main concern in the molding process cycle.

With the increasing demand for high-quality injection molded plastic parts, the production problems associated with the molding process often cause significant time delays and cost increases since it is a complex mix of injection machine variables, mold complexity, operator skills, and plastic material properties [14-15].

CAE applications in molding process design is to help designers determine the process conditions and parameters. Some of the molding conditions can be determined based on the properties of plastic materials, while some of the molding conditions and parameters need to be decided based on the characteristics of the molded part. CAE applications in process design is to conduct in-depth analysis to further provide informative simulation results for the determination of the optimal process conditions and parameters. Processes which can efficiently solve certain macro-defects, such as short shots, flash, burn marks, sink marks, poor weld line appearance, warpage, etc., would be the prime concern.

In designing the molding processes, the analyses including flow simulation, cooling simulation, part structural analysis, and fiber orientation prediction are all needed. These analyses are usually conducted during the part design stage or at the early design stages of mold design to determine the process parameters for molding machine selection or process configuration. Since individual simulation and analysis can only address certain issues, simultaneous simulations and analysis of all the above-mentioned issues are required. Through hybrid analysis of all the simulation results and the iterative analysis based on the different process categories and settings, the designers can have a thorough understanding of the tacit and explicit phenomena in the molding process cycle and design the most feasible molding processes.

5.4.5 CAE in Product Quality Assurance

Plastic moldings may contain macro-defects generated due to poor design of the molding activities and the plastic melt flow pattern. To avoid the defects occurrence and improve the product quality, the root causes of these problems should be investigated and explored. CAE analysis tools can be used to achieve

CAE Applications in Mold Design

this goal and help to avoid or eliminate the following common product quality defects to a good extent [16-18].

5.4.5.1 Weld and Meld Lines

Weld lines are generated when two or more cooling polymer melt flow fronts from opposite directions meet together within the mold cavity and create a hairline features on a molding. Weld lines occur when the plastic melt flow has been split around an obstacle in the flow path and then rejoins on the other side. Weld line is actually one of the defects generated during the molding process due to an irrational flow pattern.

Meld lines are also flow defects created in the molding process, and they are formed when two flow fronts flow and meet together in the same direction. The meld lines are also caused due to the irrational flow pattern. Both weld and meld lines could reduce the mechanical properties of the molding at the point where they are located. Ideally, they should be shifted to the least sensitive areas if they cannot be eliminated.

CAE software tools can be used to identify the root causes of these defects for complex moldings and corresponding solutions can be recommended based on the causes identified.

5.4.5.2 Shrinkage

Shrinkage occurs as the polymer melt cools in the mold due to density variation of the polymer from the molding temperature to the surrounding temperature. At the molecular level, this is caused by the polymer chains relaxing or the so-called recoiling and aligning polymer chains with their neighboring chains. For highly crystalline polymers, they have greater shrinkage due to the formation of denser crystal structures. For thicker cross section parts, sink marks may occur in some areas where the variation of the cross sections of the molding is large and the plastic melt is slower to cool. Besides, the variation in shrinkage in a molding induces residual stress which could warp the molding or make it crack, if the residual stress exceeds the strength limit of the molding. Usually, at higher molding and mold temperatures, the plastic melt would shrink more since it has greater molecular energy and subsequent ability to recoil. Higher packing pressures may compensate for this shrinkage.

CAE applications in shrinkage analysis and prediction can identify the causes which make shrinkage occur. First of all, they reveal the fiber orientation to predict the maximum shrinkage direction and identify appropriate measures to be adopted to avoid or eliminate the shrinkage. They also identify the non-uniform temperature distribution and unbalanced cooling of a molded part, which could cause differential shrinkage. Further, a high molding pressure contributes greatly to the variation of the polymer density, which, in turn, further contributes to the non-uniform shrinkage of the molded parts. To predict and eliminate shrinkages, simultaneous analysis of heat transfer and compressible fluid flow coupled with a mold-cooling simulation is required [16].

5.4.5.3 Warpage

Warpage is a distortion which may occur in molded parts due to molecular chain orientation. In the plastic molding process, the plastic melt flows along small cavity channels and the molecular chains become deformed, stretched and aligned in the flow direction. When the plastic cools, the molecules try to relax to their original coiled state, but the deformed and stretched molecular chains have no time to relax to their original preferred coiled state and maintain the uncoiled state due to the quick cooling process. After molding, the molecular chain still tries to recoil, resulting in part warpage. On the other hand, warpage can be considered a result of differential shrinkage. Variation in shrinkage can be caused by molecular and fiber orientation, temperature variations, variable packing, and different pressure levels in the molding.

CAE application in warpage analysis can identify the causes of warpage based on simultaneous fiber orientation prediction, cooling simulation, filling and packing analysis, and part structural analysis. From these analyses, the warpage location and degree can be identified, and appropriate measures to eliminate or decrease the warpage can be recommended. Solutions can be obtained through improving uniformity of flow and cooling to reduce the shrinkage variation in the molding.

5.4.5.4 Air Traps

Air traps are the air accumulation in a plastic molding. During the molding process, the air in the plastic melt can move during the flow of the plastic melt. When the air is not sufficiently vented, they can accumulate and form pockets of compressed air in the molding, which are called air traps.

CAE flow simulation can analyze the forming process and locations of the air traps in the molding based on melt-front advancement results. Usually, if the air traps are near the parting surface, the air can be vented easily by including venting holes in the mold design. If the air traps are inside a molding, they will block the plastic melt from entering into the space occupied by them. These air traps can result in local over-heating and cause burn marks in the molding. CAE analysis can identify the optimal gate location to avoid the air traps.

5.5 APPLICATION EXAMPLES

The application of CAE analysis in injection and casting molds design has been in use for more than two decades. There are a lot of successful application cases. In this section, the cooling analysis of an injection mold is conducted as a CAE application example for injection mold design. Besides, the simulation of the casting process to assist the mold and process design is another example for casting mold design. These cases show the capabilities of CAE applications in assisting injection and casting molds design.

5.5.1 Injection Mold Cooling Analysis

In this application case, the cooling analysis of the mold of a photo frame part built with a hot runner system is conducted [19]. To obtain the actual data, two thermal sensors and a pressure sensor are installed in the mold assembly. Cooling analysis based on the mold assembly is performed using ABAQUS, which is a commercial FEA analysis software. The simulations are adjusted with the experimental results to find out the heat input from hot runner and its influence on mold cooling. Based on the analysis results, the cooling system is re-designed.

5.5.1.1 Methodology

In FEA analysis, the model generation is critical. In model generation, it is necessary to make sufficient assumptions to idealize the practical engineering problem. In this case, to model the mold assembly for simulation, the following assumptions are made to simplify the problem and reduce computational time:
- Filling times are normally very small compared to cooling time; the temperature of polymer melt is thus assumed constant during the filling period.
- The gap resistance between the mold and the part interfaces is ignored and a perfect thermal contact between the two systems is presumed.
- The temperatures of surfaces contacted with hot runner nozzle are supposed to be constant based on the measurement of the sensor.
- Natural convection between ambient air and the exterior mold surface is less than 5% of the overall loss in most injection molding applications and can thus be ignored [16].
- Heat transfer between the mold and the coolant in the cooling channels is assumed to be steady.

The overall energy balance equation for a full 3D analysis is as follows:

$$\rho \cdot C_p \cdot \frac{\partial T}{\partial t} = K(\frac{\partial^2 T}{\partial x^2} + \frac{\partial^2 T}{\partial y^2} + \frac{\partial^2 T}{\partial z^2}) \quad (5.1)$$

where ρ is the density, C_p is the specific heat, T is the temperature, t is the time, and K is the conductivity. After the simplifications, the boundary conditions can be defined as:
- Constant temperatures of surfaces contacted with hot runner nozzle;
- Injection temperature of polymer during the filling stage, which is also simplified as cyclic, constant, average injection temperature;
- Forced convection between coolant and cooling channels surface.

The latter is governed by:

$$-K_m \cdot \frac{\partial T}{\partial n} = h_c(T - T_c) \tag{5.2}$$

where subscripts m and c represent the mold and the coolant, h is the heat transfer coefficient and $\partial/\partial n$ is the outward normal derivative to the boundary surfaces. In order to remove heat efficiently, a turbulent flow is required. The dimensionless Reynolds number R_e is a measure of the flow that is defined as

$$R_e = \frac{v_c \cdot D \cdot \rho_c}{\mu_c} = \frac{Q_c \cdot D \cdot \rho_c}{A_c \cdot \mu_c} \tag{5.3}$$

where v is the velocity, D and A_c are the diameter and the cross-sectional area of the cooling channels, respectively. μ is the absolute viscosity, and Q is the flow rate of coolant. A desirable and suggested value of Reynolds number for mold cooling is $R_e > 4000$ as the flow is turbulent, and is much better if $R_e > 10000$. The heat transfer coefficient h_c, based on the Dittus-Boetler correction [20], is applied to the forced convective heat transfer by turbulent flow in a circular pipe,

$$h_c = 0.023 \cdot \frac{K_c}{D} \cdot R_e^{0.8} \cdot P_r^{0.4} \tag{5.4}$$

The Prandtl number Pr is defined as

$$P_r = \frac{\mu_c}{\rho_c \alpha_c} = \frac{\mu_c \cdot (C_p)_c}{K_c} \tag{5.5}$$

where the thermal diffusivity α is defined as

$$\alpha = \frac{K}{\rho \cdot C_p} \tag{5.6}$$

The initial condition is such that the mold temperature is equal to the coolant temperature. When injection cycle starts, the mold will be heated up. It takes at least ten cycles times to reach the steady-state temperature. The mold temperature distribution at the end of the previous cycle is used as the initial condition for the new cooling cycle. Since the coolant temperature difference between the inlet and the outlet is small, the forced convection between coolant and channels can be simplified, throughout the cycle, as heat transfers to a sink of constant temperature

CAE Applications in Mold Design

of coolant with heat transfer coefficient calculated using equation (5.4). The time increment in analysis follows the mechanism of ABAQUS.

5.5.1.2 Analysis Conditions

The analyses are performed using ABAQUS analysis software. The cooling analysis, defined as transient uncoupled heat transfer analysis by ABAQUS, consists of several heat transfer steps to represent the cycles, two steps for each cycle, till the cycle of steady state. Each odd step represents the filling stage in which all the three boundary conditions are applied. Each even step represents the cooling stage in which the boundary condition of average temperature of hot polymer melt is removed.

The mold assembly with hot runner to be simulated for cooling analysis is shown in Fig. 5.3. As shown in the figure, a thermal sensor is installed to measure the plastic melt temperature from the core side; another thermal couple is to take the temperature of lateral surface of hot runner nozzle; and a pressure sensor is also installed to record the pressure of the core impression for the optimization of injection conditions.

Fig. 5.4 and Fig.5.5 show the original cooling designs of the core and cavity.

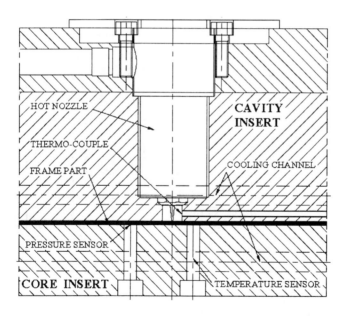

Fig. 5.3 The mold assembly with hot runner [19].

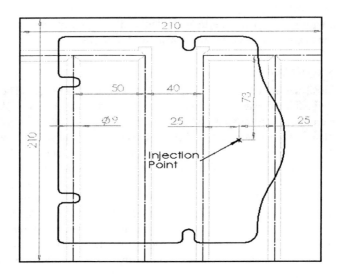

Fig. 5.4 The cooling channels layout of the core [19].

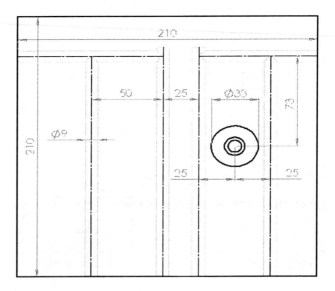

Fig. 5.5 The cooling channels layout of the cavity [19].

CAE Applications in Mold Design

The distance between the centerline of cooling channels and the mold impressions is 12 mm. It can be seen that the distances between cooling channels, d, are wide and not constant. The recommended d is 3 – 3.5 times the diameter of the cooling channel. Uniform cooling effect can only be achieved with the constant d. Therefore, the designs could be modified and further improved. The dimension of the frame part is $157 \times 177 \times 1.8$ mm with the volume of 45907 mm^3. The dimension of the assembly of core and cavity is $210 \times 210 \times 95$ mm. The material properties and manufacturing conditions are listed in Tables 5.1 and 5.2, respectively.

The analyses are based on assembly. The 3D tetrahedral meshes are automatically created using ABAQUS. Different element sizes are assigned. The important surfaces where boundary conditions applied are finer and the others are coarser. This scheme ensures the accuracy of simulation with fewer elements. However, the numbers of nodes and elements are still large, as listed in Table 5.3.

Table 5.1 The cooling channels layout of the cavity [19]

Material	Water (30°C)	Steel	Plastic (HDPE)
Density (Kg/m^3)	996	7800	950
Specific heat ($J/Kg \cdot K$)	4177	460	2200
Conductivity ($W/m \cdot K$)	0.6155	36.5	0.50
Viscosity (mm^2/s)	0.801	-	-

Table 5.2 Molding conditions of the part [19]

Conditions	Value
Average injection temperature (°C)	240
Average ejection temperature (°C)	80
Average mold temperature (°C)	70
Coolant temperature (°C)	30
Temperature of surfaces contacted with hot runner nozzle (°C)	100
Filling time (s)	1
Cooling time (s)	16
Flow rate of coolant (l/min)	10
Heat transfer coefficient (W/m$^2 \cdot$K)	11621

Table 5.3 No. of nodes and elements used in simulation [19]

	Frame part	Cavity	Core
No. of nodes	1,755	10,130	8,914
No. of elements	4,982	45,516	38,758

5.5.1.3 Analysis of Results

The temperature distribution at the end of the 10th cycle is shown in Fig. 5.6 and Fig. 5.7. It is found that the temperatures are scaled from 50°C to 80°C. Temperatures beyond this range are in the same color as the adjacent boundary temperature. The simulation result is found to correspond well with the experimental result. Obviously, the areas without the passing through of cooling channels and the areas in between the cooling channels where d is 40 and 50 mm are hotter. Besides, as the heat input is not removed instantaneously, the left-hand side of the core is getting hotter and hotter. These unbalanced temperature distributions would result in warpage of the part and further prolonging of the cycle time. Poor cooling design also causes higher injection pressure, as measured by the pressure sensor. Normally, the injection pressure drops during filling stage. Due to the unbalanced temperature distribution of mold impression, there is rebound when the melt front flows from the hotter to the cooler area.

The plane and thin-wall products are often very sensitive to uneven cooling. It is necessary to decrease the distance d and eliminate the hot areas to reduce the cycle time and improve the quality of molded part. Fig. 5.8 shows the modified cooling design of the cavity. The cooling design of the core is also modified accordingly. The simulation results are shown in Fig. 5.9 and Fig. 5.10. For the modified design, the cooling time is reduced from 16 to 10 seconds, while the temperature distributions are more uniform than the original one.

Since the hot runner nozzle needs to remain at constant temperature during injections, its heat input is found to be different from different cooling design. It can be observed that the influence of the hot runner nozzle is well confined within shorter cooling time by the modified design.

5.5.2 Simulation of the Casting Process

Similar to injection molding, casting is a high productivity process and integral to many of today's mainstream manufacturing processes. In the die casting process, a metal melt is injected into the space between the core and cavity blocks to produce a "near-net-shape" product. Squeeze casting is also one of the widely used casting operations that combines both of closed die forging and pressure die casting. Semi-solid metal casting is another popular casting process that molds the metal in semi-solid conditions.

CAE simulation of the cast molding process is an invaluable approach for mold and process designers since it offers the improvement of process and product

CAE Applications in Mold Design

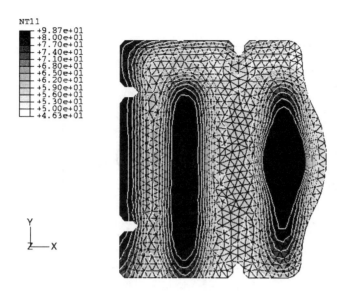

Fig. 5.6 Temperature distribution of frame part [19].

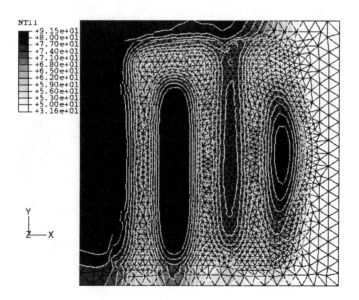

Fig. 5.7 Temperature distribution of the core [19].

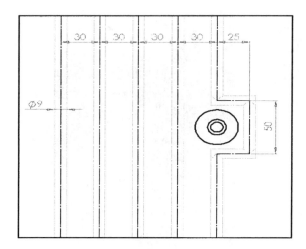

Fig. 5.8 Modified cooling channels layout [19].

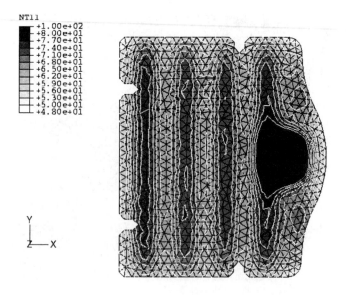

Fig. 5.9 Improved temperature distribution of the frame part [19].

CAE Applications in Mold Design

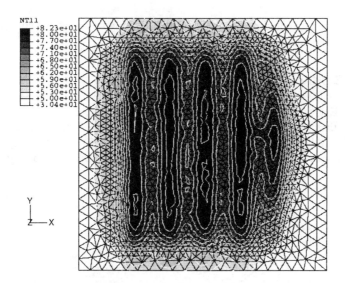

Fig. 5.10 Improved temperature distribution of the core [19].

quality, assistance in mold design, shortening of development lead times, and enhancement of designer motivation [21-26]. In product quality assurance area, CAE simulation helps to predict the product quality such as the formation and distribution of macro-segregation, prediction of microstructure including solidification of grain structure, and product porosity analysis. The simulation helps to explore the heat conduction in-between the mold and metal melt and the temperature distribution. It also helps to verify the process parameters and mold structure through simulation of the filling process. The main objective of these efforts is to finally improve product quality, decrease production cost and shorten product development lead-time.

To illustrate the CAE applications in the casting process, Fig. 5.11 shows a drum part fabricated by the squeeze casting process [25]. The material of the drum part is an aluminum alloy. The molding filling process is simulated using a commercial software called MAGMAsoft, which is a simulation tool developed by MAGMA for the optimization of castings and molding processes. The objective of the simulation in this case is to investigate and evaluate the feasibility of fabrication of the part by squeeze casting. The simulation results shown in the figure reveal the filling process and the laminar flow obtained during the squeeze casting process of the drum part.

Another casting case is shown in Fig. 5.12 [25]. The part is a small computer actuator fabricated by a semi-solid metal casting process. Through a series of simulation process, the temperature distribution in the metal and the minimum die temperature that should be kept are revealed. The figure shows the simulation results for the temperature distribution in the part.

5.6 CAE CHALLENGES IN MOLD DESIGN

CAE software tools for injection mold design have been in the marketplace for more than two decades, and they are now widely used in the molding industry that users, hitherto, have not doubted the critical roles CAE technology plays in the molding industry. However, CAE solutions for the molding industry may not

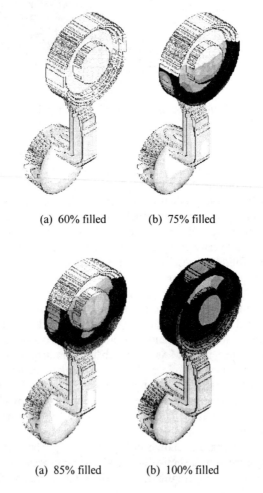

(a) 60% filled   (b) 75% filled

(a) 85% filled   (b) 100% filled

Fig. 5.11 Casting filling process in squeeze casting of a drum part [25].

CAE Applications in Mold Design

be acceptable if the desirable solution has to be topologically robust, computationally efficient, quantitatively accurate and predictable [27]. There do exist some limitations and challenges faced by CAE solutions in the current application scenario. Therefore, the mold-making engineers should be fully aware of this to avoid any mis-interpretation and over-confidence of the CAE functionalities [27-29].

To address the issues, the following discussion is only limited to the modeling problems and the complexity of the molding process. From the modeling point of view, the mold filling process is the most difficult to model. The physical phenomena likely to be present in a typical mold filling process include:

- Unsteady flow of multiple, immiscible fluids: In a typical mold filling scenario, there are at least four "fluids", each separated from the other by a discernible interface. They are mold material, gas in mold cavities, plastic melt filling the mold cavity, and the solidified plastic [27].
- Dynamic and kinetic interfaces: In a typical molding process, the interfaces in-between the above-mentioned four "fluids" possess topologies that are not only irregular but also dynamic, undergoing changes including merging, tearing, and filamenting due to the flow and interface physics such as surface tension and phase change [27].
- Dynamic heat transfer: In the mold filling and cooling processes, the heat transfer is dynamic. Not only the heat is convective, diffusive and radiative transfer from the plastic melt to the mold components and further to the coolant in the cooling system, new heat is also generated by the friction and movements between the interfaces among the above four "fluids" and inside the plastic melt, which would join the original heat transfer.
- Complex flow phenomena: During the filling process of mold cavities, the plastic melt in different locations could have different rheological characteristics such as PVT data and mechanical properties due to the non-uniform velocity and temperature distribution. The variations of rheological characteristics and material properties in the plastic melt would further affect the flow phenomena, which would, in turn, affect the other physical phenomena. For thick-wall moldings, some flow phenomena including turbulence, eddies, and swirling backflow could occur, which may conflict with the current CAE assumption that the plastic flows in a sheet-like approach with the wall thickness being small compared to the overall dimensions of the moldings.
- Difficult to control product quality: The various physical phenomena in the mold filling process would finally affect the product quality. The molding quality is easily affected by the individual physical phenomenon and any combination thereof. Finally, there exist many quality defects such as residual stress, shrinkage, warpage, weld and meld lines, air traps, etc., in the moldings. All these quality defects are difficult to control and eliminate.

For high-fidelity simulation of mold filling, developing models to simulate each phenomenon mentioned above is a formidable task. Concurrent simulation of the simultaneous occurrence of all the above phenomena and taking into consideration their interplay with each other would be extremely difficult, if not impossible. Much research is still needed before a fundamental understanding of these complex physical phenomena and their interplay can be obtained.

On the other hand, the molding process is becoming more complex. The new molding processes, novel and mixed materials, increasing complexity of part shape and bulky parts, etc., would make the modeling process more complicated than ever before. Taking the two-shot molding as an example [29], modeling and simulation of this type of molding process would be most challenging. While the first shot is modeled as a normal part, the second part has to take into account the previously solidified molding. The first molding can be considered as part of the mold steel. Modeling the mixed mold material scenario—one is a good conductor and the other is an excellent insulator—is a non-trivial issue. When the mold is transferring the heat out of the cavity, the first shot is keeping it in. All of these make the modeling more complex and conflict with some basic assumptions such as isotropic physical behavior of mold and its components. On the other hand, from material data point of view, exact material data is needed to accurately

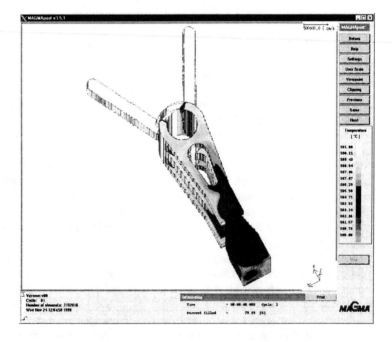

Fig. 5.12 Casting filling simulation [25].

describe the plastic behavior. Usually, these data can be obtained from the polymer suppliers. However, their data is often not of the best quality. Besides, material additives can dramatically impact the polymer behavior in the molding process. These additives are usually purchased from different manufacturers and mixed in at the molding facility. The effect of blending additives on the change of the physical properties of the base material is difficult to calculate and estimate [29]. If the CAE simulation and analyses are based on the behavior data of the base material, the final results would be doubtful. The difficulty in handling all the above-mentioned issues also presents challenges for robust CAE applications.

Although current computational techniques and software technologies provide advanced enabling technologies to support software development for high-fidelity simulation, the complexity of above physical phenomena during molding and post-molding processes would present real challenges for CAE applications in mold design. The current practice is to simplify the modeling and process complexity. To simplify the modeling, assumptions are needed. Therefore, the final CAE analysis results should be used more in a qualitative basis, although CAE tools provide detailed quantitative results. Despite all these, CAE simulation technology is an extremely valuable tool for plastic molded part designers, mold makers and molding engineers. For engineers to use CAE tools in part, mold, and process design, they should be fully aware of the limitations and weaknesses of the CAE tools. This would be very critical in the analysis and use of CAE simulation results.

5.7 SUMMARY

In injection molding activities, CAE, CAD, and CAM are three important enabling technologies. CAD technology provides solutions for representation of design and enables the design to be more automatic. CAM solution helps to physically realize the design and virtually verify the manufacturability of the design solutions. CAE technology, however, helps to generate, verify, validate, test, and optimize the design solutions. In this chapter, the CAE applications in mold design are summarized. Foremost, the CAE applications in four aspects, viz., part design, mold design, molding process determination, and product quality assurance are depicted. The CAE analysis procedures and functionalities are then presented. In mold development life cycle, how CAE details in design processes is also articulated. To illustrate the practical applications of this technology, two case studies, namely, the cooling analysis of an injection mold design and the casting process simulation for casting mold design, are presented. The challenges and limitations of CAE technology in mold design are finally summarized to end the chapter.

REFERENCES

1. M.W. Fu, J.Y.H. Fuh and A.Y.C. Nee, "A core and cavity generation method in injection mold design", International J of Production Research, Vol. 39, No.1, pp.121-138, 2001.
2. V. Travaglini, "Optimizing part and mold design using CAE technology", Proceeding of ANTEC'1998, pp.36-40.
3. Advanced CAE Technology, Inc., "C-MOLD reactive molding user's guide", on-line documentation, 1997, http://www.scudc.scu.edu/cmdoc/re_doc/06_log.frm.html.
4. Moldflow Corporation, "Moldflow Plastics Advisers^{TM} design-for-manufacture analyses: Solutions for part designers", on-line documentation, 2000, http://ni.bitpipe.com/-detail/RES/997269894_183.html.
5. Moldflow Corporation, "Moldflow Plastics Advisers^{TM} design-for-manufacture analyses: Solutions for mold designers", on-line documentation, 2000, http://ni.bitpipe.com/-detail/RES/997270202_871.html.
6. B.H. Kim and M.C. Ramesh, "Automatic runner balancing of injection molds using follow simulation", J of Engineering for Industry, Vol. 117, pp.508-515, 1995.
7. C.C. Lee and J.F. Stevenson, "Runner design with minimum volume for multicavity injection molds: Part 1: Runner sizing, Proceeding of ANTEC'1997, pp.376-382.
8. C.C. Lee, "Runner design with minimum volume for multi-cavity injection molds: Part 2: Runner layout, Proceeding of ANTEC'1997, pp.382-387.
9. B.H. Lee and B.H. Kim, "Automated design for the runner system of injection molds based on packing simulation, Proceeding of ANTEC'1996, pp.708-713.
10. B.H. Lee and B.H. Kim, "Automated selection of gate location based in designed quality of injection molded part", Proceeding of ANTEC'1995, pp.554-560.
11. K. Himasekhar, K.J. Lottey and K.K. Wang, "CAE of mold cooling in injection molding using a three-dimensional numerical simulation", ASME J of Engineering for Industry, Vol. 114, pp.213-221, 1992.
12. K.J. Singh, "Design of mold cooling system", Injection and Compression Molding Fundamentals, A.I. Isayev, ed., Marcel Dekker, New York, 1987, pp. 567-605.
13. L.Q. Tang, K. Pochiraju, C. Chassapis and S. Manoochehri, "A computer-aided optimization approach for the design of injection mold cooling systems", ASME J of Mechanical Design, Vol. 102, pp.165-173, 1998.
14. Moldflow Corporation, "Moldflow Plastics Xpert®, Process Automation and Control System", on-line documentation, 2001, http://www.moldflow.com/pdf/mpx/mpx_white paper.pdf
15. S.L. Mok, C.K. Kwong and W.S. Lau, "An intelligent hybrid system for initial process parameter setting of injection molding", International J. of Production Research, Vol.38, pp.4565-4576, 2000.
16. H.H. Chiang, K. Himaskhar, N. Santhanam and K.K. Wang, "Integrated simulation of fluid flow and heat transfer in injection molding for the prediction of shrinkage and warpage", J of Engineering Materials and Technology, Vol. 115, pp.37-47, 1993.
17. B.H. Lee, "Optimization of part wall thickness to reduce warpage in injection molded part based on the modified complex method", Proceeding of ANTEC'1996, pp.692-697.
18. A.G. Bakelite, "Defects on molded parts", developed by Paul Thienel, Bernhard Hoster and Christian Kurten, On-line documents, 2003, http://www.bakelite.de/eng/pdf/kvl-pt2.pdf.

19. Y.F. Sun, K.S. Lee, C.H. Tan and A.Y.C. Nee, "The comparison between finite element analysis and experimental results of the cooling of injection molds with hot runner system", Proceedings of the 10th Manufacturing Conference in China, 10-12 October 2002, Xiamen, China.
20. R.W. Jeppson, Analysis of flow in pipe networks, Butterworth Publishers, UK, 1976.
21. N. Ahmad, H. Combeau, J.L. Desbiolles, T. Jalanti, G. Lesoult, J. Rappaz, M. Rappaz and C. Stomp, "Numerical simulation of macro-segregation: a comparison between finite volume method and finite element method predictions and a confrontation with experiments", Metallurgical and Materials Transactions A, Vol. 29A, pp.617-630, 1998.
22. M. Rappaz, C.A. Gandin, J.L. Desbiolles and P. Thevoz, "Prediction of grain structures in various solidification processes", Metallurgical and Materials Transactions A, Vol. 27A, pp.695-705, 1996.
23. C.A. Gandin, J.L. Desbiolles, M. Rappaz, and P. Thevoz, "A three-dimension cellular automation-finite element model for the prediction of solidification grain structures", Metallurgical and Materials Transactions A, Vol. 30A, pp.3153-3165, 1999.
24. K. Davey, S. Hindula and L.D. Clark, "Optimization for boiling heat transfer determination and enhancement in pressure die casting", ed. by A.Y.C. Nee, Special Issue on Molds and Dies, Proceedings of the Institution of Mechanical Engineers, Part B, Vol. 216, pp.1589-1609, 2002.
25. B.H. Hu, K.K. Tong, C.M. Choy, T. Muramatsu and S.X. Zhang, "State-of-the-art casting and forging technologies for industry", Material Technology and Advanced Performance Materials, Vol.17, No.4, pp.217-223, 2002.
26. J.C. Gelin, T. Barriere and B. Liu, "Improved mold design in metal injection molding by combination of numerical simulations and experiments", ed. by A.Y.C. Nee, Special Issue on Molds and Dies, Proceedings of the Institution of Mechanical Engineers, Part B, Vol. 216, pp.1533-1547, 2002.
27. D. Kothe, D. Juric, K. Lam and B. Lally, "Numerical recipes for mold filling simulation", The Proceedings of the 8th International Conferences on Modeling of Casting, Welding and Advanced Solidification Process, 1998, San Diego, CA.
28. T.A. Osswald and P.J. Gramann, "Polymer processing simulations trends", On-line documentation, 2003, http://www.madisongroup.com/Publications/PPtrends.pdf.
29. G. Engelstein, "Misuse and abuse of plastic process simulation", Proceeding of ANTEC'1994, pp.520-523.

6
Computer-Aided Die and Mold Manufacture

6.1 INTRODUCTION

The functional part geometry of a mold (or simply referred as a mold in this chapter), such as in the core and cavity, often includes sculptured surfaces. Traditionally, milling of sculptured surfaces in a mold was done by copy milling. However, as the cutting tool engagement condition is unfavorable, this method can result in tool deflection and machining inaccuracies. Instead, Computer Numerical Control (CNC) milling and Electrical Discharge Machining (EDM) are the two most important machining methods in die and mold manufacturing. In general, a die or mold is usually machined by the more efficient CNC machines first. If there are regions that are too intricate for a milling cutter or the material is very difficult to machine, EDM is used.

The efficiency offered by using a CNC machine is realized only when the coded instructions (tool paths) are efficient. As a result, tool path generation is an important step in manufacturing the products. During the past decades, NC tool path generation has achieved significant improvement, and hundreds of commercial CAM software systems are now available. However, manual user interactions of cutting tool selection and cutting parameters setting are still necessary. Moreover, as the shape of a mold may be very complex, to machine the complete workpiece, a large number of electrodes may be needed; thus electrode design becomes very tedious and time consuming. Therefore, what the mold manufacturing industry really needs is a CAM system that can analyze the part data, set the machining parameters, select cutting tools, design electrodes, and generate tool paths with little or no user interactions.

As shown in Fig. 6.1, mold machining can be classified into five steps: process planning, CNC machining, electrode generation, EDM machining, and surface finishing. In process planning, the workpiece geometry, the blank geometry, the machine tool specifications, cutting tools, etc., need to be studied. The machine tools (CNC and EDM), cutting tools and the machining parameters are planned in this process. Confined by the depths and shapes of the profile, the machining of a mold with complex surfaces consists of the *roughing* and *finishing* processes. Roughing is to remove excess material from a raw stock, while finishing is to remove residual material along the surfaces after roughing is applied to achieve the specified tolerance and surface finish. Generally, roughing is performed using CNC milling, while finishing by CNC or EDM, or a combination of the two processes, according to the following factors: reduction in lead-time, reduction in cost, increase in flexibility, and the improvement in quality.

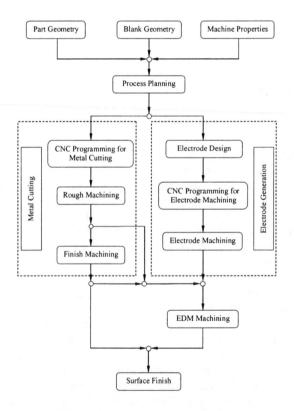

Fig. 6.1 Process of mold machining.

Computer-Aided Die and Mold Manufacture

In order to shorten the product design and manufacturing lead-time, many companies start die and mold design and manufacturing before the product is completely finalized. Subsequently, when the product design is refined, the die and mold design will need to be modified accordingly. In practice, if a die or mold design is changed after the tool path has been calculated, a new tool path must be regenerated for the entire workpiece, and the modification region is usually very small in comparison with the entire mold. This is highly unproductive and very time-consuming. A new tool path regeneration approach that can make use of the generated CL points for die and mold design modification will be very useful and is presented in this chapter. Many examples and industrial case studies will be illustrated to demonstrate the approach used.

6.2 INTERFERENCE-DETECTION IN MOLD MACHINING

As one of the most critical problems in CNC tool path generation and cutting tool selection, interference detection has been studied by many researchers. However, since mathematical results which would simplify the interference detection task are not available, most researchers detected interference from the viewpoint of tool path generation, and their methods can hardly be used in complex mold machining. The need for research on the mathematical foundation of 3-axis machining of sculptured surfaces has been reported by Choi et al. [1].

A new methodology of interference detection is described in this section. With this method, the workpiece surfaces are analyzed and classified according to their shapes. Local interference is detected by comparing the extreme curvature values of the concave and saddle surfaces with that of the cutter. Global interference is detected from the surface boundaries, and the region that is enclosed by an interference-free boundary is proven to be free of global interference. Theorems on global interference detection are proposed and tested. With this method, surface interference can be detected efficiently without calculating curve/surface offsets or generating tool paths.

6.2.1 Machining Interference

According to the shape of a profile to be machined, machining interference can be categorized into three types (see Fig. 6.2): curvature interference (Fig. 6.2a), bottleneck interference (Fig. 6.2b), and deep profile interference (Fig. 6.2c). Among them, the curvature interference (also referred to as local interference) occurs when the signed surface curvature value is greater than that of the cutter. Bottleneck interference (also referred to as global interference) occurs when the distance between a point on the offset surface to be machined and a point on another surface is less than the cutter radius. The deep profile interference occurs when the profile depth is greater than the cutter length.

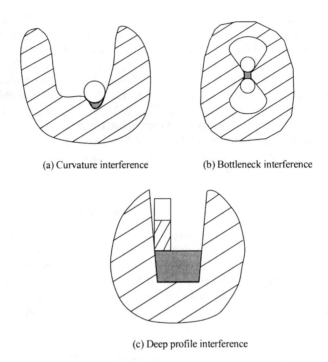

Fig. 6.2 Three types of machining interference.

Among them, the deep profile interference can be easily detected by comparing the cutter length with the profile depth. Therefore, only local and global interference will be studied in this chapter, and since in the machining of complex molds with sculptured surfaces, ball end-mills are most widely used in finish machining, and the interference detection is focused on ball end-mills.

6.2.2 Methods of Interference Detection

According to tool path generation methods, interference detection can be categorized into five types: curve offset, surface offset, CC-point and z-map methods, surface polyhedral method, and direct detection methods.

6.2.2.1 Curve Offset Method

The curve offset approach is usually used in pocket machining. With this method, the workpiece is sliced by a set of planes that are perpendicular to the z-axis, and a set of planar curves are created. Tool paths are then generated based on these intersection curves. An offset curve is defined as the locus of points traced by the unit normal vector of a planar generator curve multiplied by an offset distance [2]. When a planar contour is machined, the tool path (tool center) is the offset curve from the contour by the tool radius R. In generating tool path with the curve offset method, interference occurs if and only if there is self-intersection or intersection between the offset curves. Therefore, interference detection for curve offset method becomes finding the intersection and self-intersection of the offset curves.

As shown in Fig. 6.3, there are four types of offset curve intersections that may occur. First, an offset curve intersects itself when the curve's curvature radius is less than that of the cutter (Fig. 6.3a). Second, two offset curves of the pocket profile intersect (Fig. 6.3b). This occurs when the distance between the two curves is less than the double of the offset distance, and a shape of an '8' is formed. Third, a pocket offset intersects with an island offset curve (Fig. 6.3c). Fourth, two island offset curves intersect (Fig. 6.3d). In tool path generation of pocket profile, loops form when offsets intersect. Therefore, detecting the interfering loops is the key issue in the curve offset method.

Tiller and Hanson [3] were among the first researchers to study the interference problem. They developed an algorithm to remove the interfering loops without islands. With this algorithm, the self-intersection point of the loop is first detected, and the offset curve is then split at the intersection point. The tool path was generated by removing the interfering loop.

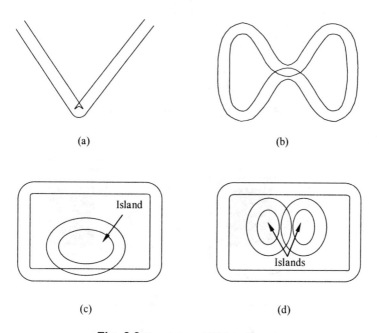

Fig. 6.3 Four types of 2D interference.

In detecting the interfering loops, the difficulty lies in determining which of the loops interfere with the workpiece, especially when there is an island inside the pocket. For example, Fig. 6.4(a) shows two loops L_1 and L_2. Loop L_1 causes no interference, and should be part of the final path. Loop L_2 causes interference and should be removed. Suh and Lee [4] solved this problem by sequencing each offset loop so that they are in clockwise order. The path was formed by following loop curves until an intersection point was met. For most situations, this discriminatory technique works satisfactorily. Problems occur when a "nested-loop" forms, where an offset crosses a loop that interferes with the workpiece (see Fig. 6.5).

Hansen and Arbab [5] solved this problem by finding the "interference indices" of the loop intersection points. All intersections of the offset chains were found first, and the offset curves were then split at the intersections. The indices of the simple chains were identified, and the interfering chains were discarded. Since at each intersection, the interference index is the same for all incident simple chains, and only one chain needs to be used for establishing the interference index.

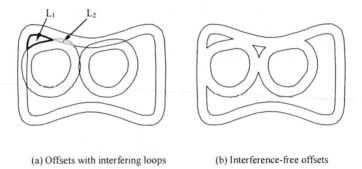

(a) Offsets with interfering loops (b) Interference-free offsets

Fig. 6.4 Interfering and non-interfering offset loops.

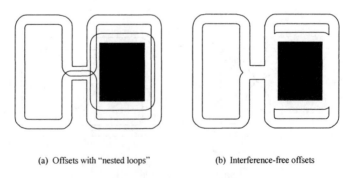

(a) Offsets with "nested loops" (b) Interference-free offsets

Fig. 6.5 Interference with "nested loops."

The curve offset method is widely used in rough machining where planar tool paths are planned, and interference detection for this approach is quite easy. One disadvantage of this method is that it is difficult to generate tool paths for other kinds of cutting tools other than flat end-mills. When machining with other cutters, the cutting point on the cutting tool changes when the surface slope changes, and the cutter's effective radius also changes, which makes the curve offset calculation very difficult. Another problem is that it is difficult to control the surface finish, which is related to the height between the two succeeding slice planes. In addition, this method needs to calculate the intersection curves, which is expensive in computation time.

6.2.2.2 Surface Offset Method

The surface offset method is usually used in machining sculptured surfaces with ball end-mills. An offset surface is a surface where every point has a distance equal to the cutter radius from the original surface [6]. The points on the offset surface are the center points of the cutter. With this approach, the offset surface is calculated first; the portion that will cause interference is then eliminated.

As in detecting interference from the curve offset, the key problem in detecting interference from the surface offset is to identify the intersection and self-intersection from the offset surfaces. Many researchers have studied this problem [7,8,9,10], and it is complex and time-consuming to calculate the intersection and self-intersection from the offset surfaces. Therefore, very few researchers detected interference by identifying the intersection and self-intersection from the offset surfaces directly [11,12]. To avoid calculating the intersection of offset surfaces, most researchers detected interference from the planar curves which were generated from the intersections of planes and the offset surfaces.

Lai and Wang [13] generated tool path with the surface offset method. In their approach, the offset surface was generated first. Planar tool paths were then generated by intersecting the offset surface with planes. Interference was detected by identifying the loops of the planar tool paths (see Fig. 6.6). As it is easy to identify the interference from curves; this method was also used by Seiler et al. [14]. However, as pointed out by Seiler et al., this method might miss the interference regions if the detection starts from an interference point. Therefore, other methods must be applied to detect the boundary points.

Tang et al. [15] developed a surface offset method to detect the interference inside one surface as well as along surface edges. In their approach, the offset surfaces and the curves of the surface boundaries were generated first, then the offsets intersected with drive planes that were parallel to z-axis. The interference-free tool paths were generated from the upper envelope of the intersection curves. By detecting interference along the boundaries of the offset surface, interference caused by G^1 (surface normal vector) discontinuity or surface gaps could be detected (Fig. 6.7).

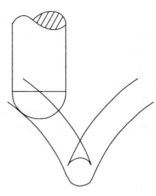

Fig. 6.6 Interference caused by the offset loops.

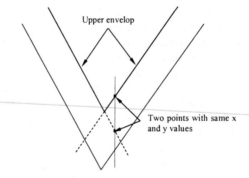

Fig. 6.7 Tool path generated from the upper envelop of the offset.

With the surface offset approach, as the offset surface can be calculated accurately, the machining accuracy can be easily controlled, and the scallop height of the machined surface can also be controlled. However, this method needs to calcu-late surface offset and to detect the possible intersection and self-intersection of the offset surfaces, which is computationally expensive. In addition, the offset surface calculation for other types of cutters other than ball end-mills is difficult, and the interference detection for these kinds of cutters is even more difficult.

6.2.2.3 *Cutter Contact (CC-) Point and Z-Map Methods*

The CC-point method can be used for all cutter types. With this method, a set of cutter contact (CC) points are planned on the compound surface, and all CC points are then offset along the surface normal vectors to compensate the effect of the cutter size. The required end positions of the cutter are thus obtained, which are called cutter location (CL) points. The final tool paths are calculated by removing the interference CL points.

Choi and Jun [16] were among the first who systematically studied interference detection for CC-point tool path generation approach. In their research, the CC points were first converted into a triangular mesh. Interference was then detected by calculating the distance between the cutter center and the mesh facet. In general, the CC-point method is computationally efficient, and the interference area can be identified directly. However, as the CC data is different from the CL data, this method can hardly be used in the circumstances where the direction of the tool path is to be defined (e.g., in rough milling, planar tool paths that are perpendicular to the cutter axis are often required). This problem can be solved with the z-map method.

The z-map, also called z-buffer method, is similar to the CC-point method. A z-buffer is a collection of z coordinate values, computed at the sampled grid points on the xy-plane, in a domain of interest. It can be obtained by the intersection of surfaces and vertical lines passing through the grip points. Tool paths are generated from the z-map using the inverse offset method. As the highest z value at each sampled point is accepted as the CL point, the final tool path is interference-free. Generating interference-free tool paths with this approach can be found in [17,18 19].

With the above two approaches, cutters other than ball or flat end-mills can be used easily. The disadvantage is that a great number of points must be determined to satisfy the machining tolerance, and extensive computation is needed to determine one tool position.

6.2.2.4 Surface Polyhedral Method

To avoid the complicated 3D surface offset calculation and to reduce the extensive point calculation arising in CC-point and z-map approaches, some researchers generated tool paths using the surface polyhedral approach. With this method, the surface is first approximated with a set of polyhedral facets, CL points are then generated based on these facets. The final tool path is generated by removing the interfering CL points.

Duncan and Mair [20] approximated the sculptured surfaces of a part model by a set of triangular facets, and for each triangular facet, explicit equations were obtained. Tool path was calculated by determining the highest z values of the cutting tool that intersected with these polyhedral faces on the domain, and interference can thus be avoided automatically.

Hwang [21] and Hwang and Chang [22] also converted the compound surface into a set of triangular polyhedron. Offset surface was then generated from these triangular facets. Interference occurs when one offset triangular polyhedron is overlapped by others. By intersecting the offset surface with a set of vertical lines, the point that intersects with the offset surface with a maximum z value is the interference-free CL point. The interference area in the original surface was then identified by detecting the cutting error between the two interference-free cutter points. If the error is too large, the area between these two points is the interference region.

As it is easy to calculate tool paths and identify interference with planes, the surface polyhedral approach is robust and reliable. However, this method needs to define equations for all the approximating planes, and the data size and computational time increase rapidly when a high resolution of surface approximation is desired. In addition, as the surface accuracy degrades in the surface approximation, a tool path generated with this method is less accurate than that of the offset method.

6.2.2.5 Direct Detecting Methods

All the above approaches detect interference from the viewpoint of tool path generation. With these methods, the interference area cannot be obtained until the offset or the tool path is calculated. If the number of available cutters is large, the computation time will be very long. Therefore, it is difficult to apply these methods in complex mold machining. To solve this problem, some researchers detected interference directly from the design data.

Yang and Han [23] detected interference with two steps: the curvature interference was first detected by comparing the surface curvature radius with that of the cutter; the global interference was then detected along the boundary of the curvature interference region. With this method, the interference region could be identified directly without tool path generation. However, large number of points are needed to be detected in global interference detection.

Pottmann et al. [24] and Glaeser et al. [25] studied the interference problem theoretically. The relation between the local interference and the offset surface's singularity, and the global interference-free condition related to local interference were studied in their researches. They solved some interference detection problems theoretically. However, the theory of global interference detection is based on the assumption that the surface is local interference free. In addition, no interference detection algorithm was given in their papers.

6.3 3-AXIS END-MILL INTERFERENCE DETECTION

In summary, many researchers have studied the interference detection problem, while most of them detected interference from the viewpoint of tool path generation or based on curve/surface offset, which was very time consuming. More theoretical research on the mathematical formulation of 3-axis machining of sculptured surfaces is therefore necessary.

To machine a surface with a ball end-mill, local interference-free is a necessary condition, but this does not ensure that the machining is gouging-free. For example, in Fig. 6.8, surfaces r_1, r_2 and r_3 are C^1 continuous, and surfaces r_1 and r_3

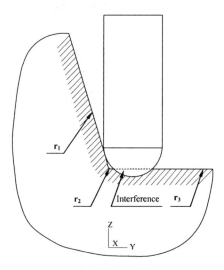

Fig. 6.8 Two local interference-free surfaces with global interference [27].

are free of local interference, but there is global interference between these two surfaces. Therefore, both local and global interference should be detected.

As described in the following sections, the novel interference detection approach [26] is performed in two steps (see Fig. 6.9): the local interference detection and the global interference detection. The input is the surface geometry of the workpiece part and the dimension of the cutting tool; the output is the detected interference regions.

In local interference detection, the surface shape is first identified, and then the extreme curvature values (κ_{max} and κ_{min}) of the concave and saddle surfaces are calculated. By comparing the surface extreme curvature values with that of the cutter, the local interference-free surface ($\kappa_{max} < (1/R)$), and the surface with all points being local interference ($\kappa_{min} > (1/R)$) are identified. The boundaries of the local interference regions are then identified in the surfaces of ($\kappa_{max} > (1/R) > \kappa_{min}$).

The global interference regions are then detected for the local interference-free surfaces, and the detection is started from the surface boundaries. If all points on the boundaries of a surface are global interference-free, the entire surface is free of global interference. If only some of the points on the surface boundaries are global interference points, the loci of global interference-free points are identified with the starting point at the surface boundary. The region that is surrounded by the boundary with all points being global interference is then detected to identify the possible interference-free region. The theories and algorithms of interference detection are described below.

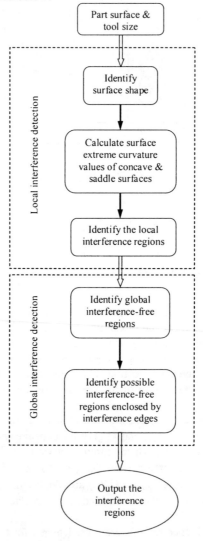

Fig. 6.9 Process of interference detection [26].

6.3.1 Local and Global Interference

In 3-axis milling with ball end-mills (the most commonly used machining method), local interference and global interference have some special properties,

Computer-Aided Die and Mold Manufacture

which will be studied in this section. Based on the theories on interference detection, an efficient interference detection algorithm will be introduced in the next section. It has been pointed out by Choi and Jun [16] that local interference occurs if and only if the signed surface maximum principal curvature value is greater than that of the cutter. Since there is no local interference inside a convex surface, if the surface shape has been identified, only the concave and saddle surfaces need to be checked. By comparing the extreme curvature values of the concave and saddle surfaces with that of the cutter, it is easy to find whether there is local interference inside a surface.

As shown in Fig. 6.10, r is the surface to be detected and e represents all of its boundaries. r_o is the offset surface of r and e_o represents all boundaries of surface r_o. The offset distance equals to the cutter radius R. r_i represents all surfaces around r. r and r_i are C^0 and piecewise C^2 continuous. It is assumed that all surface normal vectors point to the outside of the workpiece. Assuming the cutter axis T is parallel to z-axis, let $T = (0, 0, 1)$. If the surface normal vector n points to the $-z$ direction, the surface cannot be machined by a 3-axis milling machine. Therefore, it is assumed that all surface normal vectors point to the $+z$ direction, i.e., $(T \cdot n) > 0$.

Global interference occurs when the distance between a point on the offset surface r_o and surface r_i is less than the cutter radius R. Hence, global interference detection can be treated as a distance problem, and there are some special properties that can be used for global interference detection. Ding et al. [27] have proved the following theories on interference detection.

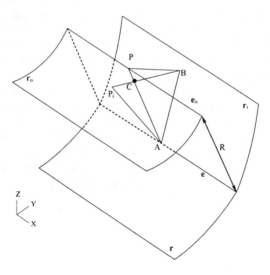

Fig. 6.10 Surface boundary and offset.

6.3.1.1 Theories on Interference Detection

Theorem 1 : Surface r is free of local interference, r_i represents all surfaces around r, and r and r_i are C^0, piecewise C^2 continuous. If all points on the boundaries of surface r do not interfere with surface r_i, then none of the points on surface r interferes with surface r_i.

Proof: To show this, the theorem proven by Pottmann et al. [24] is restated below:
Let Φ be a surface, which is represented as the graph $z = f(x, y)$ of a compactly supported, C^0, piecewise C^2 function $f: \Re^2 \to \Re$. If Φ is everywhere locally millable with a strictly convex C^0, piecewise C^2 cutter Σ with z-parallel axis, then Φ is globally millable with Σ.

In general, the surfaces of a ball end-mill are strictly convex C^0, piecewise C^2. Since surfaces r and r_i are C^0 and piecewise C^2 continuous, they can be represented with a surface Φ that is C^0 and piecewise C^2 continuous. If surface r_i is everywhere local interference-free, from Pottmann's theorem, no point on surface r interferes with surface r_i.

Now prove that the theorem is correct when there is a local interference region in surface r_i. Imagine that a ball with radius slightly greater than the cutter radius rolls on surface r_{ib}. A new surface r_n can be generated, which is the envelope of the ball on the r_i side (see Fig. 6.11). The new surface r_n is free of interference, and it is "outside" the original surface r_i. Therefore, if there is a point on surface r that interferes with surface r_i, the point will interfere with surface r_n. Except the local interference region, surface r_i and r_n have the same offset surface, and the tiny portion caused by the difference of cutter radius can be ignored in practice. Therefore, all points on the boundary of surface r do not interfere with surface r_n. Surfaces r and r_n can thus be represented as a C^0 and piecewise C^2 continuous surface Φ, and surface Φ is free of global interference.

Therefore, if all points on the boundary of surface r do not interfere with surface r_i, no point on surface r interferes with surface r_i, no matter whether there is local interference region inside surface r_i. With this theorem, the global interference can be detected from the surface boundaries, and the interference-free regions can be identified by checking the points on the surface boundaries. This gives an efficient method to identify the global interference-free regions. As shown in Fig. 6.12, surface r can be identified as global interference-free easily since all points on its boundaries are free of global interference.

Computer-Aided Die and Mold Manufacture

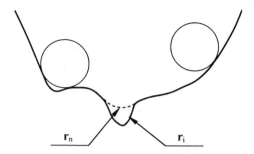

Fig. 6.11 Surfaces r_n and r_i.

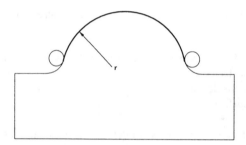

Fig. 6.12 Surface r is free of interference.

Theorem 1 gives the global interference-free condition. The following two theorems will study the surface which is surrounded by boundaries with all points on them interfering with other surfaces.

Lemma 1: Surface r is free of local interference, e is its boundary, and P is a point on its offset surface r_o. e_p and P_p are the projections of e and P on xy-plane, respectively (see Fig. 6.13). If the minimum distance between P_p and e_p is greater than the cutter radius R, then the point on surface r corresponding to point P is global interference-free.

Proof: Extrude the surface boundary e in z direction, a z-parallel cylindrical surface C_y is generated, and surface r_i is outside of C_y. Since the minimum distance between point P_p and surface C_y is greater than R, the distance between point P and surface r_i is greater than R too. Therefore, the point on surface r corresponding to point P is the global interference-free point.

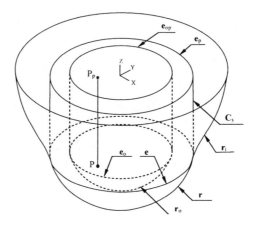

Fig. 6.13 The region with interference-free point.

In practice, it is not always easy to identify whether point P_p is inside the projection of the offset surface r_o. With the following theorem, it can be shown that if the point is inside the projection of surface r, there is also an interference-free point on the surface.

Theorem 2 : Surface r is free of local interference, P is a point on surface r, and e represents all of its boundaries. P_p and e_p are the projections of P and e on xy-plane respectively. If the minimum distance between P_p and e_p is greater than the cutter radius R, then point P is global interference-free.

Proof: Let r_p be the projection of surface r, and r_{op} is the projection of the offset surface r_o. From Lemma 1, if P_p is inside r_{op}, point P is free of global interference. Assume that point P_p is outside of r_{op}, and is between e_o and e_{op}. As surface r is free of local interference, the distance between P_p and e_p cannot be greater than the cutter radius R. Therefore, point P_p can only be inside r_{op}, and point P is an interference-free point.

Theorem 2 gives a methodology to identify the possible global interference-free point on a surface which is surrounded by boundaries with all points on them being global interference points. As it is in fact a 2D problem, it is easier to identify the interference-free point. If the interference-free point is found, the global interference-free region can be detected from this point.

Theorem 3 : r_1 and r_2 are two C^1 continuous surfaces, both of them free of local interference. For any two surface normal vectors n_1 and n_2 of surfaces r_1 and r_2 (on the same or different surfaces), $(n_1 \cdot n_2) \neq 0$, and none of the surface normal lines of the two surfaces are parallel. If all points on the boundaries of surface r_1 interfere globally with surface r_2, then all points on surface r_1 interfere globally with surface r_2.

Proof: To show this, the theorem proven by Sederberg et al. [28] is restated below:

> If two non-singular surface patches, S_1 and S_2, intersect in a closed loop C, then there exists a line that is perpendicular to both S_1 and S_2 if the following conditions are met:
> - The dot product of any two normal vectors (on the same patch or on different patches) is never zero;
> - S_1 and S_2 are everywhere tangent continuous.

Let r_{o1} and r_{o2} be the offset surfaces of r_1 and r_2 respectively. Since r_1 and r_2 are free of local interference, the normal vectors of r_{o1} and r_{o2} are the same at the corresponding points as r_1 and r_2, and they are C^1 continuous as r_1 and r_2. As all points on boundaries of surface r_1 interfere with surface r_2 globally, if there is a global interference-free region inside surface r_1, the offset surfaces r_{o1} and r_{o2} will intersect and form a closed loop. Since for any two surface normal vectors, $(n_1 \cdot n_2) \neq 0$, and the surface normal lines of the two surfaces are not parallel, surfaces r_{o1} and r_{o2} cannot intersect and form the closed loop. Therefore, there is no interference-free region in surface r_1, and every point on surface r_1 will interfere with surface r_2.

In a plastic mold, a large number of narrow pockets may be formed by the ribs of the plastic part (see Fig. 6.14). Most of the rib surfaces are planes, and these surfaces usually have tap angles for molding process. The two faced surfaces of the pocket can thus be easily identified as fulfilling the conditions of Theorem 3. If all points on their boundaries interfere with the opposite surface, these surfaces can be identified as the entire interference region.

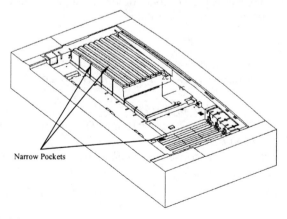

Fig. 6.14 Narrow pockets in a mold formed by ribs of the plastic part.

6.3.1.2 Interference Detection Algorithms

Based on the interference detection theories, the interference detection algorithms can be developed. As the theorems on global interference detection are based on the assumption that the detected surface is free of local interference, local interference should be detected first. Global interference is then detected among the local interference-free surfaces.

(1) Local interference detection

Local interference is detected by comparing the surface extreme curvature values with that of the cutter. Since there is no local interference in convex surfaces, surface shapes can be identified first. Only the concave and saddle surfaces' extreme curvature values need to be identified. For any point on the surface, there are two principal curvature values κ_1, κ_2. While in local interference detection, only the larger one needs to be considered. To simplify, the following equation is used:

$$\kappa_1 = K_m + \sqrt{K_m^2 - K_g} \qquad (6.1)$$

where K_m is the surface mean curvature and K_g the Gaussian curvature. Let κ_{max} be the maximum κ_1 value of the surface, and κ_{min} be the minimum κ_1 value of the surface. Let R be the cutter radius, it is obvious that if $\kappa_{max} < (1/R)$, then there is no local interference for the entire surface. If $\kappa_{min} > (1/R)$, every point on the surface is local interference point. If $\kappa_{min} < (1/R) < \kappa_{max}$, some points on the surface are interference points, while others are interference-free. By identifying the κ_{max} and κ_{min} values of the surfaces, the surfaces that are entirely interference or interference-free can be identified directly, which can shorten the detection time in cutting tool selection where more than one cutter needs to be investigated.

The algorithm and its pseudo codes of identifying the surface shape and detecting local interference are listed as follows:

LI (S, R, SI, SN)
   /* *Given all surfaces S of the workpiece and the cutter radius R, identify the local interference (SI) and non-interference (SN) surfaces* */
    {
       for (i = 0; i < number_of_S; i++)
       { sample (m × n) grip points for surface S(i) and calculate matrices of E, F, G and L, M, N;
       identify the convex/concave/saddle regions with K and K_m;
       calculate κ_{max} and κ_{min} in concave and saddle regions;
       if ($\kappa_{min} > 1/R$)   /* *every point in the region is local interference point* */
       SI ⇐ S(i);

```
        else if (κ_max < 1/R)      /* the surface is free of local interference */
          SN ⇐ S(i);
        else {
          identify the boundary of local interference region;
          trim the surface along the boundary of local interference region;
          SI ⇐ interference region;
          SN ⇐ non-interference region;
}}}
```

In local interference detection, the surface points are sampled first, and the L, M, and N values are calculated at these points. The number of points to be sampled is decided by the required accuracy. If more number of points are sampled, the detection is more accurate. The sign of Gaussian curvature value K_g is the same as that of K $(= (LN - M^2))$[29]. By computing the value of K, the saddle shape can be identified ($K < 0$). Concave and convex surfaces can then be identified with the mean curvature value K_m in the ($K \geq 0$) regions.

Then, the κ_1 values of concave and saddle surfaces are calculated, and the maximum (κ_{max}) and minimum (κ_{min}) values of κ_1 are identified in these regions respectively. By comparing κ_{max} and κ_{min} values with ($1/R$), the local interference and non-interference regions can be identified. If the surface includes both interference and non-interference regions, a set of points are sampled on the surface, and the number of the sampled points are decided by the given tolerance. The principal curvature values of these points are then calculated, and the local interference region is identified by comparing the curvature values with that of the cutter. When the interference region is identified, the surface is trimmed along the boundary of the local interference region. All interference surfaces are stored in *SI*, and the non-interference surfaces are stored in *SN*. Global interference will then be detected for the *SN* surfaces.

(2) Global interference detection

Based on the global interference features, a global interference detection algorithm *GI* is developed as following (see Fig. 6.15):

GI (SN, S, R, SG)
```
  /* Given the local interference-free surfaces (SN), and all the surfaces (S) to be
     machined, and the cutter radius R, identify the surfaces SG with all points on
     their boundaries being global interference points. */
    {
       for (i = 0; i < number_of_SN; i++) {
```

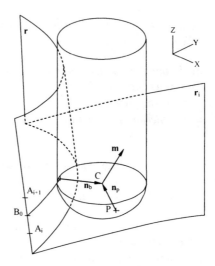

Fig. 6.15 Searching for point B_j.

```
for (j = 0; j < boundary_number_of_SN(i); j++) {
    sample m points A[m] on boundary(j);
    for (k = 0; k < m; k++) {
        n ⇐ (normalized surface normal at point A(k));
        C(k) = A(k) + R · n;   /* calculate cutter center point C */
        d(k) = 1000;           /* calculate the minimum distance between
                                  point C(k) and all the other surfaces */
        for (l=0; l<number_of_S; l++) {
            if (SN(i) ≠ S(l)) {
                d₀ ⇐ minimum distance of point C(k) with surface S(l);
                if (d₀ < d(k))
                    d(k) = d₀;
        }}
        if (d(k) > R) {
            if (d(k-1) > R)
                detect the next point;
            else
                call BD(A(k-1), A(k), SN(i), SG);
                        /* find the loci of interference-free points */
        }
        else {
            if (d(k-1) < R)
                detect the next point;
```

                    else
                        call BD(A(k-1), A(k), SN(i), SG);
            }}}
            if (all points on surface boundaries are global interference points)
                SG ⇐ SN(i);
}}

In global interference detection, the points on the boundaries of local interference-free surfaces are sampled first with the given tolerance. At each sampled point, the cutter center point C(k) is calculated, and the minimum distance d(k) between the point C(k) and all the other surfaces is then calculated. It needs to be pointed out that the measured distance is an approximation value, and the distance can be measured with the function from the CAD/CAM systems. If all points on surface boundaries are free of global interference, the entire surface is free of interference too. If all points on surface boundaries interfere with other surfaces, the surface is then further investigated with algorithm *NGI*. If some of the points on surface boundaries interfere with other surfaces while others do not, the boundary of a global interference region is then investigated with algorithm *BD* as follows:

BD (A_0, A_1, SNK, SG, S)
   /* Given the two points A_0 and A_1 on the boundary *e* of surface **SNK** with different interference conditions, and all the surfaces **S** of the workpiece, detect the region **SG** that is enclosed by global interference points. */
   {
       B(0) ⇐ Start_point (A_0, A_1, **SNK**, **S**);
                /* call sub-function Start_point to find point B(0) between points
                   A_0 and A_1 that separates interference and non-interference
                   regions */
       n_b ⇐ (normalized surface normal vector at point B(0));
       C(0) = B(0) + R · n_b;   /* calculate the cutter center point */
       i = 0;
       do {
                /* find the point P and the surface normal n_p on surface r_i
                   that has the distance of R with point C(0) */
           if (S(i) ≠ SNK) {
               d ⇐ (minimum distance between point C(0) and surface S(i));
               if ((d>(R-ε)) && (d<(R+ε))) {
                   P ⇐ (the point on surface S(i) that the distance with point
                        C(0) equals R);
                   n_p ⇐ (surface normal at point P of S(i) at point P);
               }
               else
                   i++;

```
}} while (((d<(R-ε)) or (d>(R+ε))) && (i<number_of_S));
j = 0;
do {                    /* detect the points that separate the interference
                           and non-interference regions */
    C(j) = B(j) + R · n_b;
    m ⇐ (n_b × n_p);
                        /* calculate the searching vector of the cutter center point */
    C(j+1) = C(j) + l · m;    /* search the next cutter center point */
    d ⇐ (minimum distance between C(j+1) and SNK);
    if ((d>(R+ε)) or (d<(R-ε)))
        shorten l, until C(j+1) is in the tolerance;
    B(j+1) = C(j+1) - R · n_b;    /* the point on the interference re-
                                     gion boundary */
    j++;
    if (B(j+1) reaches to surface boundary)
        search along surface boundary;
} while (B(j+1) has not reached to point B(0));
trim surface SNK along the detected boundary;
SG ⇐ interference region;
}
```

If two points A_0 and A_1 on the surface boundary change interference conditions (see Fig. 6.15), point $B(0)$ on the boundary that separates the interference and non-interference points is identified first. With $B(0)$, point P and the corresponding surface normal n_p on surface $S(i)$ that has the minimum distance of R with point $C(0)$ is identified. With n_p and the surface normal n_b of **SNK** at point $B(0)$, the searching vector m is calculated from the following equation:

$$m = \pm \frac{n_b \times n_p}{|n_b \times n_p|} \tag{6.2}$$

where the sign of m is decided by the previous searched point. If n_b is parallel to n_p, m is defined as perpendicular to n_b.

The next cutter center point $C(j+1)$ is detected by adjusting the step length l, so that the distance between point $C(j+1)$ and surface **SNK** is the cutter radius R (in the acceptable tolerance). The new point $B(j+1)$ that separates interference and non-interference regions can thus be calculated from point $C(j+1)$. The searching continues until the new searched point reaches the surface boundary. The searching is then performed along the surface boundary. When the first point $B(0)$ is searched again, the searched points form a closed boundary, which terminates the searching task. The surface is then trimmed along this boundary, and the surface with all points on its boundaries being global interference is stored in **SG**.

Computer-Aided Die and Mold Manufacture 243

The first point $B(0)$ on the boundary that separates the interference and non-interference regions is identified with the algorithm of *Start_point*. The point is searched by halving the region between the points A_0 and A_1 along the surface boundary, and the algorithm is shown below:

Start_point *(A_0, A_1, **SNK**, S);*
/* *Given the two points A_0 and A_1 on the boundary of surface **SNK**, search the point $B(0)$ on the boundary between the two points, which separates the interference and non-interference regions* */
```
{
    B(0) ⇐ (middle point of A₀ and A₁ along the surface boundary);
    n ⇐ (surface normal at point B(0));
    C ⇐ B(0) + R · n;          /* cutter center point at point B(0) */
    d ⇐ (minimum distance between point C and all other surfaces);
    if ((d<(R-ε)) or (d>(R+ε))) {
       do {
          n₀ ⇐ (surface normal at point A₀);
          C₀ ⇐ A₀ + R · n₀;    /* cutter center point at point A₀ */
          d₀ ⇐ (distance between point C₀ and other surfaces);
          if (d < (R-ε)) {
             if (d₀ < (R-ε))
                A₀ ⇐ B(0);
             else
                A₁ ⇐ B(0);
          }
          else if (d > (R+ε)) {
             if (d₀ < (R-ε))
                A₁ ⇐ B(0);
             else
                A₀ ⇐ B(0);
          }
          B(0) ⇐ (middle point of A₀ and A₁ along the surface boundary);
          n ⇐ (surface normal at point B(0));
          C ⇐ B(0) + R · n;    /* cutter center point at point B(0) */
          d ⇐ (minimum distance between point C and all other surfaces);
       } while ((d<(R-ε)) or (d>(R+ε)));
    }
    return (B(0));
}
```

When a region of a surface is enclosed by a boundary with all points on it being global interference points, it is further detected with algorithm *NGI* to identify the possible interference region. The detection algorithm *NGI* is as follows:

NGI(SI, SN, SG, S)
/* *Given surfaces **SG** with all points on their boundaries being global interference points and all surfaces **S** of the workpiece, identify the interference (**SI**) and non-interference (**SN**) regions.* */
{
   if (all points on the boundary of **SG** interfere with another surface r_i, and their normal lines are neither perpendicular nor parallel)
        SI $\Leftarrow$ *SG*;   /* *all points on **SG** interfere with surface r_i* */
   else {
       project the boundary of **SG** onto XY plane;
       if (the distance between P_p and e_p is greater than R /* *see Fig. 6.13* */
       {
          P $\Leftarrow$ *(the point on surface **SG** corresponding to point P_p)*;
          (u, v) $\Leftarrow$ *(surface parameter at point P)*;
          Δu $\Leftarrow$ *(step-over size of parameter u)*;
          B(0) $\Leftarrow$ ***SG**(u+Δu, v)*;   /* *the point on surface **SG** with the parameter being (u+Δu, v)* */
          d $\Leftarrow$ *(minimum distance between point B(0) and other surfaces)*;
          do {
             /* *find the first point B(0) on surface **SG** which separates the interference and non-interference regions* */
             $u = u + \Delta u$;
             B(0) $\Leftarrow$ ***SG**(u+Δu, v)*;   /* *the next point on surface* */
             d $\Leftarrow$ *(minimum distance between point B(0) and other surfaces)*;
          } while $((d > (R+\varepsilon))$ or $(d < (R-\varepsilon)))$;
          identify the global interference-free region from point B(0);
       }
       else {
          sample grip points on **SG**, identify the interference condition on these points;
          if (all points are interference points)
             SI $\Leftarrow$ *SG*;
          else, identify the interference-free region;
}}}

If all points on the boundaries of surface **SG** are global interference points, the surface is further investigated. If all points on surface boundaries interfere with another surface $S(i)$, and none of the surface normal lines are perpendicular or

Computer-Aided Die and Mold Manufacture

parallel for surfaces SG and $S(i)$, then all points on surface SG interfere with surface $S(i)$.

Otherwise, the boundaries of surface SG are projected onto xy-plane, and a point P_p inside the region of the projected boundaries e_p is investigated (refer to Theorem 2). To reduce computation time, the points are investigated along the curve normal line, and only a few points are sampled (e.g., the mid point and the end points of the projected curves). If P_p exists, the global interference-free region is investigated from this point. If none of the above conditions is satisfied, the grip points on surface SG are sampled and investigated. The regions of global interference are stored into SI, and the non-interference surfaces are stored into SN. That ends the interference detection for all workpiece surfaces.

6.3.2 Illustrative Example

Fig. 6.16 shows an example of an industrial part. The dark color regions in the figure show the detected interference regions when the cutter radius is 10mm. The local interference surfaces are the cylindrical surfaces and the blend surfaces at the corners of the large pocket. In global interference detection, there are some rib surfaces with all points of the surface boundaries interfere with another surface, and none of the surface normal lines are parallel or perpendicular. Therefore, all points on these surfaces are interference points. All points on the boundaries of surface A interfere with other surfaces. As it is easy to identify a point inside the projection of the surface boundaries, with the distance between the point and the projected boundary being greater than the cutter radius, which is the interference-free point. The interference-free region can thus be identified from this point.

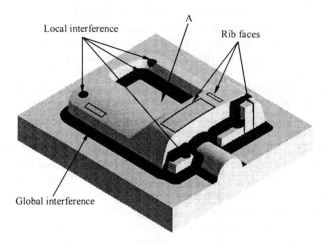

Fig. 6.16 An example of interference detection.

6.4 OPTIMAL CUTTER SELECTION

Cutter selection is a difficult task in mold machining owing to the complex constraints of surface geometry. A larger cutter may cause interference in tool movement. On the other hand, a smaller cutter needs longer machining time and causes unnecessary tool wear. Both cases increase the machining cost. In most CAM systems, the cutters are usually selected by the user according to his/her experience. In machining complex molds with sculptured surfaces, it is very difficult for an engineer to select the appropriate cutter size. Therefore, it is necessary to develop a methodology to select cutters automatically.

Cutter selection is a key decision in planning the machining operation of sculptured surfaces, and many researchers have studied this problem. Selection of an optimal set of cutting tools requires the consideration of geometric constraints as well as process time, and most researchers studied the problem from one or both of these two points.

6.4.1 Previous Work

Bala and Chang [30] studied the problem of multiple cutting tool selection for 2D pocket machining. Their approach was restricted to the selection of two cutting tool sizes only. The smaller cutter size as equal to the smallest corner radius of the profile was first selected. The larger one was then selected such that the unmachined area that remained after its use could be removed by the smaller cutter with one pass along the boundary of the finish cutter. As pointed out by the authors that for certain types of pocket geometry, the two cutter sizes that were chosen by using this approach might not be very different. In addition, the total processing time was not taken into account in their approach.

Lee et al. [31] generated tool paths by slicing the workpiece surfaces with a set of planes that were perpendicular to the z-axis. Appropriate cutters for each 2D profile were identified with the method proposed by Bala and Chang [30]. To reduce unnecessary cutter changes, some planes were merged. The cutters were then chosen based on the merged planes. However, if there is a narrow area in which only a small cutter can be used, then only the small cutter can be used for these merged planes. In addition, as in [30], the machining efficiency was not considered in the research.

Yamazaki et al. [32] introduced a system to select cutters and identify the areas that needed EDM. The surfaces to be machined were classified into "core" and "cavity" areas by identifying the surface inclination. Three cutters were selected for rough, semi-finish, and finish machining operations, respectively. The largest cutter that could machine the geometry without interference was selected for each operation. If the smallest cutter could not mill the entire workpiece without interference, it was selected, and the interference region was left for EDM. The unmachined region was identified with cutting simulation.

Since only geometric constraints were taken into account by the above researchers, the selected cutters might not be the optimal, and the number of selected cutters was usually restricted to one or two for each operation. To solve these problems, some researchers considered both geometric constraints and machining time in their approaches.

Lee et al. [33] selected cutting tools on the basis of the geometry of the final product and the raw material. In their approach, the solid model to be machined was first generated with Boolean operations. An octree representation of the solid model was then created and became the initial target of the machining. Large tools were used to machine the major portion of the octree and the smaller tools were used to machine the complex portion that was left. To improve the computing efficiency, the feature to be machined was approximated with hexahedrons, and the corners of the octants might not be removed with this method. In their research, the cutters were selected based on the volume of octree, and feed-rates for different cutting tools were not considered. In addition, the non-cutting tool path and tool change times were not considered in their research.

Veeramani and Gau [34] described a two-phase methodology for selecting an optimal set of cutting tool sizes to machine a 2D pocket that resulted in the minimum processing time. The tool paths and machining times for all the available cutters were first calculated with the Voronoi mountain concept, and the best set of cutters were selected based on the total machining time. Both cutting and non-cutting times were considered in machining time calculation. However, only rough machining with flat end-mills was studied in their research and, in addition, it was assumed that the workpiece could be machined with an available cutter that was small enough.

Chen et al. [35] presented a methodology of cutter selection and machining plane determination for pocket machining. The cutting tools were selected with two steps: in the first step, a series of hunting planes, which were parallel to xy-plane and with different z-values, were used to extract the geometric constraints and to identify all feasible cutters for each hunting plane. In each hunting plane, a set of candidate cutters and their corresponding machining times were evaluated. Cutter selection and machining plane determination were performed in the second step to minimize the total machining time. Two methods, the integer programming (IP) method and the dynamic programming (DP) method were proposed to solve the cutter selection and machining plane determination problems.

Mizugaki et al. [18] introduced a genetic algorithm for cutting tool selection. The area that could be machined by a cutter was first identified with the z-map method, the machining times for the available cutters were then calculated. The cutters were selected such that the total machining time and the unmachined area were minimal. However, the machining time calculation method was not introduced in the paper, and how to calculate the index F for each genotype was not clearly discussed.

To reduce the number of candidate tools for machining, Kyoung et al. [36] determined the range of tool sizes with geometrical constraint analysis first. The largest tool size was obtained from the maximum offset distance, while the smallest tool size was determined from the round radius specified at a concave vertex, or the minimum distance between convex vertex and other surfaces. Tool paths for different cutters were then generated, and the machining times for these cutters were calculated. The machining times for different cutter combinations were then calculated, and the set with minimum total machining time was selected as the optimal cutters.

Yang and Han [23] studied the selection of different cutter sizes for finishing machining. The total area to be machined was calculated first. The feasible and infeasible areas were then identified with interference detection algorithms. The machining time was estimated by considering tool size, scallop height, and accessible surface area. The combination of tools that possessed the minimum overall machining time was selected as the optimal tool sizes. As the machining time was calculated based on the machining area without tool path generation, this method can be used to select optimal cutters from the large number of available cutters in complex mold machining, and a similar method was used in this research.

In summary, many researchers have studied the cutter selection problem. Most of them concentrated on pocket machining, and the cutters were selected from the viewpoint of geometric constraints, the total tool path length, or the material removal rate. While in finish machining of complex molds, surface finish is another important factor in cutting tool selection besides geometric constraints. In addition, the time for tool path generation of finish machining is very long, it is impossible to generate tool paths for all the available cutters. Therefore, more research on cutter selection for finish machining is necessary.

6.4.2 Criteria of Selection

The factors that influence the decisions on cutter selection can be categorized as follows: geometric constraints, production cost, machining quality, machining accuracy, cutting tool life, and machine tool performance. Among the six factors, as the material left for finishing is relatively small and the required machining power is low, it is assumed that the CNC machine tool has enough power and torque for all the available cutters and the given feed-rates. Therefore, the machine tool performance will not be considered. Since mold machining frequently consists of a single piece or a small batch size, tool life is not a primary consideration. It is also assumed that all the available cutters are able to perform the entire machining task without tool changing. Furthermore, as the workpiece is milled using a CNC machine, the machining accuracy is decided by the CNC machine, and is not a critical problem in cutting tool selection. Hence, the formulation of optimal cutter selection for finishing is focused on achieving the required machining quality with minimum production cost while satisfying the geometric constraints.

Computer-Aided Die and Mold Manufacture

Machining quality is an important concern in finishing, and it has a direct effect on machining time. Besides machining accuracy, machining scallop height is the most important factor that decides the machining quality. As shown in Fig. 6.17, in surface machining, there is a small distance between two adjacent tool paths which is called the step-over size (g), the unmachined material between the two tool paths is called the cusp or scallop, and the upper limit height of the scallop is called the scallop height (h).

For a given profile, machining scallop height is decided by the cutting tool size, the tool path step-over size, and the surface shape. As shown in Fig. 6.18, although the distances between the two adjacent tool paths are the same, the scallop height values are not the same for different cutter sizes. To achieve the same machining scallop height, the step-over size for a small cutter should be smaller than that of the large one, which will lead to a longer machining time.

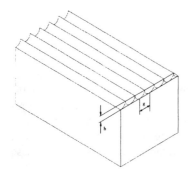   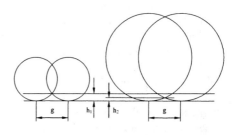

Fig. 6.17 Machining scallop height h and step-over size g.

Fig. 6.18 Different scallop height because of different cutter size.

In mold machining, shorter machining time reduces the total production cost, and an increase in cutter size would bring about a reduction in machining time. An obvious approach would be to use the largest possible cutter. But this may give rise to problems, since a large region of the profile may be left unmachined if a large cutter is used, that needs to be machined with smaller cutters which may lead to longer total production time. Hence, the selection of an optimal set of cutting tool sizes requires the consideration of processing time as well as the geometric constraints. Therefore, the problem addressed in cutter selection is as follows: given a mold workpiece and a set of available cutters (specifically, ball end-mills), find an optimal set of cutter sizes that can machine the workpiece with minimum total production time while satisfying the geometrical constraints and the surface finish.

6.4.3 Machining Time and Machining Area

Supposing for the ith cutter, the total length of the tool path is L_i, the feed-rate is f_i, and the machining time T_{mi} for this cutter is:

$$T_{mi} = \frac{L_i}{f_i} \tag{6.3}$$

In finishing, the feed-rate f_i is decided by the CNC machine tool and the cutter size, and it can be considered as a constant for a given cutter in cutting tool selection. A feed-rate database can be constructed for the cutters available in workshop. For a given feed-rate, the machining time T_{mi} is decided by the length of the tool path. The tool path length is a function of the machining area A_i, the tool path topology and the tool path step-over size g_i, which will be studied in the following subsections, respectively.

Considering geometrical constraints, the available region, the feasible region, and the unmachined region for each cutting tool size can be identified. The available region is defined as the region that needs to be machined by the cutting tool. The feasible region is defined as the region that can be machined by the cutting tool without causing interference. The unmachined region is defined as the available region minus the feasible region. The corresponding areas of these regions are defined as the available area A, the feasible area A_f, and the unmachined area A_u, respectively. The feasible and unmachined regions can be identified with interference detection algorithms as introduced in Chapter 4, and the areas of these regions can thus be calculated.

In calculating the machining area with multiple cutters, besides the feasible region that is identified by interference detection, the machining process should also be taken into consideration. If there is discontinuity between tool paths, or when a profile is machined with different cutters, there is usually some unmachined material left at the common boundary of the two tool paths (see Fig. 6.19). This is usually caused by the calculation tolerance, and it will lead to a poor surface finish. To solve this problem in practice, when a smaller cutter is used to machine the region left by a previous tool, the tool path for the small cutter is usually generated to overlap with the previous tool path generated by the larger cutter (see Fig. 6.20). The overlap distance is usually small, and it can be assumed to be 2 mm for all the cutters in most places. Hence, the area of the overlap region can be approximately calculated as:

$$A_{oi} = 2 S_{oi} \tag{6.4}$$

where A_{oi} = the overlap area of the ith cutter
S_{oi} = the boundary length of the interference region that is left by the previous cutter

Computer-Aided Die and Mold Manufacture

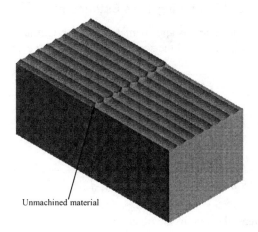

Fig. 6.19 Unmachined material because of discontinuous tool paths.

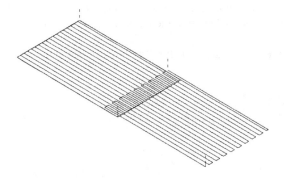

Fig. 6.20 Overlap of the two tool paths.

Although there is no overlap problem for the first cutter, in practice, the machining region for the first cutter is usually generated in such a way that it is a bit larger than the actual one to assure that the entire region is machined. It is assumed that the area of this region is the same as that for tool path overlap. The machining area A_{mi} for the ith cutter can thus be calculated as:

$$A_{mi} = A_{fi} + A_{oi} = A_{fi} + 2S_{oi} \tag{6.5}$$

where A_{fi} is the feasible area for the ith cutter.

6.4.4 Step-Over and Machining Time Estimation

When the surface curve is approximated by a straight line, the step-over size g_i for a cutter with radius R_i and the scallop height h can be calculated as (see Fig. 6.21):

$$g_i = 2\sqrt{R_i^2 - (R_i - h)^2} = 2\sqrt{2R_i h - h^2} \qquad (6.6)$$

Therefore, the machining time for the *i*th cutter can be calculated as:

$$T_{mi} = \frac{A_{mi}}{2f_i\sqrt{2R_i h - h^2}} + \frac{S_{ii} + S_{oi}}{2f_i} = \frac{Af_i + 2S_{oi}}{2f_i\sqrt{2R_i h - h^2}} + \frac{S_{ii} + S_{oi}}{2f_i} \qquad (6.7)$$

where A_{fi} = the feasible area for the *i*th cutter
f_i = the feed rate of the *i*th cutter
R_i = the cutter radius of the *i*th cutter
h = scallop height
S_{ii} = the boundary length of the unmachined region for the *i*th cutter
S_{oi} = the boundary length of the unmachined region for the $(i-1)$th cutter

Therefore, the total machining time for N cutters can be expressed as:

$$T = \sum_{i=1}^{N} \left(\frac{A_{fi} + 2S_{oi}}{2f_i\sqrt{2R_i h - h^2}} + \frac{S_{ii} + S_{oi}}{2f_i} + t_c \right) \qquad (6.8)$$

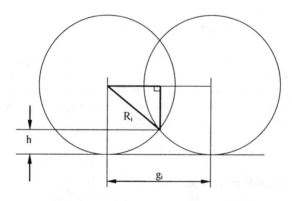

Fig. 6.21 Step-over size calculation.

6.4.5 Cutter Selection Algorithms

In machining a given surface area, the larger the cutter, the less the machining time needed will be. The best strategy of cutter selection is to choose the largest possible cutter under the geometric constraints. Therefore, the smallest cutter to be selected is the largest one among the available cutters that can machine the entire workpiece without interference. If none of the cutters can mill the entire workpiece without interference, the smallest available cutter is selected. In addition, it is assumed that the machining task is performed from the largest cutter to the smaller ones. The cutter selection algorithm and its pseudo codes are listed as follows:

Select_cutter *(R, N, f, h)*
   /* *Given N cutters with their radii R and feed-rates f, and the required surface scallop height h, select the optimal set of cutters that can machine the workpiece with the minimum time* */
   {
      $i = 0$;
      $S_{o0} \Leftarrow$ *(the boundary length of the entire region to be machined)*;
      do {
         detect interference for cutter R_i;
         calculate A_i & S_i for cutter R_i;
         $i++$;   /* *number of available cutters* */
      } while $((i<N)$ && *(the cutter cannot machine the entire workpiece))*;
      $T = 10^{10}$;   /* *large enough initial minimum machining time* */
      for $(k = \underline{0}; k <i; k++)$ {
         $(t_k, R_k) \Leftarrow$ *(calculate the minimum machining time and identify the corresponding cutters for the given i cutters)*;
         if $(t_k < T)$
            $(T, R_s) \Leftarrow (t_k, R_k)$;   /* *the minimum machining time T and the corresponding cutter combination* */
   }}

In selecting cutting tools, the input includes the given workpiece surfaces, the N available cutters with the radii R, the corresponding feed-rates f, and the surface scallop height h. It is assumed that the cutters are arranged in the order of the cutter dimensions, i.e., from the largest cutter to the smaller ones. The outputs are the selected optimal cutters.

The length S_{o0} of the entire region to be machined is calculated first. The interference detection algorithms introduced in Chapter 4 can be used to detect the interference regions for a specific cutter. The feasible regions, the unmachined regions and the boundaries of the interference regions are identified. Since the feasible region for a large cutter is also feasible for a smaller one, only the infeasible regions left by the previous cutter need to be investigated for the following cutters. The corresponding area A_i and the boundary length S_i are then calculated. Here, the area A_i is defined as the feasible area under the assumption that the entire workpiece is machined with the ith cutter.

The above steps are repeated until the smallest available cutter is detected, or the cutter that can machine the entire workpiece, and this cutter is accepted as the smallest selected cutter. If none of the cutters can finish the entire workpiece, the un-machined region will be machined with electrodes, which will be introduced in the next section.

For a given number of tool changes, the minimum machining time and the corresponding cutter combination are then identified. In determining the machining time, the feasible area A_{fk} and the boundary length S_{ok} are calculated as follows:

$$A_{fk} = A_k - A_{k-1} \qquad (6.9)$$
$$S_{ok} = S_{i(k-1)} \qquad (6.10)$$

where k and $(k-1)$ represent the kth and the $(k-1)$th selected cutter. With the values of A_{fk}, S_{ik} and S_{ok}, the machining time for the given cutters are estimated with Eq. (6.3). Comparing the machining times for different cutter combinations, the cutters R_k that can machine the workpiece with minimum time t_k are identified. The optimal cutters R_s are selected such that the cutters can machine the workpiece with minimum time T.

6.5 COMPUTER-AIDED ELECTRODE DESIGN AND MACHINING

6.5.1 Introduction

In the machining of molds with sculptured surfaces, die-sinking or plunge-type EDM plays an important role. Although the process is relatively slow when compared with CNC milling, EDM is accurate, and suitable for producing deep and narrow cavities. It can operate unmanned, machine profiles that no CNC milling cutter can reach, and, furthermore, productivity is dependant on workpiece conductivity rather than hardness. Therefore, most mold manufacturing companies use EDM in finish machining of hardened material and the regions where no cutter can mill.

Fig. 6.22 shows the schematics of EDM process. In the EDM process, material is removed by a series of discrete electrical discharges (sparks) that occur in the machining gap between the electrode and the workpiece. When the fluid between the two closed points of the electrode and the workpiece becomes ionized, the dielectric fluid creates a path for the discharge. The dielectric surrounding the column of electrical conduction is vaporized and decomposed by the discharge energy. As conduction continues, the diameter of the discharge column expands and the current increases. The small area in which the discharge occurs is heated to

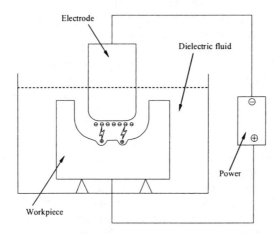

Fig. 6.22 Schematics of EDM process.

an extremely high temperature, and a small portion of the workpiece material is elevated above its melting temperature and is removed.

In EDM, the machining property is determined by surface roughness, over-cut, metal removal rate (MRR), and electrode wear. Among them, the first four parameters can be controlled using voltage, current, frequency, capacity, and fluids. Surface roughness is a function of the current divided by the frequency, and an increased discharge frequency can improve the surface finish. Over-cut equals to the length of the sparks that are discharged, and is determined by the initiating voltage and the discharge energy. The metal removal rate depends on the volume of metal removed by each spark and by the frequency of the discharge. The volume of metal removed per discharge is a function of the discharge energy and is increased by increasing the current. It also varies with the melting point of the workpiece.

Electrode wear is determined by the electrode material and the energy of the discharge. Materials having good electrode wear characteristics are the same as those that are difficult to machine. These are materials that require large amounts of energy to melt a given volume and usually have high melting temperatures. As one of the most widely used materials, graphite does not melt but goes directly to the vapor phase in EDM. For a given particle size, the vaporization requires a greater amount of energy than the melting of a similar particle of a metallic electrode and accounts for the favorable wear characteristics of graphite.

EDM researchers have mainly concentrated on achieving faster and more efficient material removal rates, coupled with a reduction in electrode wear, and improved surface integrity/finish. Mohri et al. [37] studied how the electrode wear changed in the beginning and stationary stages. Lonardo and Bruzzone [38] studied the influences of electrode material, dimension, flushing, depth of cut, and planetary motion on EDM performances, which included the workpiece removal rate, the electrode wear, accuracy, and surface texture.

Some researchers studied the problem of EDM process planning. Kruth [39] introduced a CAD/CAM system for mold design and manufacturing. The integration of CAD/CAM and EDM, including EDM process planning, electrode manufacturing, and on-line control of the EDM process were considered in the system. Lauwers and Kruth [40] introduced a process planning system for EDM operations. The system evaluated the process plans and selected the optimal process plan based on three cost factors: machining cost, electrode cost, and workpiece setup cost. However, how to design electrodes from sculptured surfaces was not studied in the system.

Some researchers studied how to simulate the EDM process with computers. Dauw [41] focused on the geometrical simulation of the EDM die-sinking process. The simulation showed how the electrode wear and workpiece geometry progressed during machining. Kunieda et al. [42] introduced a reverse simulation method for die-sinking EDM. The research studied how to obtain the appropriate electrode shape for achieving the desired final workpiece shape.

Bayramoglu and Duffill [43] introduced a "frame tool" method to machine a 3D profile based on the idea of wire EDM machining. With this method, the electrode was designed with a 3D frame shape. In the machining, the electrode feeds along the normal direction of the frame, so that the raw material can be removed in block inside the frame. Although this method could improve the EDM material removal rate, the machining of the electrode was difficult, and only simple shape surfaces could be machined with this method.

Yu et al. [44] applied 2-1/2D CNC pocketing idea in machining a 3D profile with a flat end cylindrical shape electrode. In their approach, the electrode was considered as a flat end cutter, and the workpiece was machined with the layer-by-layer method. The tool paths were planned in such a way that for two adjacent layers, the tool path directions and the start and end points were reversed. Therefore, uniform wear at the end of the electrode was realized, and the electrode wear can be compensated for in the depth direction. With this method, a 3D profile could be machined with a simple shape electrode accurately. However, as pointed out by the authors, the machining efficiency of this method was low and, in addition, the profile accuracy was also decided by the electrode length tolerance and should be properly controlled.

In summary, these researchers mainly concentrated on how the EDM control parameters or electrode shapes affected the machining efficiency and accuracy, or on how to machine some specific profiles with simple electrodes. However, how to find the region to be removed by EDM and design the electrode that can machine any profile from a 3D workpiece automatically was not considered. Further research would need to be conducted in this direction.

This section introduces a methodology and implementation of designing electrodes according to the workpiece geometry. With the defined electrode

Computer-Aided Die and Mold Manufacture

boundaries, the system creates an electrode automatically through curve trimming, solid body extrusion, and Boolean operations. An algorithm of sharp corner interference detection for electrode design is also developed and introduced in this chapter.

6.5.2 Principles of EDM Electrode Design

An electrode usually includes two parts, i.e., the electrode tool and the holder (see Fig. 6.23). The electrode tool is used to form the shape of the workpiece geometry. The electrode holder is used to hold the electrode tool during EDM. The electrode tool and the holder are often designed and constructed in one piece. In the following sections, an electrode tool will be called an electrode. The electrode has different position coordinate systems in CNC (Fig. 6.23) and EDM (Fig. 6.24). To avoid confusion, the electrode coordinate system in EDM machining status will be considered as the normal position. That is, the top face of the electrode tool connects with the electrode holder.

Since a large number of electrodes are used in complex mold manufacturing, to shorten the mold manufacturing time, an electrode is usually machined with CNC machines. Therefore, in this system, the electrode is assumed to be machined to its near net-shape with a CNC milling machine. In addition, the electrode design module can form part of the mold design system whose output of the designed mold is a 3D solid model, and the workpiece is therefore assumed to be a solid part.

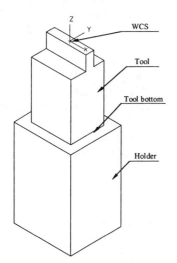

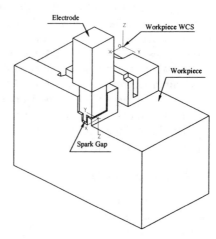

Fig. 6.23 An EDM electrode.   **Fig. 6.24** An electrode and the workpiece.

An electrode is used to machine the workpiece using the EDM process. To do so, the electrode itself must first be machined to shape before it can be used. A unique feature of electrode design is that the electrode must be designed based on the workpiece geometry to be produced. The following principles of electrode design are described:

(1) An electrode is used to form the shape of workpiece geometry, and it can be considered as a form-shaped cutting tool. According to the CNC and EDM machining process requirements, two or more electrodes can be united into one, and one electrode can be separated into several electrodes.
(2) To improve the machining efficiency, roughing, finishing, and even semi-finishing electrodes are used in EDM. The spark gaps for these electrodes are different, hence their geometries are different from each other. But these electrodes differ from each other only in the surfaces' offset values (EDM spark gaps). By setting different stock values (equal to the spark gaps) when generating the CNC milling programs for machining the electrode, roughing, semi-finishing, and finishing electrodes can be generated from the same electrode model. Therefore, only one electrode model will be used for the roughing, semi-finishing, and finishing, and the dimension of the designed electrode is the same as that of the workpiece.
(3) Unmachined material left on an electrode will cause over-cut to the workpiece in EDM, thus the electrode should be designed such that it can be machined completely with CNC milling.
(4) The electrode material is easy to cut using high speed milling machines that are widely used in electrode machining. The CNC milling time is short when compared with the electrode setting time; therefore, an electrode should be designed such that it can be machined in one CNC setting.
(5) The surface region that needs EDM may be very small, but the electrode that undergoes the CNC and EDM machining processes cannot be too small. Raw material of an electrode should be of a standard size if possible.
(6) When an electrode is used in the EDM process, the positional relationship between the electrode and the workpiece becomes very important. A clear reference point on an electrode is necessary. In practice, the origin point of the electrode working coordinate system (WCS) is often used as the reference point on the electrode design (see Fig. 6.23).

6.5.3 Electrode Tool Design

An electrode tool often contains sculptured surfaces, and these sculptured surfaces need to be machined with CNC machines. Therefore, the designed electrode tool should be a surface or solid model. Two methods can be used to obtain the electrode geometries/surfaces; one method is to extract the surfaces from the workpiece and construct the electrode from these surfaces. However, since an electrode often consists of several surfaces, it is difficult to obtain these surfaces.

Computer-Aided Die and Mold Manufacture

In addition, the region that needs an electrode may only be part of a surface, the surfaces to be used to form the electrode need to be trimmed to the electrode boundary, and it is difficult and time-consuming to perform the surface trimming operation automatically.

The second method is to generate the electrode tool through the electrode boundary with Boolean operations. The idea of this method is as follows:
  i) Define the boundaries that need an EDM electrode, and create curves from them;
  ii) Extrude the curves in the z direction to create a solid body; and
  iii) Subtract the extruded body by the workpiece using the Boolean operation, and a solid electrode tool with the reverse shape of the workpiece is generated.

Using this method, the surfaces that need EDM can be identified and trimmed automatically through the Boolean operation, and the generated electrode is a solid model. Therefore, the second method was used in a reported system [45]. As shown in Fig. 6.25, the electrode tool design module includes workpiece selection, electrode boundary identification, curve creation and trimming, solid body extruding, and Boolean operation, etc., and they will be introduced in the following subsections.

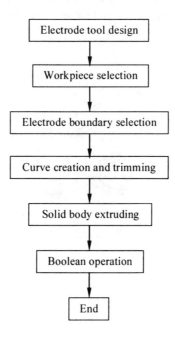

Fig. 6.25 Flow chart of the electrode tool design system.

6.5.3.1 Workpiece and Electrode Boundary Selection

A workpiece is selected first. Its extreme z values (z_max and z_min) are calculated. These values are then used to control the height of the extruded body. Generally, EDM is only used when the workpiece cannot be machined using CNC milling cutters, and the electrode boundary can be determined after cutter selection. With this method, after interference detection and cutter selection, the region that no cutter can reach without interference is detected, and the boundaries of these regions are identified. As the boundary of an interference region is a closed one, it can be used directly in electrode design. But for some processes (e.g., the material is very difficult to machine or the machining task needs to be performed in the same EDM setting), an electrode may be needed even when the surface profile can be machined using a milling cutter. Then the electrode boundary should be defined and selected by the user. Since identifying interference area boundary has been introduced in Section 6.3, this section will focus on the latter case, where the electrode boundary is determined and selected by the user.

The electrode boundaries define the region of the workpiece that requires an electrode to machine (see Fig. 6.26). They are used to generate an extruded body, and they should form a sequenced closed loop. A closed loop/boundary means that the surface boundaries or curves connect end to end, and there is no gap or overlap between them. To make electrodes generation more convenient, both curves and surface boundaries can be selected as the electrode boundary. The algorithm and its pseudo codes of selecting curves and surface boundaries as electrode boundaries are given below:

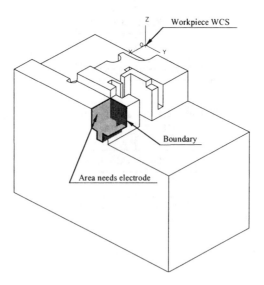

Fig. 6.26 Electrode boundary identification.

Computer-Aided Die and Mold Manufacture

Select_curve_boundary (W, c, m, b, n)
  /* Given the workpiece W and the m available curves c, select curves and
     surface boundaries as electrode boundaries **b**, and the number of **b** is n */
{
    (e, l) ⇐ (search all l surface boundaries **e** of the workpiece W);
    (bd, k) ⇐ ((c, m) + (e, l));   /* put all surface boundaries and curves
                                                         to bd, k is the number of bd */
    b(0) ⇐ bd(i);   /* user selects the first boundary */
    n = 1;
    if (**b**(0) is not closed) {
        bc = 0;   /* **b** is not closed */
        P_1 ⇐ (one endpoint of **b**(0));
        display arrow on **b**(0) which points to P_1;   /* define the searching
                                                                                   direction */
        if (the arrow is reversed by the user);
            P_1 ⇐ (the other endpoint of **b**(0));
        P_0 ⇐ (the endpoint other than P_1 of **b**(0));
        do {
            if (selected by user) {
                b(n) ⇐ (selected boundary/curve);
                n++;
            }
            else if (user create curve) {
                call functions to generate the curve;
                b(n) ⇐ (generated curve);
                n++;
            }
            else {
                j = 0;   /* the number of curves/surface boundaries that
                                                  connects with **b**(n) */
                for (i=0; i<k; i++) {
                      if (bd(i) is not **b**(n)) {
                          d ⇐ (the distance between point P_1 and bd(i));
                          if (d < ε) {   /* ε is small enough */
                              ec(j) ⇐ bd(i);   /* the curves/surface boundaries
                                                        connecting with **b**(n) */
                            j++;
            }}}
                for (i=0; i<j; i++) {
                    highlight ec(i);   /* for user to accept or reject the
                                                    curve/boundary */
                    if ((ec(i) is the wanted curve/boundary) {/* user decides */
                        b(n) ⇐ ec(i);

```
                n++;
}}}
P₁ ⇐ (one endpoint of b(n-1));
display arrow on b(n-1) which points to P₁;
if (the arrow is reversed by the user);
        P₁ ⇐ (the other endpoint of b(n-1));
        P₂ ⇐ (the endpoint other than P₁ of b(n-1));
        d ⇐ (the distance between P₀ and P₂);
        if (d < ε)
                bc = 1;                                  /* b is closed */
} while ((bc==0) && (user does not terminate the selection));
}}
```

6.5.3.2 Curve Creation and Trimming

Since it is difficult to operate with surface boundaries in trimming and projection, if the selected electrode boundary is a surface boundary type, a curve needs to be created automatically to represent the surface boundary. The curves are then projected onto the *xy*-plane (see Fig. 6.27).

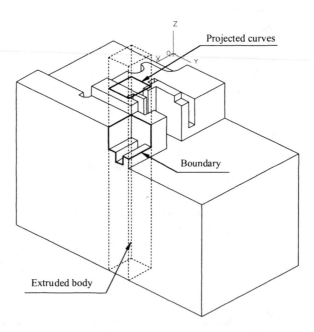

Fig. 6.27 Curve projection and solid body extruding.

Computer-Aided Die and Mold Manufacture 263

When the workpiece part file is converted from another CAD/CAM system, the surface boundaries that should connect end by end may not connect well and a gap or overlap between the surface boundaries may be found. The curves created from them may not connect well. To generate a solid body with these curves, they need to be trimmed to make them connected end to end. As shown in Fig. 6.28, there are five types of curve relations. In Fig. 6.28a, the curves form a closed loop perfectly. In Fig. 6.28b, the curves' endpoints are inside their intersection point, and the curves must be extended. In Fig. 6.28c, the curves' endpoints are outside their intersection point, and the portion outside the intersection point must be trimmed off. In Fig. 6.28d, a curve is entirely outside the intersection point. The curve must be extended first, and the outside portion must be trimmed off subsequently.

In Fig. 6.28e, two curves do not intersect, and there is an overlap or gap between them. Obviously, these curves cannot be connected by trimming or extending. One way to connect them is to construct a line from the two curves' endpoints. However, this will form sharp corners between curves, which will cause problems in CNC machining. Another method is to move one of the curves' endpoint until it touches the required endpoint. Since the gap or overlap is usually caused by the system tolerance, the distance between the two curves' endpoints is relatively small. Hence, the error caused by moving the curve endpoint is acceptable. The algorithm of curve trimming is as follows:

Curve_trim (C, n)   /* Given n curves **C**, trim them to form a closed loop */
{
    for (i=0; i<(n-1); i++) {
        d ⇐ the distance between C_i and C_{i+1};
        (P_i, P_{i+1}) ⇐ (two endpoints on C_i and C_{i+1} with minimum distance);
        if (d > ε) {   /* ε is small enough */
            find the intersection point **P** of C_i and C_{i+1};
            if (**P** does not exit)
                move P_i to P_{i+1};
            else if (**P** is outside C_i and C_{i+1})
                extend curves C_i and C_{i+1} to **P**;
            else
                cut off the portion of curves C_i and C_{i+1} outside **P**;
    }}
    trim C_0 and C_{n-1};
}

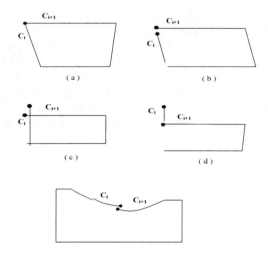

Fig. 6.28 Conditions of curve trimming/extension.

6.5.3.3 Solid Body Extrusion and Boolean Operation

Extruding the trimmed curves in the z direction, a solid body with the shape of the curves is generated (see Fig. 6.27). To prevent an electrode holder from sparking the workpiece, the extruded body is made higher than the workpiece's highest point for a value that is greater than the largest possible spark gap (e.g., 5mm). The extruded body's lowest point has a value of the workpiece's z_min value. Subtract the extruded body by the workpiece, an electrode tool with the reverse shape of the workpiece is created (see Fig. 6.29).

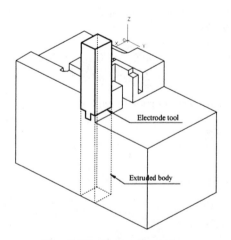

Fig. 6.29 Boolean operation.

Computer-Aided Die and Mold Manufacture

6.5.4 Electrode Holder Design

To make it easy to set the reference point in CNC and EDM, the electrode holder is designed with a rectangular shape. Standard-sized electrode holders should be used whenever possible, so that standard fixtures can be used, which can greatly reduce the electrode setting time in CNC and EDM. A library containing standard sizes of electrode holders can be constructed for use. When the electrode holder is generated, it is positioned at the top of the electrode tool while the center coincides with each other in x and y directions (see Fig. 6.29).

The WCS of the electrode is used in CNC and EDM. It reflects the relative position of the electrode and the workpiece, and should be easy to handle in CNC and EDM processes. Here, the electrode WCS for CNC and EDM is set to the same. The axes of the WCS are set parallel to the workpiece's WCS corresponding axes. The origin of the WCS is set at the center of the electrode holder in x- and y-directions, and it is positioned at the bottom of the electrode tool. The steps of electrode holder generation and WCS setting are given as follows:

Create_holder_WCS(T)
        /* *Given designed electrode tool(s) **T**, create the electrode holder **H** and set the coordinate system WCS for the electrode* */
{
    T $\Leftarrow$ *(select the electrode tool(s));*
    (X, Y) $\Leftarrow$ *(the dimensions of **T** in x and y directions);*
    if ((X $\leq L_{max}$) && (Y $\leq L_{max}$))   /* *the dimension is less than the largest standard size of the electrode holder* */
        (X, Y) $\Leftarrow$ *(select the size fit for the electrode tool(s) from the standard size database);*
    else
        (X, Y) $\Leftarrow$ *(integral multiple of 5 that fits the electrode);*
    *create **H** and put it at the top of the electrode tool(s) **T**;*
    *unite **H** with **T**;*
    set the WCS for the electrode;
}

6.5.5 Sharp Corner Interference Detection

As discussed previously, milling interference can be categorized into three types according to the shape of a profile to be machined: *local interference*, *global interference*, and *deep profile interference*. Generally, these three types of interference can be avoided by changing the dimensions of the cutter. While in electrode design, a "new" type of interference: *sharp corner interference* may occur. As shown in Fig. 6.30, sharp corner interference occurs when two surfaces form a sharp corner that cannot be machined by any cutter at their common surface boundary. By definition, sharp corner interference is in fact a special type of global interference. If this kind of interference occurs, the interference region for a given cutter can be identified with the global interference detection algorithms, as introduced in Section 6.3.1.

Since the sharp corner interference is caused by the geometry of the workpiece, when this type of interference exists, the material cannot be removed by the same type of the cutter. In addition, the electrode material is relatively easy to machine and the size of an electrode is usually small, thus the cutter which is used to machine the electrode can be very small. For example, cutters as small as 0.5mm in diameter may be used in machining an electrode. In addition, high speed machining is widely used in machining electrodes, thus cutters with a large ratio of length/diameter are often used. Hence, only sharp corner interference where no cutter can reach without interference is considered in electrode design.

An electrode is usually machined using a CNC milling machine. Two types of milling cutters are usually used in electrode machining: *flat end-mill* and *ball end-mill*. The main difference between them is that a flat end-mill has a sharp edge at its bottom, while a ball end-mill has a spherical face at its bottom. If a profile cannot be machined by either flat or ball end-mill, it cannot be machined by any other types of cutters. Therefore, in this section, only sharp corner interference for ball end-mill and flat end-mill is discussed.

6.5.5.1 Terminologies

Before detecting sharp corner interference, some terminologies used in this section are given below (see Fig. 6.31):

(1) Common boundary e: the surface boundary that the two connected surfaces share, $e = r_1(s, t) \cap r_2(u, v)$.
(2) Boundary normal plane P_n: the plane that contains the point (P_t) on e and is normal to the tangent vector t_e of e at P_t.

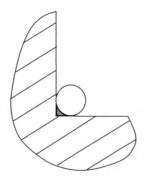

Fig. 6.30 Sharp corner interference.

(3) Surface tangent vectors t_{s1}, t_{s2}: P_n intersects with the two connected surfaces r_1 and r_2 to create two intersection curves C_1 and C_2. The tangent vectors t_{s1} and t_{s2} of the two curves at their common vertex point (P_t) are defined as the surface tangent vectors at that point.
(4) Surface tangent angle α_i: P_n intersects with a plane, that contains point P_t and is parallel to the *xy*-plane, to create a line l_p. The angle between t_{si} and l_p is the angle α_i. If t_{si} points to the positive direction of *z*-axis, $\alpha_i > 0$, otherwise $\alpha_i \leq 0$. In Fig. 6.31, $\alpha_1 < 0$ and $\alpha_2 > 0$.
(5) Angle θ: the angle between t_{s1} and t_{s2}, it is measured in the side without workpiece material, and $0° \leq \theta < 360°$.

6.5.5.2 *Sharp Corner Interference*

When two connected surfaces form a sharp corner at their common boundary, interference may occur. The interference condition is decided by the surface boundary vector t_e and the angles of α_i and θ, i.e.,

- If θ ≥ 180° (Fig. 6.32a), no interference occurs at the common boundary of the two surfaces for ball and flat end-mills;
- If θ < 180° (Fig. 6.32b), interference will occur for ball end-mill. For flat end-mill, if the boundary tangent vector t_e is parallel to the *xy*-plane, the surface tangent angles α_1 and α_2 must be checked:

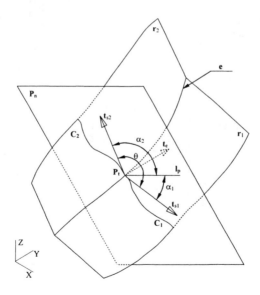

Fig. 6.31 Surface angles and directions.

- If $\alpha_1 \leq 0°$ and $\alpha_2 \geq 90°$ (Fig. 6.32c), there is no sharp corner interference; otherwise,
- Interference occurs (see Fig. 6.32d);

• If t_e is not parallel to the *xy*-plane (see Fig. 6.32e), interference will occur.

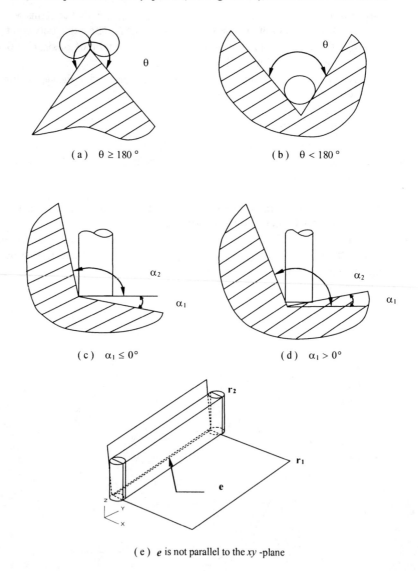

Fig. 6.32 Sharp corner interference for flat and ball end-mills.

Computer-Aided Die and Mold Manufacture

The sharp corner interference conditions are summarized in Table 6.1. It can be found from the table that if there is a sharp corner interference for the flat end-mill, no other types of milling cutter can mill the surface boundary without interference, the electrode must then be redesigned.

6.5.5.3 Algorithm

Based on the above discussions, a sharp corner interference detection algorithm is listed below:

Sharp_corner_interference_detection (e, n)
/* Given surface common boundaries *e* and the number *n* of *e*, find the interference boundaries SP_EDG, and the number *m* of SP_EDG */
{
    for (i=0; i<n; i++) {
        $(r_1, r_2) \Leftarrow$ (surfaces sharing e(i));
        $(P_t, k) \Leftarrow e(i);$ /* sample *k* points P_t on e(i) */
        j = 0;
        interference_sign = 0;   /* no interference is detected */
        do {
            $t_e \Leftarrow$ (boundary vector at point $P_t(j)$);
            $P_n \Leftarrow$ (surface normal plane at point $P_t(j)$);
            $(C_1, C_2) \Leftarrow (r_1, r_2$ and $P_n)$;   /* intersection curves */
            $t_{s1} \Leftarrow$ (surface tangent vector of r_1);
            $t_{s2} \Leftarrow$ (surface tangent vector of r_2);
            $(\alpha_1, \alpha_2, \theta) \Leftarrow$ (calculate angles between t_{s1} & t_{s2});
            if ($\theta < 180°$) {
                if (t_e parallel to *xy*-plane) {
                    if ($\alpha_1 > 0°$ && $\alpha_2 < 180°$)
                      interference_sign = 1;   /* interference detected */

Table 6.1 Conditions of interference determined by the common edge of two surfaces

Surface tangential angle α_1, α_2 and θ	θ	Interference for flat and ball end-mill		
		Ball	Flat	
	$\geq 180°$	No	No	
	<180°	Yes	t_e parallel to *xy*-plane	$\alpha_1 \leq 0°$ & $\alpha_2 \geq 90°$	No
				All else	Yes
			t_e not parallel to *xy*-plane	Yes	

```
        }
        else
            interference_sign = 1;
    }
    if (interference_sign == 1)
        highlight e(i);  /* prompt the user to redesign the electrode */
    j++;
} while ((interference_sign == 0) && (j <k));
}}
```

In detecting sharp corner interference, all surface boundaries of the electrode are identified. One surface boundary is then analyzed, and the surfaces r_1 and r_2 that share the surface boundary are identified. A set of points on the surface common boundary are sampled, and the number of the points is decided by the detecting accuracy. The points are sampled in such a way that they all have equal curve length along the surface boundary.

The angles at the sampled points are then calculated. A surface boundary vector t_e at the sampled point $P_t(j)$ is calculated, and the surface boundary normal plane P_n is constructed. P_n intersects with the surfaces r_1 and r_2, and the intersection curves C_1 and C_2 are constructed. Surface tangent vectors t_{s1} and t_{s2} are calculated for curves C_1 and C_2 at their common vertex point $P_t(j)$. The angles α_1, α_2 and θ are then calculated with surface tangent vectors t_{s1} and t_{s2}, and the sharp corner interference is detected with these angles. If there is no sharp corner interference at that point, the next point is investigated until all points of interference on the surface boundary are investigated. If the sharp corner interference is detected, the electrode must be redesigned. Generally, the electrode needs to be separated into two or more electrodes along the interference boundary.

6.5.6 Illustrative Examples

The computer-aided Cutter Selection and Electrode Design System [45] (CSEDS) has been reported to select cutting tools and design electrodes automatically with little or no user integration for a 3-axis mold machining. As shown in Fig. 6.33, CSEDS works as follows: the workpiece and the available cutters are first selected by the user; the interference is then detected automatically for these cutters; and the interference regions are then identified. The machining times for these cutters are also calculated, and the set of cutters that can machine the workpiece with minimum machining time is identified as the optimal cutters. Electrodes are designed for the regions where no cutter is able to machine. The electrode boundaries can be designed with boundaries defined by the user. With the selected cutters and the designed electrodes, tool paths are generated automatically. The system includes five main parts: the interface, the database, the interference detection module, the cutter selection module, and the electrode design module. The inputs of the system are the workpiece geometry, the dimensions and

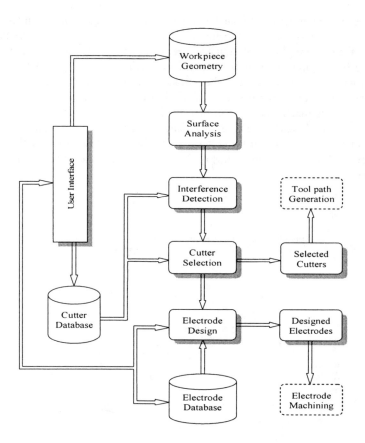

Fig. 6.33 Architecture of the CSEDS [45].

feed-rates of the available cutting tools, and the surface scallop height requirement. The outputs are the selected optimal cutting tools and the designed electrodes.

Fig. 6.34 shows an example of a hand phone cover core to be machined. Besides the bottom and the four side surfaces, there are 87 other surfaces to be machined. The total surface area to be machined is 28244.6 mm^2 and the boundary length of the region to be machined is 695.4 mm. There are 52 convex surfaces and 35 concave surfaces. The smallest curvature radius of these surfaces is on the "ear hole" surfaces (0.516 mm). The dark color regions in Fig. 6.35a shows the detected interference regions when the cutter radius is 20 mm. Since the curvature radius of the ear hole surfaces is less than the cutter radius, these surfaces are local interference surfaces. For global interference detection, some surfaces have global interference-free edges (e.g., the upper large surface). Therefore, global interference needs only to be investigated along these surface edges, and no investigation needs to be performed inside the edges of the surface. Some points on the flashing surfaces' edges are global interference-free, while others are not. The global interference-free region can thus be detected by finding the loci of interference-free points with the starting point at the surface edge. Fig. 6.35b shows the detected interference regions when the cutter radius is 2 mm. It can be found that the interference area is much smaller than that of Fig. 6.35a. The cutters' radii, the feedrates, the feasible and infeasible areas, and the lengths of infeasible regions are listed in Table 6.2. None of the cutters can machine the entire mold without interference, and the cutter with radius of 2 mm is therefore the smallest cutter to be selected.

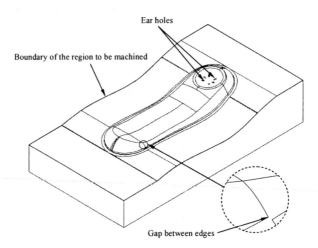

Fig. 6.34 A hand phone cover mold to be machined.

Table 6.2 Areas of feasible/infeasible regions and lengths of the infeasible regions

	R_0	R_1	R_2	R_3	R_4	R_5	R_6
R_i (mm)	20	15	12	10	8	5	2
A_{ui} (mm^2)	4927.8	4264	3614.2	3193.6	2760.5	2321.3	1026.9
A_i (mm^2)	23316.8	23980.6	24630.4	25051	25484.1	25923.3	27217.7
S_{ii} (mm)	796.5	772.1	751.1	766.4	748.3	719.6	690.5

Computer-Aided Die and Mold Manufacture

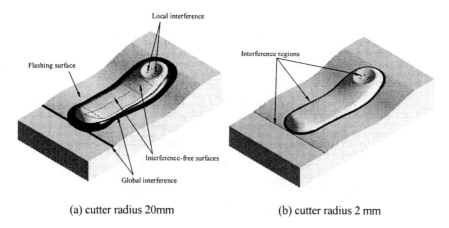

(a) cutter radius 20mm   (b) cutter radius 2 mm

Fig. 6.35 Interference regions with different the cutter sizes.

Table 6.3 lists the minimum machining times for different number of tool changes. It can be found from the table that among the different cutter combinations, the cutters of (R_0, R_6) can machine the mold with minimum time (106.6 minutes). The two cutters are therefore selected as the optimal cutters. In this example, the minimum machining time is less than 1/3 of the maximum machining time (330.8 minutes). Therefore, optimal cutter selection can greatly shorten the mold machining time. Since the available cutters cannot machine the entire workpiece without interference, electrodes are needed to machine the unmachined regions. Fig. 6.36 shows some electrodes designed automatically with the system. As the interference regions of the "ear holes" are very small, the electrodes are combined and constructed into one. On the contrary, although the boundaries of the interference region on the flashing surfaces form a closed loop, in practice, it is not optimal to design a large electrode to machine this region. The electrode can be separated into a number of smaller ones.

Table 6.3 Minimum machining times for different number of cutter changes

Number of tools	Machining time t_i (min)	Cutter combinations
1	330.8	R_6
2	**106.6**	R_0R_6
3	110.5	$R_0R_4R_6$
4	120.6	$R_0R_2R_4R_6$
5	145.8	$R_0R_1R_2R_5R_6$
6	164.8	$R_0R_1R_2R_4R_5R_6$
7	183.2	$R_0R_1R_2R_3R_4R_5R_6$

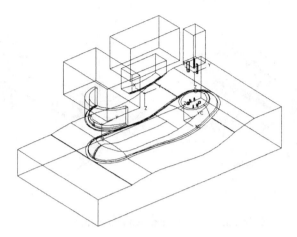

Fig. 6.36 Some automatically designed electrodes.

Fig. 6.37 shows another example of some designed electrodes for machining the core mold of a different hand phone cover. To machine this mold completely, more than 40 sets of electrodes are needed. It may take over 16 hours to design these electrodes with the general CAD/CAM functions. Using the above-discussed electrode design methodology, the electrodes can be designed in a much shorter time.

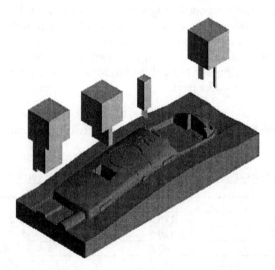

Fig. 6.37 Electrodes designed for EDM a hand phone cover mold [45].

6.6 MODIFICATION OF MOLD DESIGN AND TOOL PATH REGENERATION

6.6.1 Introduction

In the last several decades, many researchers have studied tool path generation methodologies [19,46,47] and proposed a unified CAM system architecture for die and mold machining, but how to efficiently modify the tool path based on the previously generated tool paths and the frequently modified die and mold design has not been studied. In this section, the problem of die and mold design changes and how it affect reprogramming and regeneration of NC tool paths will be addressed. At the time of writing this book, it appears that there is no reported research on tool path regeneration of sculptured surfaces based on existing tool paths, and none of the commercial CAM systems has considered reuse of the cutter location (CL) points in tool path regeneration.

To reduce the time required for regenerating tool path, this section presents a new tool path regeneration approach [48] that can make use of the generated CL points for die and mold design modification. With this approach, the affected CL points which cannot be reused are identified and removed first, new CL points are then calculated and added to replace the removed ones, and those unaffected CL points are maintained in the new CL data. Hence, only the affected tool path points need to be recalculated; the new tool paths for the modified workpiece can be regenerated efficiently. The algorithm of tool path modification has also been developed and tested with several industrial parts.

Since most dies and molds are machined using 3-axis CNC machines and ball end-mills for finish machining sculptured surface, this section focuses on regenerating tool path for this configuration. It is assumed that tool paths have been generated for the original workpiece, while the modified region and its boundary due to design change have already been identified. In addition, the sizes of the cutters that are used to machine the original and the modified workpiece are assumed to be the same.

6.6.2 Basic Concepts and Notations

Before studying tool path regeneration, some concepts and notations of tool path generation are introduced. To regenerate tool path efficiently, the affected CL points must be identified first. Since the identification of the affected CL points is related to the tool path topology, it will also be studied here.

6.6.2.1 CC Point, CL Point, and Cutter Center Point

The surface normal n at an arbitrary point $P(u, v)$ is expressed as

$$n = \frac{\mathbf{r}_u \times \mathbf{r}_v}{|\mathbf{r}_u \times \mathbf{r}_v|} \tag{6.11}$$

where r_u and r_v are the derivatives along the u and v directions on surface r at point P.

For a ball-end mill, let P_{cc} be the cutter contact (CC) point, R be the ball-end cutter radius, n be the surface normal, $T = (0, 0, 1)$ be the cutter axis that is parallel to the z-axis, the cutter center point (P_c) is on the offset surface of the workpiece and is given by

$$P_c = P_{cc} + R_n \qquad (6.12)$$

The cutter location (CL) point (P_{cl}) is calculated using

$$P_{cl} = P_c - RT \qquad (6.13)$$

It can be found from Eq. (6.13), a CL point has the same x and y values as the corresponding cutter center point. In 3-axis machining, if the surface normal vector n points to the $-z$ direction, the surface cannot be machined. As such, it is assumed that all the surface normal vectors point to the $+z$ direction, i.e., $(T \cdot n) > 0$.

6.6.2.2 Tool Path Generation

As described previously, for a 3-axis CNC machining, the tool path generation methods can be categorized into five types: the iso-parametric method [49], the constant scallop method [50], the parallel tool path method [46], the CC-point method [16] and the z-map method [19]. Among them, it is difficult to generate continuous tool path for multiple surfaces with the iso-parametric and the constant scallop methods. With the parallel tool path method, the CC-point method and the z-map method, the tool path lines are usually parallel to a line or a plane, and the tool path is continuous for multiple surfaces. Therefore, these methods are widely used in die and mold machining. In this section, it is assumed that the tool path has been generated using one of these three methods.

6.6.3 Proposed Algorithm

The idea of identifying the affected CL points is that if the boundary C of the modified region is interference-free, the corresponding CL points form a closed boundary C_o. All the affected CL points are enclosed by C_o, while the CL points out of C_o are not affected by die or mold modification.

To make it easier to implement, the tool path is generated using the parallel line method, and the tool path lines are parallel to the x-axis, i.e., the y values for all the CL points in the same tool path row are then the same.

Computer-Aided Die and Mold Manufacture   277

6.6.3.1 Framework of Tool Path Regeneration

Fig. 6.38 shows the algorithm of tool path regeneration. The input is the boundary C of the modified region, the dimension of the cutting tool, the workpiece and the CL file f_{old}, which stores the CL points before modification. The output is the CL file f_{new}, which stores the new CL points for the modified workpiece. The affected CL points' boundary C_o is calculated based on C, and the extreme y values (y_{min} and y_{max}) of C_o are also calculated. CL points read from f_{old} are then identified. If the CL point is not affected, it is output to f_{new} directly. Otherwise, it is replaced with new CL point. The CL points in f_{old} are identified one by one, until all the CL points are identified.

6.6.3.2 Create C_o from C

To make sure that all the affected CL points are enclosed by C_o, C must be interference-free. However, in practice, there may be interference points on C. Therefore, interference must be detected for C. If there are interference points on it, a new interference-free boundary C' needs to be created by enlarging C. Since many researchers have studied the problem of identifying interference, we will not further elaborate it here. For example, Ding [26] algorithm can be used to detect interference and identify the interference-free boundary. After interference detection, C_o is then calculated based on the interference-free boundary as in the following:

1) Identify the surface that each curve of C' belongs to;
2) Sample points along each curve with given tolerance;
3) Calculate surface normal **n** at each point;
4) Calculate CL points; and finally
5) Connect the CL points to form C_o;

The boundary of the modified region may be more than one, and these curves may belong to different surfaces. To calculate the CL point for a point on a curve, the surface that contains this must be identified first. With a given tolerance, points are sampled along each curve. The unit surface normal vector **n** at each point is calculated with Eq. (6.11), and the corresponding CL point is calculated using Eq. (6.13). Connecting all the CL points, C_o is created.

6.6.3.3 Identifying the Affected CL Points

Calculate the extreme y values (y_{min} and y_{max}) of C_o and compare the y values of a CL point with them. It is very easy to identify whether the CL point is in the affected tool path row ($y_{min} < y < y_{max}$) or not ($y > y_{max}$ or $y < y_{min}$). If the CL point is not in the affected tool path row, it can be output to f_{new} directly.

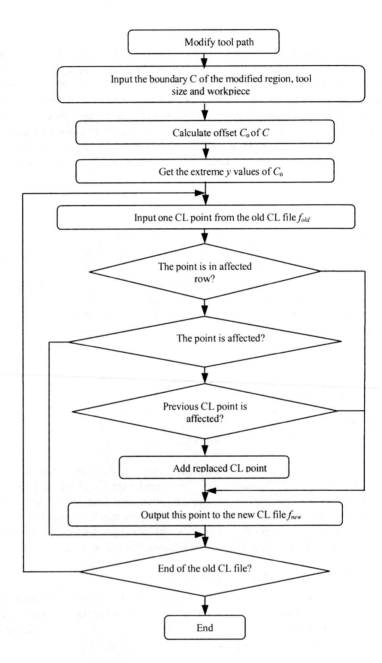

Fig. 6.38 Tool path modification flow chart.

Computer-Aided Die and Mold Manufacture

As shown in Fig. 6.39, for an affected tool path row, it is likely that only some of the CL points are affected and the affected CL points are enclosed by two extreme points. Since the *y*-values for all the CL points in the same row are the same, these two extreme points can be calculated by intersecting C_o with the plane $Y = y$, where *y* is the *y*-value of that tool path row. Moreover, for a gouge-free tool path, there can only be one *z*-value for the given values of (x, y). As a result, the *x*-values (x_{min} and x_{max}) of the two intersection points can be used to identify whether the CL point is affected. If the *x*-value of a CL point is in the range of (x_{min}, x_{max}), it is an affected CL point and must be removed. Otherwise, it can be used to machine the modified workpiece.

When a CL point in an affected row is identified as unaffected, there exist two cases: (1) affected CL points have been identified in that row, and (2) no affected CL points have been detected. The first case indicates that all the affected CL points in that row have been detected and removed (e.g., point *A* in Fig. 6.39), and new CL points need to be added before the current CL point.

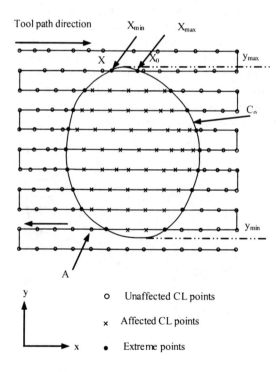

Fig. 6.39 Extreme points and effected points.

The second case can be further classified into two situations by comparing the absolute values of $|x - x_0|$ and $|x_{max} - x_{min}|$, where x is the x-value of the current CL point, x_0 is the x-value of the previous CL point in that row which has been identified as unaffected and is nearest to the current point. If $|x - x_0| < |x_{max} - x_{min}|$, no CL point needs to be added before the current CL point. On the contrary, if $|x - x_0| \geq |x_{max} - x_{min}|$, new CL points need to be added. The latter case occurs when the profile of the workpiece before modification is relatively flat, and there are no CL points between the two intersection points of the ($Y = y$) plane and C_o.

6.6.3.4 Replace the Affected CL Points

When the affected CL points are removed, they must be replaced with new CL points. The new CL points have the same y-value as the removed CL points. Since the profile of the modified region is different from that of the original workpiece, the x-values of the newly generated CL points may not be the same as the deleted ones. They must be calculated according to the machining tolerance.

With the given x and y values, the z-value of the new CL point is calculated with the z-map method, i.e., moving the cutter along the z-axis, until the cutter surface is tangent to the workpiece. As shown in Fig. 6.40, when the point (x_0, y, z_0) is calculated and accepted as a CL point, the z-values of z_1 and z are calculated using x_1 (= $x_0 + \delta$) and x (= $x_0 + \delta/2$), respectively. The linear center point's z-values (z') of the two points (x_0, y, z_0) and (x_1, y, z_1) can be then worked out ($z' = (z_0 + z_1)/2$). If the difference between z and z' is within the given tolerance ε, the point (x_1, y, z_1) is accepted as the new CL point. Otherwise, the step-size δ is shortened by half until the new point meets the given tolerance.

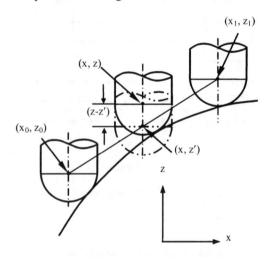

Fig. 6.40 Difference between z and z'.

6.6.4 Illustrative Examples

The tool path modification algorithm has been tested using several industrial parts. Fig. 6.41a is a mold of riser. There are 219 trimmed sculptured surfaces in the original mold, and the surface area to be machined is 191,215 mm^2. Some materials are added to the mold (Fig. 6.42a), and the area of the added surfaces is 907 mm^2. Fig. 6.41b shows the tool path for the original riser mold that was generated with a commercial CAM software, Uunigraphics [51]. The machining parameters were as follows: the radius of the ball end-mill was 5 mm, the machining tolerance of the tool path was 0.01 mm, and the scallop height was 0.03 mm. The tool path was generated with the parallel tool path method. It took about 10 minutes to generate the tool path on a HP C240 workstation. Fig. 6.42b shows the new tool paths with replaced points for the modified mold using the developed tool path modification algorithm. The new tool path used the same machining parameters. It took

Fig. 6.41a Riser mold before modification.

Fig. 6.41b Tool paths of riser mold before modification.

Fig. 6.42a Modified riser mold.

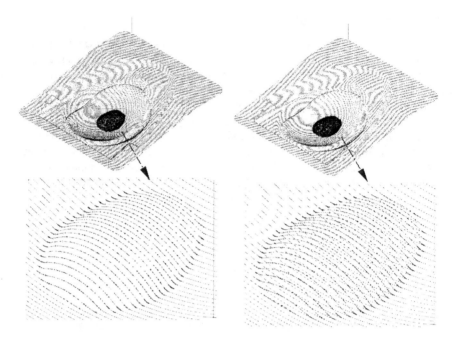

Fig. 6.42b Regenerated tool paths, points replaced after modification.

Fig. 6.42c Tool paths before and after modification with replaced points.

about 3 seconds to identify the affected and the unaffected CL points with the same computer. It can be found from the enlarged view in Fig. 6.42b that the new tool path is gouging-free and smooth. Fig. 6.42c shows the tool paths with replaced points before and after modification for comparison.

6.7 SUMMARY

The machining of complex molds with sculptured surfaces is a difficult task and has been addressed by many researchers. However, most commercial CAD/CAM systems cannot generate tool paths unless the user has selected the proper cutters. This chapter describes the methodology that can detect interference, select cutters and design EDM electrodes automatically for mold manufacturing. Theories, algorithms, and pseudo codes for implementation are discussed in details. Several industrial parts are used to illustrate the usefulness of the reported approaches in complex mold machining.

The novel algorithm to tool path regeneration for die and mold design modification based on the previously generated tool paths is also described in the end. With this algorithm, when a die or mold profile is modified, the new tool path for the modified workpiece can be generated efficiently by calculating only the CL points on the modified region and thus greatly reduce the chance of NC reprogramming errors. It is shown that the reported approach is able to regenerate gouge-free tool paths efficiently for die and mold design modifications.

REFERENCES

1. B. K. Choi, D. H. Kim, and R. B. Jerard, "C-space approach to tool path generation for die and mold machining", Computer-Aided Design, 29, pp.657-669, 1997.
2. T. Maekawa, "An overview of offset curves and surfaces", Computer-Aided Design, 31, pp.165-173, 1999.
3. W. Tiller and E. G. Hanson, "Offsets of two-dimensional profiles", IEEE Computer Graphic Application, 4(9), pp.36-46, 1984.
4. Y. S. Suh, and K. Lee, "NC milling tool path generation for arbitrary pockets defined by sculptured surfaces", Computer-Aided Design, 22, pp.273-284, 1990.
5. A. Hansen and F. Arbab, "An algorithm for generating NC tool paths for arbitrarily shaped pockets with islands", ACM Transactions on Graphics, 11, pp.152-182, 1992.
6. T. Maekawa and N. M. Patrikalakis, "Interrogation of differential geometry properties for design and manufacture", The Visual Computer, 10, pp.216-237, 1994.
7. Y. J. Chen, and B. Ravani, "Offset surface generation and contouring in computer-aided design", Journal of Mechanisms, Transmissions and Automation in Design, ASME Transactions, 109(3), pp.133-142, 1987.
8. S. Aomura, and T. Uehara, "Self-intersection of an offset surface", Computer-Aided Design, 22 (7), pp.417-422, 1990.
9. Y. Wang, "Intersection of offsets of parametric surfaces", Computer-Aided Geometric Design, 13, pp.453-465, 1996.
10. T. Maekawa T., W. Cho and N.M. Patrikalakis, "Computation of self-intersection of offsets of Bezier surface patches", Transactions of the ASME, J of Mechanical Design, Vol.119, 275-283, 1997.
11. K.K. George and N.R. Babu, "On the effective tool path planning algorithms for sculptured surface manufacture", J of Computers and Industrial Engineering, 28, pp.823-838, 1995.
12. K.I. Kim. and K. Kim, "A new machine strategy for sculptured surfaces using offset surface", Int'l J. of Production Research, 33, pp.1683-1697, 1995.
13. J.Y. Lai and D.J. Wang, "A strategy for finish cutting path generation of compound surfaces", Computers in Industry, 25, pp.189-209, 1994.
14. A. Seiler, V. Balendran and K. Sivayoganathan, "Tool interference detection and avoidance based on offset nets", Int'l J of Machine Tools and Manufacturing, 37, pp.717-722, 1997.
15. K. Tang, C.C. Cheng and Y. Dayan, "Offsetting surface boundaries and 3-axis gouge-free surface machining", Computer-Aided Design, 27, pp.915-927, 1995.
16. B.K. Choi and C.S. Jun, "Ball-end cutter interference avoidance in NC machining of sculptured surfaces", Computer-Aided Design, 21, pp.371-378, 1989.
17. Y. Takeuchi, M. Sakamoto, Y. Abe, and R. Orita, "Development of a personal CAD/CAM system for mold manufacture based on solid modeling techniques", Annals of the CIRP, 38(1), pp.429-432, 1989.

18. Y. Mizugaki, M. Hao, M. Sakamoto, H. Makino, "Optimal tool selection based on genetic algorithm in a geometric cutting simulation", Annals of CIRP, 43, pp.433-436, 1994.
19. A.C. Lin and H.T. Liu, "Automatic generation of NC cutter path from massive data points", Computer-Aided Design, 30 (1), pp.79-90, 1998.
20. J.P. Duncan and S.G. Mair, "Sculptured surfaces in engineering and medicine", Cambridge University Press, Cambridge, 1983, pp. 3-126.
21. J.S. Hwang, "Interference-free tool-path generation in the NC machining of parametric compound surfaces", Computer-Aided Design, 24, pp.667-676, 1992.
22. J.S. Hwang and T.C. Chang, "Three axis machining of compound surfaces using flat and filleted end-mills", Computer-Aided Design, 30, pp.641-647, 1998.
23. D.C.H. Yang and Z. Han, "Interference detection and optimal tool selection in 3-axis NC machining of free-form surfaces", Computer-Aided Design, 31, pp.303-315, 1999.
24. H. Pottmann, J. Wallner, G. Glaeser, and B. Ravani, "Geometric criteria for gouge-free three-axis milling of sculptured surfaces", Transactions of the ASME, J of Mechanical Design, 121, pp.241-248, 1999.
25. G. Glaeser, J. Wallner and H. Pottmann, "Collision-free 3-axis milling and selection of cutting tools", Computer-Aided Design, 31, pp.225-232, 1999.
26. X.M. Ding, Computer-aided cutter selection and electrode design for mold machining, PhD dissertation, National University of Singapore, 2001, pp. 47-73.
27. X.M. Ding, J.Y.H. Fuh and K.S. Lee, " Interference detection for 3-Axis mold machining", Computer-Aided Design, 33, pp.561-569, 2001.
28. T.W. Sederberg, H.N. Christiansen and S. Katz, "Improved test for closed loops in surface intersections", Computer-Aided Design, 21, pp.505-508, 1989.
29. T.J. Willmore, "An introduction to differential geometry", Clarendon Press, Oxford, 1959, pp. 31-94.
30. M. Bala, M. and T.C. Chang, "Automatic cutter selection and optimal cutter path generation for prismatic parts", Int'l J of Production Research, 29, pp.2163-2176, 1991.
31. Y.S. Lee, B.K. Choi and T.C. Chang, "Cutter distribution and cutter selection for sculptured surface cavity machining", Int'l J of Production Research, 30, pp.1447-1470, 1992.
32. K. Yamazaki, Y. Kawaharh, J.C. Jeng, and H. Aoyama, "Autonomous process planning with real-time machining for productive sculptured surface manufacturing based on automatic recognition of geometric features", Annals of the CIRP, 44, pp.439-444, 1995.
33. K. Lee, T.J. Kim and S.E. Hong, "Generation of toolpath with selection of proper tools for rough cutting process", Computer-Aided Design, 26, pp.822-831, 1994.
34. D. Veeramani and Y.S. Gau, "Selection of an optimal set of cutting-tool sizes for 2.5D pocket machining", Computer-Aided Design, 29, pp.869-877, 1997.
35. Y.H. Chen, Y.S. Lee and S.C. Fang, "Optimal cutter selection and machining plane determination for process planning and NC machining of complex surfaces", J of Manufacturing Systems, 17, pp.371-388, 1998.
36. Y.M. Kyoung, K.K. Cho, and C.S. Jun, "Optimal tool selection for pocket machining in process planning", J. of Computers & Industrial Engineering, 33, pp.505-508, 1997.
37. N. Mohri, M. Suzuki, M. Furuya, and N. Saito, "Electrode wear process in electrical discharge machining", Annals of the CIRP, 44, pp.165-168, 1995.
38. P.M. Lonardo and A.A. Bruzzone, "Effect of flushing and electrode material on die sinking EDM", Annals of the CIRP, 48(1), pp.123-126, 1999.

39. J.P. Kruth, "Steps towards an integrated CAD/CAM system for mold design and manufacture: Anisotropic shrinkage, component library and link to NC machining and EDM", Annals of the CIRP, 35, pp.83-87, 1986.
40. B. Lauwers and J.P. Kruth, "Computer aided process planning for EDM operations", J of Manufacturing Systems, 13, pp.313-322, 1994.
41. D.F. Dauw, "Geometrical simulation of the EDM die-sinking process", Annals of the CIRP, 37, pp.191-197, 1988.
42. M. Kunieda, W. Kowaguchi and T. Takita, "Reverse simulation of die-sinking EDM", Annals of the CIRP, 48(1), pp.115-118, 1999.
43. M. Bayramoglu and A.W. Duffill, "Manufacturing linear and circular contours using CNC EDM and frame type tools", Int'l J of Machine Tools and Manufacturing, 35, pp.1125-1136, 1995.
44. Z.Y. Yu, T. Masuzawa and M. Fujino, "Micro-EDM for three-dimensional cavities: development of uniform wear method", Annals of CIRP, 47, pp.169-172, 1998.
45. X.M. Ding, J.Y.H. Fuh, K.S. Lee, Y.F. Zhang, and A.Y.C. Nee, "A computer-aided EDM electrode design system for mold manufacturing", Int'l J of Production Research, 38, pp.3079-3092, 2000.
46. J. E. Bobrow, "NC machine tool path generation from CSG part representations", Computer-Aided Design, Vol.17, pp. 69-76, 1985.
47. B. K. Choi, Y. C. Chung, J. W. Park and D. H. Kim, Unified CAM-system architecture for die and mold manufacturing, Computer-Aided Design, Vol. 26, pp.235-243, 1994.
48. L.P. Zhang, J.Y.H. Fuh and A.Y.C. Nee, "Tool path regeneration for mold design modification", Computer-Aided Design, Vol. 35, 813-823, 2003.
49. G. C. Loney and T. M. Ozsoy, NC machining of free form surface, Computer-Aided Design, vol.19, pp.85-90, 1987.
50. K. Suresh and D.C.H. Yang, Constant scallop-height machining of free-form surfaces, ASME J of Engineering for Industry, Vol.116, pp.253-258, 1994.
51. Unigraphics on-line documentation, Version 15.0, UG Solutions Inc., Maryland Heights, 1999.

7
Computer-Aided Process Planning in Mold Making

7.1 INTRODUCTION

Process planning is the function within a manufacturing facility that establishes the processes and process parameters to be used in order to convert a piece-part from its original form to a final form that is specified on a detailed engineering drawing. For the manufacturing domain, process planning includes determination of the specific operations (machines, cutters, fixtures, etc.) required and their sequence. In some cases, the selection of depth of cut, feed rate, and cutting speed for each machining operation is also included in the process planning for a part. Due to the advent of computer technology and the need for shorter production lead time, there has been a general demand for computer-aided process planning (CAPP) systems to assist human planners and achieve the integration of computer-aided design (CAD) and computer-aided manufacturing (CAM). Recently, the emerging product design practice employing the design for manufacturing (DFM) philosophy also needs a CAPP system to generate the best process plan for a given part for manufacturing evaluation. On the other hand, in order to meet today's ever-tighter delivery schedule from the customers (especially for the tooling industry), the production planner needs a powerful and flexible CAPP system to accommodate the dynamic scheduling needs.

Over the last two decades, there have been numerous attempts to develop CAPP systems for various parts manufacturing domains, such as rotational and prismatic parts. The levels of automation among the reported systems range from interactive, variant, to generative [1]. The discussion in this chapter focuses on the generative type as it is the most advanced and preferred. In general, the approaches employed by those reported generative CAPP systems, in terms of the various decision-making activities in process planning, can be summarized as follows:

(1) Machining features recognition – The workpiece, both the raw material and finished part, is generally given in a CAD solid model representation. The difference between the finished part and the raw materials represents the volumes that need to be removed. The total removal volumes are extracted and decomposed into individual volumes, called machining features (e.g., holes and slots), which can be removed by a certain type of machining process (e.g., drilling and milling).
(2) Process selection – For each machining feature, a machining process (in name) or a set of processes (e.g., roughing and finishing) is selected.
(3) Machines and cutters selection – For each process, a machine and a cutter are assigned to it for execution, based on the available shop floor resources.
(4) Set-up plan generation – A set-up refers to any particular orientation of the workpiece on the machine table together with a fixture arrangement where a number of processes can be carried out. The tool approach direction (TAD) for each process is firstly identified. A set-up is determined based on the commonality of TAD and fixture arrangement for several processes.

The output, in general, is a set of sequenced process methods (machine, cutter, and fixture arrangement). Ideally, the activities in (2)-(4) should be carried out simultaneously (the way that a human expert normally does), in order to find a feasible solution. This is because these decision-making problems are inter-related rather than independent of each other. On the other hand, process planning is also an optimization problem due to the fact that for each decision-making problem in (2)-(4), the feasible solutions are most likely many but one. The overall solution space is sufficiently large that a search algorithm is required to find a good solution. Over the years, most reported CAPP systems handle these decision-making activities, to a large extent, in a linear manner and each is treated as an isolated deterministic problem. As a result, they found little industrial acceptance as the plans generated are often non-executable and/or far from being optimal. Recently, it has been noticed that some CAPP approaches have been designed to tackle the problems mentioned above, in various degrees, by integrating the decision-making activities [2-8]. This simultaneous approach represents a realistic direction for developing generative CAPP systems for the various job-shop types of manufacturing domains.

Mold making is an important branch of manufacturing industry. It has several unique characteristics:

(1) A mold-making workshop is typically of the job-shop type. It normally has full range of common-purpose machine tools. Production volume for a par-

CAPP in Mold Making

ticular mold is normally less than five, but the number of on-going jobs is generally large.

(2) A mold assembly has a large number of various components (see Fig. 7.1). Most components are so-called standard parts, e.g., mold-bases and ejection pins, that can be purchased directly. The components that need to be fabricated are core, cavity, sliders, and lifters.

From point (1), it is obvious that CAPP is necessary for mold-making shops, in order to increase productivity and smoothen the link between CAPP and scheduling. Based on point (2), parts that need CAPP are core, cavity, sliders, and lifters. Generally, processes involved in the fabrication of cavity and core are rough milling, finish milling, and electronic discharge machining (EDM). Rough and finish milling (both 3-axis) are used to remove most of the materials before the workpiece is hardened. EDM is then used to produce the portions which are geometrically too complex to be handled by the milling method. Therefore, the majority of machining planning for cavity and core is NC tool-path generation, which can be handled comfortably using commercially available CAM software. Little is required to conduct process selection and sequencing. On the other hand, the number and types of sliders and lifters are generally much larger, and, at the same time, requiring different processes (machine and cutters) for each. The demand for process

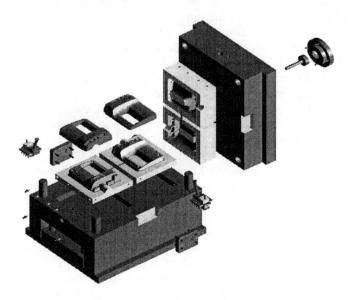

Fig. 7.1 A mold assembly (Courtesy of Manusoft Technology, Singapore).

planning in the fabrication of sliders and lifters is generally high and requires a CAPP system to assist the process planners. Therefore, the application of CAPP for mold making can be narrowed down to CAPP for fabricating sliders and lifters.

In this chapter, a unique approach that models the process planning problem in an optimization perspective is first presented. Although a more comprehensive reading on this can be found in [9], it is briefly introduced here for an easy understanding of the method used for mold making. This is followed by some observations of the general practice used in industries for the fabrication of sliders and lifters. Combining the unique CAPP approach and the general industrial practice, a hybrid approach for developing a CAPP system for manufacturing sliders and lifters has been developed. The developed prototype system is to be presented with a case study.

7.2 AN OPTIMIZATION MODELING APPROACH TO CAPP

The process planning domain for a CAPP system is defined by the spectrum of parts, and machining environment it deals with. Here, it is assumed that the geometry of parts is represented as a solid model created using a CAD software. The machining features are recognizable. As for the machining environment domain, a CAPP system should be flexible enough to handle common parts in traditional job shops. In this discussion, the job shop information, in terms of its machining capability and current status, is treated as a user input through an open architecture, to update the currently available machining resources (machines, tools) along with their technological attributes such as the limits of dimensions, achievable accuracy, and surface finish. The machines used include conventional machines (e.g., horizontal and vertical milling, drilling, and grinding) and CNC machining centers.

Before the process planning modeling method is introduced, some terms are defined first. The term *process* has two levels (high and low) of meaning. At the high level, a process is generally defined as a machining method in name that is able to produce one or a set of machining features. At this level, a process is called process type (PT). Examples of PTs are drilling (for holes) and end-milling (for planes). At the low level, a process is defined specifically as a machining method in relation with a machine and a cutter and the part (set-up on the machine table) for execution. A process at this level is called process method (PM), which can be defined by machine/cutter/TAD (M/T/TAD). In process planning, both PT and PM are involved. However, in manual process planning, the line between the consideration of PT and PM is not very clear. In the present modeling method, these two processes are involved in different stages towards the construction of the overall process plan solution space. The modeling algorithm is described as follows:

CAPP in Mold Making

Algorithm: Solution Space Construction

Input: A set of features, job shop information (available machines and cutters)
Output: Number of PTs required, all feasible PMs (M/T/TAD) for each PT, precedence relationships between PTs

(1) For each machining feature (F_i), find all the PTs that can achieve its attributes (shape, dimensions, tolerances, and surface finish), based on shop level machining capabilities. The resulting PTs for F_i can be expressed as $\Sigma PT_i = \{(PT_{1-1}, PT_{1-2}, ..., PT_{1-k}), (PT_{2-1}, PT_{2-2}, ..., PT_{2-k}), ..., (PT_{m-1}, PT_{m-2}, ..., PT_{m-k})\}$, where m is the total number of feasible set and k is the number of PTs in each set.

(2) Identify all the precedence relationships (PRs) between features as well as among the PTs in each set for a feature. These PRs can be identified based on datum dependency, fixture constraints, geometrically parent-child relationship, and good manufacturing practice.

(3) For each PT in each ΣPT, find all the combinations of machines (Ms) and cutters (Ts) with which it can be executed, based on available machine and cutters. For each combination of M and T, find all the feasible TADs.

End of Algorithm Solution Space Construction

The output from the above algorithm is a fixed number of PTs required for fabricating a part and along with each PT, one or several sets of PM(M/T/TAD). The PRs among the PTs are also explicitly represented.

The next step is to find a process plan from the solution space, i.e., a PM for each PT with an appropriate sequence among the PMs. It is obvious that there can be many feasible solutions for a given process planning problem. Therefore, there is a need to find the best plan. To achieve this, an objective function must be established first. Commonly used criteria for process plan evaluation include processing time and cost. At this stage, since the detailed information on tool paths and machining parameters is not available, the processing time and cost can only be estimated. The main factors to be considered are numbers of machine changes, set-up changes, and cutter changes, which can be accurately calculated. For the estimation of other factors such as machining time and cost, Zhang and Nee [9] provided some models.

Once an evaluation criterion is established, a search algorithm needs to be formed to find the best solution. By looking at the solution space, it is obvious that this is an NP-hard problem and an optimization algorithm is therefore needed. Zhang and Nee [9] developed two evolutionary search algorithms by applying genetic algorithm and simulated annealing respectively. Testing results showed that both algorithms are capable of finding solutions of high quality.

In the following section, this approach will be adopted to model the process planning problem for fabricating sliders and lifters. Some changes are also to be made to suit the current manufacturing practice in the mold-making companies.

7.3 CAPP FOR SLIDERS AND LIFTERS

7.3.1 Design of Sliders and Lifters

During the last two decades, the advances of computer applications in design, manufacturing, and engineering analysis have gradually changed the mold design from a complete manual process to a computer-aided process. This evolution has been propelled by the introduction of some additional functions for mold design (e.g., core and cavity creation) in some commercially available CAD tools. However, the conversion from manual to computer-aided is far from complete as many design phases of mold design are carried out interactively with the basic CAD functions. The effectiveness of using CAD in mold design is largely dependent on the user's skill. Recently, there has been much effort in developing functionally complete CAD tools for mold design. One of such CAD tools is the IMOLD®, developed by a team from the National University of Singapore. Starting with a 3D part model as input, it provides a comprehensive set of functional modules that helps the user to complete a mold design process from core and cavity creation to cooling and ejector system design. The output is a complete 3D mold assembly which components can be used either collectively or individually for downstream applications. One notable feature of IMOLD is the incorporation of the parametric modeling technique for the design of components, including sliders and lifters.

Fig. 7.2a shows a typical slider assembly. Generally, a slider assembly consists of slider body, slider head, wear plate, guide, heel block, angle pin, and stop block. The slider head forms the profile of the undercut, and it is usually created together with the slider body. Through the slider body, the angle pin moves upward and downward. Guides provide the moving pass for the slider body and are attached to the moving half of the mold base. The wear plate is a rectangular plate, which prevents the slider body from wearing out. The heel block provides the locking engagement. The stop block resists the slider body from moving out from its actual stroke length. Similarly, a lifter assembly (see

CAPP in Mold Making

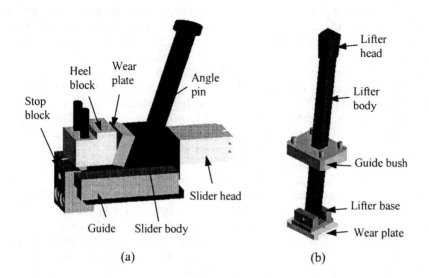

Fig. 7.2 Typical slider and lifter assemblies.

Fig. 7.2b) consists of a *lifter body, lifter head, lifter base, guide bush,* and *wear plate*. The lifter head forms the profile of the undercut, and it is usually created together with the lifter body. The lifter base supports the lifter body during ejection. Guide bush provides the moving pass for the lifter body. The wear plate is a rectangular plate that prevents the lifter body from wearing out. Among the components of sliders and lifters, usually only the slider body and lifter body are customer-made. Therefore, only these two types of components will be addressed in the development of a CAPP system.

7.3.2 A Hybrid CAPP Approach

Generally, a slider/lifter can be designed in many ways and different mold-making companies have different geometric models for sliders/lifters. However, it has been observed that only a limited number of geometric models are used for sliders/lifters in a particular company. This is mainly due to the trend in industry towards design standardization in order to improve design and manufacturing efficiency. This kind of design practice is referred to as the "quasi-standard" approach. Fig. 7.3 shows the common types of slider bodies used in a particular mold company. The decision on which type of the common models is to be used to resolve a particular undercut is mainly based on the dimensions of the undercut.

The use of quasi-standard approach in design leads to a limited number of types of sliders/lifters. This in turn leads to a rather fixed process plan for each type of slider/lifter model. For example, a typical process plan used for slider type-A (see Fig. 7.3) is shown in Fig. 7.4. If one takes a closer look at this process plan, however, one can see that it is a plan with only required PTs and their sequence; the specific PMs for each PT are yet to be decided. Based on these observations, a hybrid CAPP approach is proposed to deal with sliders/lifters. In the high level, a database is constructed in which every type of slider/lifter model has a set of sequenced PTs, called a *plan template*. During process planning, upon the identification of a particular slider/lifter type, its plan template can be retrieved. In the low level, the optimization approach introduced in section 7.2 is adopted. All the feasible PMs for the PTs are identified based on the workshop status (available machining resources). An optimization search algorithm is then used to find the optimal process plan, i.e., one PM to one PT. The general structure of the CAPP approach is shown in Fig. 7.5. It is expected that this hybrid approach is able to find process plans of high quality by extensively exploring the flexibility of process planning. On the other hand, without too much modification, this approach can be customized to different mold-making companies in a straightforward manner.

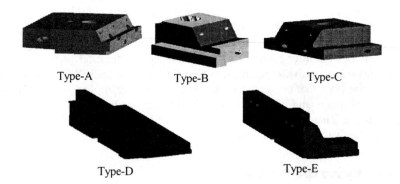

Fig. 7.3 Some commonly used slider bodies in a company [10].

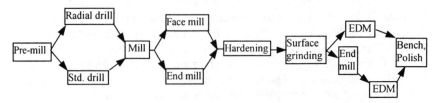

Fig. 7.4 A process plan template for a typical slider body [10].

CAPP in Mold Making

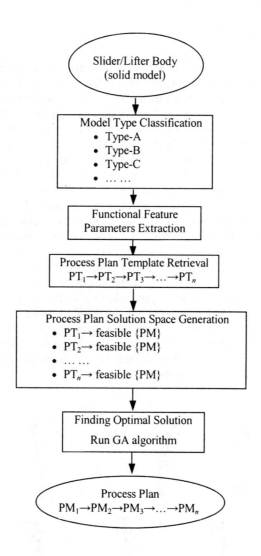

Fig. 7.5 The proposed hybrid CAPP approach.

7.3.3 Process Planning Problem Formulation

7.3.3.1 Feature Parameter Extraction

Referring to Fig. 7.5, the first step to the problem formulation is to recognize the type of the slider/lifter model and extract all the parameters of the functional features on the model. Since the slider/lifter is modeled in solid mode, feature recognition is, in general, considered a feasible approach for the above task. Looking at the basic geometric features on the slider models shown in Fig. 7.3, the basic features of a slider can be recognized without much of a problem. However, some lifter models with intersected features (see Fig. 7.6) may pose some problems. At the same time, it is also noticed that more and more solid model-based CAD tools for injection mold design have appeared in the market. In those available CAD tools, parametric design is commonly adopted for designing sliders and lifters and the parameters of the sliders/lifters are stored explicitly in a so-called *expression file*. In the present method, this expression file is considered as an input. Although this assumption does not apply to general CAD tools, it can be applied to many mold design CAD tools. Here, one of such CAD tools, IMOLD, is used to illustrate the feature parameter extraction process. In IMOLD, every slider/lifter model has a specific parametric model. Upon selecting a specific model, the user is prompted to enter the parameters in a pop-up window and these parameters are then stored in an expression file. Fig. 7.7a shows part of an expression file for the slider model shown in Fig. 7.7b. Since every variable in the expression file is explicitly defined, the parameters of all the features in a particular slider/lifter model can be identified with little difficulty.

Fig. 7.6 Some commonly used lifter bodies in a company [11].

CAPP in Mold Making

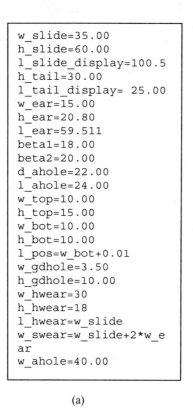
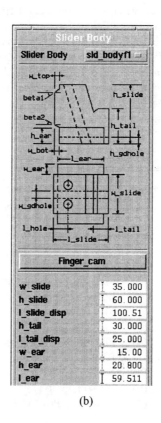

(a) (b)

Fig. 7.7 A slider body design interface and its expression file [10].

7.3.3.2 Identifying All Feasible PMs

Upon the identification of the slider/lifter model type and the extraction of its parameters, the specific process plan template can be retrieved, in which a set of sequenced PTs are present. The next step is to identify all the feasible PMs, based on technical consideration, from the available resources. The whole solution space of the process planning problem can thus be established.

A PM includes the combination of a machine tool, a cutter, and a fixture device. The identification of all the feasible PMs for a given PT is based on two criteria. The first criterion is that the machine tool, cutter, and fixture device must be available at the time of the execution of this process plan. This kind of information is based on the master schedule of the workshop. The user needs to update the database of the resources (machine tool, cutter, and fixture device) by indicating their availability. The second criterion is that the machining capability of the PM must be sufficient for the required PT and the part. The machining capability attributes include both dimensional and achievable accuracy limits. These attributes of a machine tool are first compared with the required PT's attributes and the part dimension to decide whether the machine tool is feasible. Feasible cutters and fixture devices can then be selected to complete the identification of all feasible PMs for this machine tool. The whole process for identifications of PMs is illustrated below:

Algorithm: Identification of All PMs for All PTs

Input: A set of PTs in order, $\{PT_1, PT_2, ..., PT_m\}$
Available machines, cutters, fixture devices

Output: All feasible PMs for each PT

(1) Take PT_i ($i = 1, m$)
(2) Find all machines that can perform PT_i and, at the same time, have sufficient dimensional and accuracy limits. Store these machines in $\{M_1, M_2, ..., M_k\}$.
(3) Take machine M_j ($j = 1, k$)
(4) Find all cutters (T) and fixture devices (FD) that can be used to perform PT_i on M_j. Store these combinations in cluster $PM_i\{M/T/FD\}$.
(5) If $j < k$, go to (3)
(6) If $i < m$, go to (1)
(7) Output all $PM_i\{M/T/FD\}$, $i = 1, m$.

The $\{PM_i\{M/T/FD\}, i = 1, m\}$ output from the above algorithm form the solution space of the process plans. Picking a M/T/FD combination from every PM_i gives a feasible process plan to the slider/lifter on hand. Therefore, it is very often that more than one feasible process plan may be available. Under such circumstances, an optimization technique to find the optimal plan is certainly most desirable.

7.3.4 Optimization Techniques for Process Planning

It is shown in the last section that process planning is a problem with multiple feasible solutions. Therefore, there is a need to find the optimal, if not the best, solution. In order to achieve this, evaluation criteria must be established. In this section, close to realistic evaluation criteria are introduced first, followed by the introduction of two optimization search algorithms.

7.3.4.1 The Process Plan Evaluation Criteria

Currently, the most commonly used criteria for evaluating process plans include shortest processing time and minimum processing cost. Considering the availability of certain information at this planning stage, in general, detailed information such as tool paths and machining parameters is not available. Therefore, the total processing time and cost have to be estimated. Zhang and Nee [9] established a set of cost indices for estimating processing cost of a process plan with only a set of ordered process methods. They classified the overall cost into two categories, i.e., the cost due to the usage of machines and tools and the cost due to set-ups (machine change, set-up change on the same machine, and tool change). For the cost due to machine/tool usage, a constant cost index is assumed every time a particular machine or a tool is used. For the cost due to set-ups, a constant cost index is assumed when a particular set-up is required. Given a feasible process plan, i.e., an ordered set of (M/T/FD), the cost resulted due to these five aspects (machine usage, cutter usage, machine change, set-up change, and cutter change) can be calculated. This cost estimation scheme has achieved certain success for the process planning of prismatic parts [9].

For mold manufacturing, one would often find the shortest processing time to be the most desirable criterion. The processing time for a part can also be classified into two categories: time due to machining and time due to set-ups. For the time due to set-ups, a constant time penalty is assumed to have incurred when a particular set-up is required. There are three types of set-ups, i.e., machine change, set-up change on the same machine, and tool change. The three time factors for a process plan can be estimated as follows:

(1) Machine change time (MCT): a machine change is needed when two adjacent PMs are performed on different machines.

$$MCT = MCTI \times \sum_{i=1}^{n-1} \Omega(M_{i+1} - M_i) \qquad (7.1a)$$

$$\Omega(M_i - M_j) = \begin{cases} 1 & \text{if } M_i \neq M_j \\ 0 & \text{if } M_i = M_j \end{cases} \qquad (7.1b)$$

where n is the total number of PMs in a plan, $MCTI$ is the machine change time index, and M_i is the ID of the machine used for PM$_i$.

(2) Fixture change time (*FCT*): a fixture change is needed when two adjacent PMs performed on the same machine need different fixture devices.

$$FCT = SCTI \times \sum_{i=1}^{n-1}((1-\Omega(M_{i+1} - M_i)) \times \Omega(FD_{i+1} - FD_i)) \qquad (7.2)$$

where *SCTI* is the fixture device change time index, and FD_i is the ID of the fixture device used for PM_i.

(3) Tool change time (*TCT*): A tool change is needed when two adjacent PMs performed on the same machine use different cutters.

$$TCT = TCTI \times \sum_{i=1}^{n-1}((1-\Omega(M_{i+1}-M_i))\times\Omega(T_{i+1}-T_i)) \qquad (7.3)$$

where *TCTI* is the tool change time index.

The different set-up time indices, i.e., *MCTI*, *SCTI*, and *TCTI*, can be provided by the mold manufacturing workshop. The set-up time (T_{set-up}) for a given process plan is simply the sum of these three factors.

The estimation of machining process time, however, is not a simple issue at this planning stage. Since the tool path and the cutting parameters are not available yet, one would need to find a way to approximate the process time for each PM. One simplified approach is to treat the material removal rate for a process on a specific machine as a norm. For example, the material removal rate for M_i to carry out a *face-milling* process is defined as R^i_{f-mill}. As a result, each machine has a set of material removal rate for the corresponding executable processes. On the other hand, the material volume (V_j) to be removed corresponding to PT_j can be estimated. Given a process plan, the total machining process time (T_m) can be estimated as,

$$T_m = \sum_{i=1}^{n} \frac{V_i}{R^j_{PT_i}} \qquad (7.4)$$

where *n* is the total number of PTs. $R^j_{PT_i}$ is the material removal rate of M_j that is selected to carry out PT_i in the current process plan. In this estimation method, the assumption of constant material removal rate for a machine to conduct a specific process will surely result in some deviation from the true value. Based on the present observation, the difference of material removal rate for different machines (performing similar processes) in a workshop is quite significant. Therefore, it is expected that this assumption does not affect the machine tool selection significantly.

Finally, the processing time for a given process plan is simply the sum of T_{set-up} and T_m. With this established plan evaluation criterion, a search algorithm is needed to find the best solution. With all the feasible PMs for each PT, searching for the best process plan is obviously a combinatorial problem, and an intelligent search algorithm is needed. In the next section, two search algorithms based on genetic algorithm and simulated annealing, respectively, are to be introduced.

7.3.4.2 An Optimization Search Method Based on Genetic Algorithm

Genetic algorithm (GA) is a stochastic search technique based on the mechanism of natural selection and natural genetics. It starts with a set of random solutions called *population*, and each individual in the population is called a *chromosome*. The chromosomes evolve through successive iterations, called *generations*, during which the chromosomes are evaluated by means of measuring their *fitness*. To create the next generation, new chromosomes called *offsprings* are formed by two genetic operators, i.e., *crossover* (merging two chromosomes) and *mutation* (modifying a chromosome). A new generation is formed by *selection*, which means that chromosomes with higher fitness values have higher probability of being selected to survive. After several generations, the algorithm converges to the best chromosome, which hopefully represents the optimal or near-optimal solution to the problem. To apply GA to search for the optimal process plan for fabricating a slider/lifter, the general implementation procedures are described in the following sections.

(1) *Chromosome representation*
Zhang et al. [12] applied GA to the process planning of prismatic parts. Their chromosome representation for a feasible plan can be used here. A process plan is represented by a string of segments (strings), with each segment corresponding to a PM.

(2) *Initial population generation*
A feasible plan (chromosome) can be generated by randomly selecting a feasible (M/T/FD) for each PT. Similarly, other chromosomes can also be generated.

(3) *Fitness function*
The overall processing time for a process plan is treated as its fitness.

(4) *Selection*
The selection is generally carried out by utilizing two mechanisms [11,12]. Firstly, it applies the "elitism" by copying the chromosome having the lowest fitness, thus keeping the best fitness among the chromosomes non-increasing. Secondly, several selection mechanisms can be used for the reproduction of the remaining chromosomes. For example, Zhang et al. [12] used the "roulette wheel" method, while Alam [11] used the exponential ranking mechanism in order to overcome some drawbacks of the "roulette wheel" method.

(5) *Genetic operators*

In GA implementation, crossover acts as the main genetic operator means to perform a wide spread search, while mutation acts as a supplementary means to induce the smallest possible change in the search space in order to perform an intensive search.

In the GA developed by Zhang et al. [12], the crossover operator is used mainly to explore the solution space due to the sequence change among all PTs. Here, since the sequence of PTs in a plan template is fixed, the crossover can be optional. Alam [11] adopted the crossover algorithm of Zhang et al. [12] by removing the sequence correction algorithm. This algorithm is described as follows:

(i) Take two chromosomes C-1 and C-2 randomly at a time. Use a given crossover probability $P_{crossover}$ to determine whether crossover should take place. If the answer is no, stop.

(ii) Determine a cut point randomly from all the positions of a string. Each string is divided into two parts, the left side and the right side, according to the cut point.

(iii) Copy the left side of the C-1 to form the left side of the offspring-1, and the right side of C-2 forms the right side of offspring-1. Similarly, Copy the left side of the C-2 to form the left side of the offspring-2, and the right side of C-1 forms the right side of offspring-2.

This crossover action will change the PMs of certain PTs in a chromosome, but not their sequence.

The mutation operator generally changes one or several segments of a chromosome. Here, it is used to explore the solution space due to multiple feasible PMs for a PT. Since there are three items in a PM, two different mutation mechanisms can be used. The first one uses three mutation operators for the three items, respectively [12], and the second uses only one mutation operator by treating the three items as a single combination [11]. There is no basic difference between these two mechanisms. The only difference lies with the selection of the mutation rate, $P_{mutation}$. The algorithm by with a single mutation operator is described as follows:

(i) Take a chromosome. Select a PT randomly in it and use a predetermined probability (mutation rate $P_{mutation}$) to determine if the PM for this PT needs to be changed.

(ii) Randomly select a PM from all the alternative PMs of the PT to replace the current PM.

The above gives a general implementation of the GA-based search algorithm. In theory, this search algorithm is able to traverse the whole solution space by changing the PMs for the PTs, the selection of the GA parameters plays an important role in finding the global optimal solution. Based on the present experience, the setting of $P_{crossover} = 0.6 - 0.8$ and $P_{mutation} = 0.1 - 0.2$ generally gives a good convergence within a reasonable period of time [11]. On the other hand, in general, a fixed number of generations serves as a good criterion for stopping the search [12].

7.3.4.3 An Optimization Search Method Based on Simulated Annealing

Apart from GA, there are other search algorithms that are able to find the optimal solution for combinatorial problems. One of such method is search based on the principle of simulated annealing (SA). It is often incorporated into the gradient descent method. The gradient descent method starts with a random (but valid) solution and then small changes are considered to this initial solution. Only those changes leading to a better solution are accepted. This is repeated until no changes can be made that lead to a better solution. The problem with this approach is that the solution is often trapped in local minima. To solve this problem, the principle of SA can be incorporated into the gradient descent algorithm. This SA-based gradient descent search method can be applied in the present process planning method to find the optimal plan [9].

The details of applying SA to the process planning of prismatic parts are given in [9]. Here, the general approach is adopted and some changes are made in order to accommodate the characteristics of the solution space for process planning of sliders/lifters. The implementation procedure of the SA-based algorithm is briefly described as follows:

(1) Randomly generate a feasible plan, $(PM_1, PM_2, ..., PM_n)$, called the *current-plan*. The method is the same as generating a random chromosome in the GA based algorithm. Set the initial temperature T_0.
(2) Start from $T = T_0$, while not reaching the final temperature T_{final}
{
    (a) Make a random change to the *current-plan*. This is done by randomly selecting a PT from the *current-plan* and replace its PM with another alternative randomly, if there is any. Let *temp-plan* be the plan after the change.
    (b) Calculate the processing time of *current-plan* (E_1) and *temp-plan* (E_2).
    (c) IF $E_1 > E_2$
        Accept the change (let the *temp-plan* be the *current-plan*)
        ELSE
            Generate a random number, X ($0 < X < 1$)

IF $X < e^{\frac{E_1-E_2}{T}}$
    Accept the change
ELSE
    Reject the change
END IF
END IF
(d) Repeat (a) to (c) until a criterion is satisfied, e.g., the *current-plan* remains for a number of rounds.
(e) Reduce the temperature to a new T according to a cooling schedule. For example, the fast simulated annealing schedule suggested by Szu [13] can be used here as

$$\frac{T_t}{T_0} = \frac{1}{1+t} \qquad (7.5)$$

Where *t* is an integer.
}

Similar to the implementation of GA, the parameters in the SA-based algorithm, T_0 and T_{final}, must be carefully selected in order to achieve the global optimal solution. Zhang and Nee [9] proposed some basic guidelines for the selection of these two parameters. The initial temperature T_0 should be chosen sufficiently high, so that even the worst processing time increase after a random change, the possibility given by Boltzman's expression is high enough to lead to further randomisation of the plan. On the other hand, the final temperature T_{final} should be chosen low enough so that only changes leading to shorter processing time is accepted. In the present process planning problem, T_0 and T_{final} should be related to the three set-up time indices, i.e., *MCTI*, *SCTI*, and *TCTI*. According to Zhang and Nee [9], T_0 and T_{final} can be set as: $T_0 = 10 \times Max(MCTI, SCTI, TCTI)$ and $T_{final} = 0.25 \times Min(MCTI, SCTI, TCTI)$.

7.3.5 Discussions

In this section, a hybrid approach to process planning of sliders/lifters is described, together with its implementation procedures. It follows the quasi-standard approach for the design and fabrication of sliders/lifters used by mold manufacturing companies, and at the same time, aims at finding the optimal process plan by exploring the flexibility of alternative combination of machines, cutters, and fixture devices. Two optimization search algorithms, based on GA and SA, respectively, are introduced. The workshop resource database is also customizable to suit any mold manufacturing workshop. Given a slider/lifter design in a solid model, it is expected that this process planning approach is able to find the optimal process plan with the shortest processing time.

One problem with this approach is that it runs as a stand-alone system. Therefore, it tends to choose the machines of best performance all the time. Although at the start of planning, the user needs to indicate the availability of each machine, it will be very difficult to give precise information if the plan is to be executed sometime later in a master schedule. Thus, there is a possibility that the generated plan needs some modifications when executed. To resolve this problem comprehensively, the integration of process planning and scheduling is necessary. Zhang et al. [14] developed a novel integration approach that makes use of the flexibility of process planning. The process planning approach for sliders/lifters can fit into this integration approach very nicely.

7.4 SYSTEM IMPLEMENTATION AND AN EXAMPLE

In Section 7.3, the proposed hybrid approach to process planning of sliders/lifters in mold fabrication is described, together with detailed implementation procedures. The implementation requires a CAD system with injection mold design functions, so that sliders and lifters can be designed using parametric models. In the following sections, one CAPP system, called IMOLD_CAPP, that follows this approach, developed by Alam et al. [10,11] is used as an example for implementation. One example is described to illustrate the efficacy of this approach.

7.4.1 The IMOLD_CAPP System

The IMOLD_CAPP system is developed on a HP UNIX workstation, with IMOLD V4 and Unigraphics (UG) V15. Unigraphics is a commercial CAD system that offers many functionality, such as the UG/Open-API, which allows user to develop customer-based application functions. The IMOLD_CAPP system is written in C with UG/Open API language. The graphical user interface (GUI) is developed using UI Styler.

In total, there are five functional modules that have been developed, namely, *feature extraction, technological information, database management, GA,* and *graphic simulation.* Fig. 7.8 shows the main interface menu that has been added into the UG menu. Each item in the pull-down menu corresponds to a functional module, which has its own dialog interface.

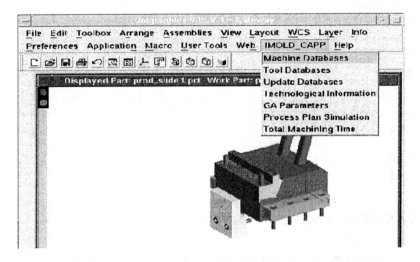

Fig. 7.8 Main menu of the IMOLD_CAPP system [11].

The *machine database* and *tool database* allow the user to view information on machining resources. By clicking these items, users can obtain information about different machines, cutters of a particular machine and their availability. Fig. 7.9 shows a dialog box and relevant information about the end mill cutters of a CNC milling machine in the database. The *update databases* allows the user to add new machining resources or indicate unavailability of some existing resources. Fig. 7.10 shows the updating database of CNC milling machines. By clicking the "CNC Milling Machines" button of the dialog box, the CNC milling database opens up and users can modify the data and save the file. The *technological information* allows the user to input technological information on surface finish and dimensional tolerances, which cannot be attached to the geometric model of sliders/lifters. Fig. 7.11a shows a dialog box that allows the input of technological parameters for a slider body.

The *GA parameters* allow the user to modify the GA parameters that are set by default. These parameters include the population size, the maximum number of generations, the crossover rates, and the mutation rates (note that mutation is applied to machine and cutter separately in this implementation). Fig. 7.11b shows a dialog box for the GA parameters.

The process planning simulation allows the user to view a graphic display of the execution of the generated plan, one step at a time. The immediate status of the workpiece is shown after each step in a plan. Fig. 7.12 shows an example of the display on the screen.

CAPP in Mold Making

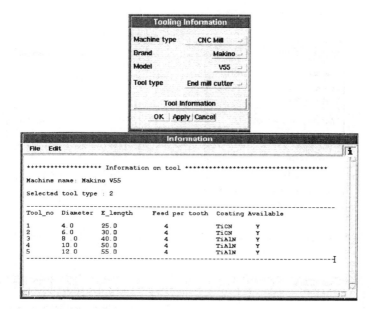

Fig. 7.9 Tool database management [11].

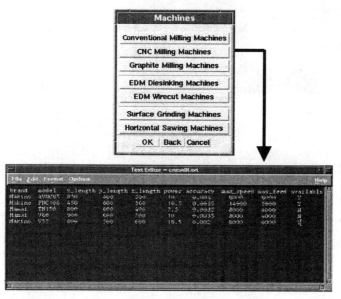

Fig. 7.10 Updating a CNC milling machines database [11].

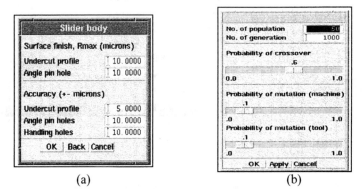

Fig. 7.11 Dialog boxes for technological information and GA parameters [11].

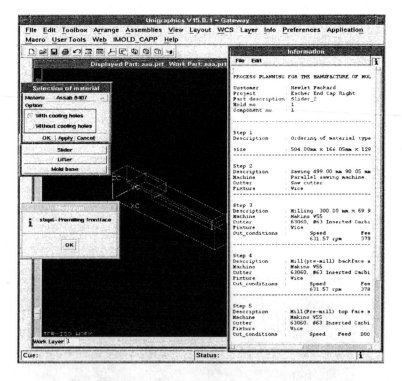

Fig. 7.12 Process plan simulation on the screen [11].

7.4.2 An Example

The slider body is shown in Fig. 7.13. This type of slider is used for small undercuts. It contains the following features: six faces, four angle faces, two handling holes that are located on face #4 and face #5 respectively, three counter-bored holes for cooling pipes, two steps that are on face #2 and face #3 respectively, two chamfers on face #5, one angle pin through-hole on the face #4, three screw holes on face #6 for wear plate, and a profile for undercut on face #1. The surface finishes are $R_{max} = 10$ μm for profile and $R_{max} = 12$ μm for angle pin hole. The accuracy is the default value, e.g., 10 μm for both feature machining.

The GA parameters selected for this case study are $P_{crossover} = 0.8$ and $P_{mutation} = 0.15$ and the population size is 50. The maximum number of generations is set at 1000. It was found that within this limit, the GA-based search algorithm is able to find the optimal solution most of the time. The final generated process plan is shown graphically in Fig. 7.14.

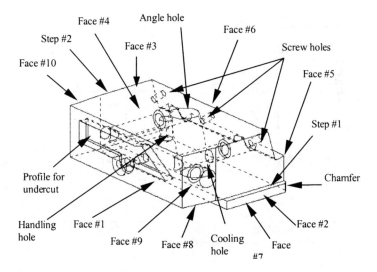

Fig. 7.13 The example slider with all the machining features [10].

Step	Process	Machine	Cutter	Fixture	Graphic Illustration
1	Pre-mill all faces with 0.2 mm allowances	Makino V55	Face mill	Machine vice	
2	Drill cooling holes on face #3 and face #5	Radial drilling machine	T-drill (Ø10 mm), T-drill (Ø12.5 mm), machine tap	Clamps	
3	Drill handling hole on face #5	Makino V55	C-drill, T-drill (Ø 8.5 mm.), tap (M10)	Machine vice	
4	Milling face #6　　Drill screw holes on face #6	Makino V55	Face mill (Ø32 mm) C-drill, T-drill (Ø5.2 mm.), tap (M6)	Angle vice	
5	Create step on face #2　　Chamfering	Makino V55	Face mill (Ø32 mm) Chamfer cutter	Machine vice	
6	Create step on face #3　　Chamfering	Makino V55	Face mill (Ø32 mm) Chamfer cutter	Machine vice	
7	Drill angle hole on face #4	Makino V55	C-drill, T-drill (Ø 11mm.), T-drill (Ø 22 mm)	Angle vice	
8	Drill handling hole perpendicular to face #4	Makino V55	C-drill, T-drill (Ø 8.5 mm), tap (M10)	Machine vice	
9	Milling face #7 (bottom angle face)	Makino V55	End mill cutter (Ø 12 mm)	Angle vice	

10	Pre-milling for the profile at face #1	Makino V55	End mill cutter (Ø 12 mm)	Machine vice	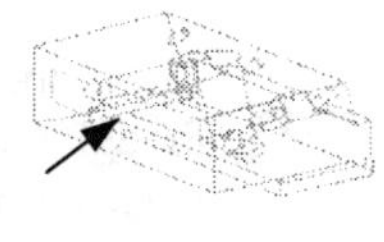
11	Grind all faces, steps to size	MHT surface grinding machine	Grinding wheel	Vice and magnetic table	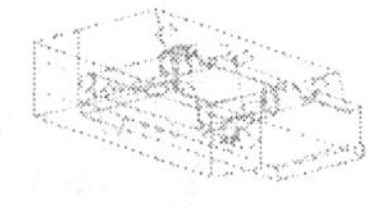
12	Milling profile on electrode	Makino SNC64 milling machine	End mill cutter (Ø 12 mm)	Erowa chucks and vice	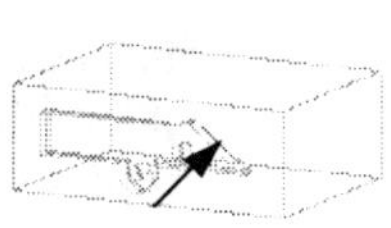
13	EDM for undercut profile on the face #1	Makino EDNC43	Graphite electrode	Magnetic table and latch clamp	
14	Polishing and benching				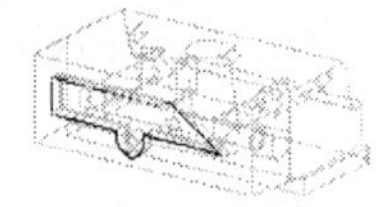

Fig. 7.14 The generated process plan for the slider [10].

7.5 SUMMARY

An injection mold contains a large number of components. Among these components, those requiring to be machined are mainly the cavity, core, sliders and lifters. Process planning for the core and cavity is relatively straightforward. Therefore, the focus of process planning effort is on sliders and lifters.

Process planning is an optimization problem with inter-related decision making activities and multiple constraints. This chapter firstly describes a unique process planning approach that integrates the tasks of routing and sequencing for obtaining a globally optimal process plan for a machined part. This process planning modeling method is then adopted for sliders and lifters. Based on the industrial practice, a hybrid approach, i.e., basic plan template for process type selection and an optimization approach for process method selection, has been developed. Intelligent search methods, such as GA-based and SA-based algorithms, are introduced for finding the globally optimal solution based on the proposed process planning model. The proposed method has been implemented for a number of slider and lifter models of a molding company. Testing results are satisfactory. The approach employed has the following advantages:

(1) The proposed approach uses a template-based plan as a starting point. Although this seems to limit the solution space for the process type and sequence selection, in reality, the negative effect is very limited due to the quasi-standard approach in design adopted by many mold-making companies. On the other hand, an optimization approach is adopted for the selection of process method, which makes the approach, to a large extent, generative in nature.

(2) The developed system is easy to be customized for another company. To achieve this, the only change needs to be done is to create a set of plan templates for the company and develop the data base for the available machine tools, cutters, and fixtures.

A limitation of this approach lies with machining feature recognition, which can only be carried out with models generated using the IMOLD system. A more general feature recognition system is under development that targets general B-rep models of sliders and lifters.

REFERENCES

1. L. Alting and H.C. Zhang, "Computer-aided process planning: the state-of-the-art survey", International J of Production Research, Vol. 27, No. 4, pp. 553-585, 1989.
2. C.L.P. Chen and S.R. LeClair, "Integration of design and manufacturing: solving set-up generation and feature sequencing using an unsupervised-learning approach", Computer-Aided Design, Vol. 26, No. 1, pp. 59-75, 1994.
3. C.C. P. Chu, C.C. P. and R. Gadh, "Feature-based approach for set-up minimization of process design from product design", Computer-Aided Design, Vol. 28, No. 5, pp. 321-332, 1996.
4. D. Yip-Hoi and D. Dutta, "A genetic algorithm application for sequencing operations in process planning for parallel machining", IIE Transactions, Vol. 28, pp. 55-68, 1996.
5. C.C. Hayes, "P^3: A process planner for manufacturability analysis", IEEE Transactions on Robotics and Automation, Vol. 12, No. 2, pp. 220-234, 1996.
6. S. K. Gupta, "Using manufacturing planning to generate manufacturability feedback", J of Mechanical Design, Vol. 119, pp. 73-80, 1997.
7. J. Chen, Y.F. Zhang, and A.Y.C. Nee, "Set-up planning using Hopfield Net and Simulated Annealing", International J of Production Research, Vol. 36, No. 4, pp. 981-1000, 1998.
8. G.H. Ma, An Automated Process Planning System for Prismatic Parts, M.Eng. thesis, National University of Singapore, 1999.
9. Y.F. Zhang and A.Y.C. Nee, "Application of genetic algorithms and simulated annealing in process planning optimization", Computational Intelligence in Manufacturing Handbook, edited by J. Wand and A. Kusiak, Boca Raton, FL: CRC Press LLC, pp. 9:1-24, 2001.

10. M.R. Alam, K.S. Lee, M. Rahman, and Y.F. Zhang, "Automated process planning for the manufacture of sliders", Computers In Industry, Vol. 43, pp. 249-262, 2000.
11. M.R. Alam, "Computer-Aided Process Planning for the Manufacture of Injection Molds", PhD dissertation, National University of Singapore, 2003.
12. F. Zhang, Y.F. Zhang, and A.Y.C. Nee, "Using genetic algorithms in process planning for job shop machining", IEEE Transactions on Evolutionary Computation, Vol. 1, No. 4, pp. 278-289, 1997.
13. H. Szu, "Fast simulated annealing", Proceedings of the American Institute of Physics Conference on Neural Computing, 1986, pp. 420-425.
14. Y.F. Zhang, A.N. Saravanan, and J.Y.H. Fuh, "Integration of process planning and scheduling by exploring the flexibility of process planning", International J of Production Research, Vol. 41, No. 3, pp. 611-628, 2003.

8

Early Cost Estimation of Injection Molds

8.1 INTRODUCTION

Early cost estimation plays an important role in the product development cycle. It has been widely reported that 70-80% of the product costs are committed during the design stage although only a small proportion of the total product costs has actually been incurred [1-3]. Therefore, there can be significant savings in product costs, as well as improved productivity and product quality, if necessary changes of the product are made in the early design stage. In order to achieve this, the product cost must be estimated fairly accurately even in the early design stage. Traditionally, the cost of a product is derived by the summation of the various cost components such as the material cost, machining cost, directly labor cost, administration, and engineering cost. However, this requires information not only on a detailed product design but also its process plan and indirect cost factors that are not available in the design stage. A product cost estimation method based on the design information is therefore a necessary step to help the designer achieve good trade-off decisions.

Early cost estimation for injection molds is particularly important for the mold makers. This can be seen from two aspects. Firstly, when a customer approaches a mold-making company with a part design, the quotation engineer needs to estimate the cost of the mold based on the given part design and some basic tooling requirements (e.g., number of cavities) in a short period of time. A quotation is then generated for the job based on the estimated cost. The accuracy of this estimation is critical to the company. If the estimated cost is much lower than the actual cost, the company faces loss if the job is awarded. On the other hand, if the estimated cost is much higher than the actual cost, the company will often lose the biding for the job. Secondly, when a mold is to be designed for a given part, the design engineer needs the estimated cost when trying different alternatives of mold design. Presently, the quotation engineer carries out cost estimation mainly based on experience, which could lead to inaccuracy and inconsistency of the estimation. In the mold design stage, the design engineer has limited knowledge of manufacturing costs since cost estimation has traditionally been the province of manufacturing engineers. Obviously, a quick and more systematic early cost estimation tool is needed for both mold quotation and design engineers.

Since the concept of concurrent engineering was introduced in early 1980, there has been a large amount of literature on cost estimation; some of them are aimed particularly at cost estimation during the early design stage. Their approaches can be basically classified into the following categories:

(1) *The simplified breakdown approach* – Cost estimation is based on the assumed optimum manufacturing methods irrespective of the process and equipment that will actually be used. Empirical equations are often required to be developed for breakdown elements. Typical examples of this approach include the work of Dewhurst and Boothroyd [4] on cost estimation models for machining and injection molding components, Boothroyd and Reynolds [5] on a cost model for turned parts, and Boothroyd and Radovanovic [6] on a cost model for machined components. A shortcoming of this approach is that it is difficult to update the cost estimate based on actual cost information.

(2) *The Group Technology (GT)-based approach* – Cost estimation is based on the similarity principle. It typically uses a basic cost value while taking into account the effects of variable cost factors such as size and complexity. Linear relationship between the final cost and the variable cost factors is assumed. One example of this approach is the cost estimation tool for injection molds developed by Poli et al. [7] in which a six-digit part design code is used. Look-up tables were used to extract the cost factors based on the code.

(3) *The cost increase function approach* – This approach is used for a specific group of products. A product's cost can be represented as a function related to the cost of the basic design for that group. The coefficients and exponents are derived through the regression analysis of historical cost data. A comprehensive description of this method is given by Wierda [8] and Hundal [9]. The problem of using cost functions lays in deducing them. With a large number of variables, the functions tend to be very complicated and difficult to present. Moreover, for each group of products, a new function has to be derived.

(4) *Activity-based costing (ABC)* – ABC is a method for accumulating product costs by determining all the cost factors associated with the activities required to produce the product. Based on historical, observed, or estimated data, the cost per unit of the activity's output is calculated. The estimated

Early Cost Estimation of Injection Molds 317

cost for a new product can be obtained according to the product's consumption of these activities. The work by Ong [10] on cost tables to support wire harness design in the assembly of electrical connections is an example of this category.

(5) *Cost function approximation using neural networks* – This approach uses a learning approach to approximate the relationship between the cost and the cost-related factors (design and manufacturing) from historical data (product vs. cost). The learning process is carried out by using a neural network. It is based on an assumption that the implication of a product's design specifications on its manufacturing processes and cost is consistent given consistent manufacturing practices. A typical example of this approach is the cost estimation model for packaging products developed by Zhang et al. [11-12].

Cost estimation for injection molds based on the part design only is a more difficult task than that for machined parts. This is due to the fact that injection mold design itself is a difficult task. It has been found that the factors affecting the cost of a mold are multi-dimensional, and the relationship between the cost and these factors is complex and highly non-linear. Therefore, it would be extremely difficult, if not impossible, to develop an accurate analytical model based on known mathematical and scientific principles. On the other hand, it is also very desirable that a fairly accurate computing model can be developed so that cost estimation can be done quickly. Considering these requirements, the cost function approximation approach is an obvious choice for developing a cost estimation tool for injection molds.

In this chapter, the approach of cost function approximation using neural networks is introduced first. This is followed by the identification of cost-related factors based on molded part design and their quantifications. The learning process using neural networks for cost function approximation is then described using historical design and cost data.

8.2 COST FUNCTION APPROXIMATION USING NEURAL NETWORKS

There are two major stages in the cost function approximation approach. The first one is the data preparatory stage in which the variables in the cost function are defined and the historical data collected. The second stage involves the function approximation by using neural network. The important steps in the first stage are described as follows:

(1) The product/part domain needs to be defined. This is often defined as a cluster of products/parts produced in a company over a period of time. The period of time should be chosen such that during which the materials costs and manufacturing technology/practice are relatively consistent.

(2) All the cost-related factors (CFs) based on design specifications need to be identified. The CFs should cover all the products/parts in the defined domain.
(3) A quantification scheme needs to be established to quantify each of these CFs. There are generally two types of CFs. The first type has continuous numerical values. A simple normalization can be used for quantification. The second type is of design alternatives, which generally do not have any values. A numerical value should be assigned to each alternative for quantification. Generally, the numerical values assigned to different states of a CF should reflect the order of contribution towards the cost, not necessarily proportionally. A necessary requirement for the quantification scheme is that all the possible states of each CF should be covered.
(4) Historical cost data is collected. It is effectively a sampling process and the selected samples will be used for function approximation. Therefore, the spectrum of the previously produced products/parts should be covered thoroughly in the final sample set. For each selected sample, its CFs are extracted and quantified and its actual cost calculated using the detailed breakdown method. The end result is a set of {CFs} vs. actual cost.

At the end of this stage, a complete set of samples are collected, and each sample points to a numerical set of $(CF_1, CF_2, ..., CF_n) \rightarrow$ cost, where a total number of n CFs is assumed.

With the set of numerical samples, there are different ways to approximate the function between $(CF_1, CF_2, ..., CF_n)$ and cost. The regression method is typically a conventional approach for function approximation, in which the coefficients and exponents are derived through the regression analysis of historical data. However, since the nature of the non-linearity relationship between the cost and CFs is not known, this has to be assumed.

Artificial neural networks (NN) is an information processing technique that simulates biological neurons using computers [13]. One of the most important applications of NN is modeling a system with an unknown input-output relation. Usually, the accurate information of the system is not available and one can only utilize a number of examples observed from the actual system and make the network learn from it. Given a fixed architecture of the network, learning is carried out through modifying the parameters of the network by the stochastic gradient decent method that eventually minimizes a certain loss function. Among various NN types, the back-propagation NN (BP-NN) is a commonly used NN for learning input-output relationship from numerical examples, which serves for function approximation very well.

A BP-NN is a multi-layer network with an input layer, an output layer, and one to several hidden layers between the input and output layers. Each layer has a number of processing units, called neurons. Fig. 8.1 shows a three-layer BP-NN with a single output. A neuron simply computes the sum of their weighted inputs, subtracts its threshold from the sum, and passes the results through its transfer function. This can be expressed mathematically as:

$$y_i = f_i(\sum_{j=1}^{n} w_{ij} x_j - s_i) \qquad (8.1)$$

where x_j represents one of the inputs, y_i the output of the neuron, w_{ij} the weight associated with x_j, s_i the threshold value of the neuron, and f_i the transfer function [14]. The value of w_{ij} determines how strongly the input influences the activity of the neuron. The magnitude of a weight can be changed during a training process. It is mainly through this mechanism that the neuron is made adaptive to new information presented to it and hence the learning process is accomplished. The BP-NN refers to their training algorithm, known as error back-propagation. The training of such a network starts with assigning random values to all the weights. An input is then presented to the network, and the output from each neuron in each layer is propagated forward through the entire network to reach an actual output. The error for each neuron in the output layer is computed as the difference between an actual output and its corresponding target output. This error is then propagated backwards through the entire network, and the weights for a particular neuron are adjusted in direct proportion to the error in the units to which it is connected. In this way the error is reduced and the network learns. This training process is also called *supervised training* since the target output for each sample's input is known.

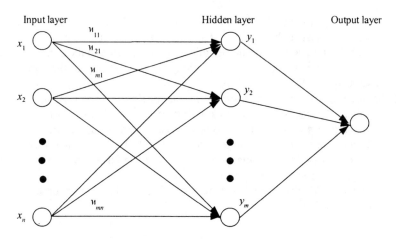

Fig. 8.1 A three-layer neural network.

In the neural network learning stage, an error tolerance should be set as the stopping criterion for training. Once training is completed, the NN, with a set of finalized weights and biases, represents the approximated cost function. Given a new product/part, the CFs are firstly extracted and quantified, and then used as the input to the NN and an estimated cost can be calculated.

8.3 COST-RELATED FACTORS FOR INJECTION MOLDS

Cost-related factors (CFs) refer to those that affect the cost of the mold. These factors include the molded part specifications as well as the conceptual mold design specifications. The former comes from the customers and the latter is based on the study of the part design. For example, part specifications include information such as part size, weight, tolerances, number and type of bosses, ribs or protrusions, as well as the types of surface finishing, engraving, and texturing of the part. Conceptual mold design specifications include the number of cavities, shape of parting line, and the number of external and internal undercuts (corresponding to the number of sliders and lifters required).

The identification of CFs was carried out mainly through consultations with experienced engineers in the fields of quotation, mold design, and manufacturing. As a result, a total of 19 CFs have been identified. These CFs are then quantified and a value between 0 and 1 is assigned to each CF based on its specific states. It is worth noting that the CFs and normalization factors are based on a specific mold making company reported by Lim [15] and Sharifuddin [16]. The CFs and their quantification scheme are described as follows:

(1) *Number of cavities* (N): This is decided during the conceptual mold design stage. The larger the number of cavities, the more expensive the mold is. $CF_1 = N/16$ (The largest number of cavities ever made in the company is 16).
(2) *Weight of part* (W g): $CF_2 = W/2000$ (The heaviest part in the company is assumed as 2000 g).
(3) *Projected area of the part along the line of draw* (A cm^2): A larger A will increase the cost of core and cavity plates and mold base. $CF_3 = A/30,000$ (The largest A of a part in the company is assumed as 30,000 cm^2).
(4) *Height of the mold in the line of draw* (H cm): A larger H will increase the cost of cavity and core plates. $CF_4 = H/30$ (The largest H of a part in the company is assumed as 30 cm).
(5) *Type of mold*: There are four types of molds and their corresponding CF_5 is given as follows:
    a. Two-plate cold runner $CF_5 = 0.1$
    b. Three-plate cold runner $CF_5 = 0.13$
    c. Hot-tip single drop $CF_5 = 0.2$
    d. Hot-tip (2-6 drops) $CF_5 = 0.5$

(6) *Parting surface types*: Parting surface separates the cavity and core. The more complex the parting surface is, the more cost of machining it will occur. The parting surface can be classified into six types and their corresponding CF_8 is given below:
   a. Flat parting plane $CF_6 = 0$
   b. Plane with a single step $CF_6 = 0.3$
   c. Simple steps or simple curved $CF_6 = 0.5$
   d. Greater than 4 steps $CF_6 = 0.63$
   e. Complex curved $CF_6 = 0.75$
   f. Complex curved with steps $CF_6 = 1$

(7) *Number of sliders on cavity* (NS_{cavity}): Sliders are used to release the external undercuts along the line of draw on the molded part. Sliders incorporated into the cavity need independent hydraulic actuation for releasing, thus increasing cost. $CF_7 = NS_{cavity}/2$ (The largest number of sliders on cavity among the molds ever produced is 2).

(8) *Number of sliders on core* (NS_{core}): Having sliders on the core costs much less than having them on the cavity as cam and spring actuation can be used to release the former. However, increasing the number of sliders will increase the cost of the mold. $CF_{10} = NS_{core}/12$ (The largest number of sliders on core among the molds ever produced is 12).

(9) *Number of large lifters* (NL_{large}): Lifters are used to release the internal undercuts along the line of draw on the molded part. Adding lifters into the mold will definitely increase the cost. $CF_9 = NL_{large}/4$ (The largest number of larger lifters among the molds ever produced is 4).

(10) *Number of small lifters* (NS_{small}): $CF_{10} = NS_{small}/8$ (The largest number of small lifters among the molds ever produced is 8).

(11) *Surface finish*: Increasing surface finish requirement will certainly increase the cost of the mold. The types of surface finish requirements and their corresponding CF_{11} are given as follows:
   a. Not critical $CF_{11} = 0.1$
   b. Opaque standard (SPE #3) $CF_{11} = 0.15$
   c. Opaque high gloss $CF_{11} = 0.2$
   d. Transparent high gloss (SPE #2) $CF_{11} = 0.3$
   e. Highest gloss (SPE #1) $CF_{11} = 4$

(12) *Engraving*: There are generally three types of engraving for molds: *normal lettering*, *raised letters or logos*, and *molded-in labels/icons*. The quantifications of CF_{12} is given as follows:
   a. No engraving $CF_{12} = 0.1$
   b. Normal lettering $CF_{12} = 0.2$
   c. Raised letters/logos $CF_{12} = 0.5$
   d. Molded-in labels/icons $CF_{12} = 0.7$

(13) *Texturing on the core*: If there is texturing on the core side, $CF_{13} = 0.2$; otherwise, $CF_{13} = 0.1$.

(14) *Texturing on the cavity*: If there is texturing on the cavity side, $CF_{14} = 0.2$; otherwise, $CF_{14} = 0.1$.

(15) *Part dimensional tolerance*: The part dimensional tolerance requirement has a big impact on the manufacturing processes used in mold making, thus the cost of the mold. This is because the mold maker is required to work within a small portion of the part tolerances in order to leave the remainder of the tolerance bands to cover variations in the molding process. The quantification of CF_{15} is given as follows:

    a. Greater than ±0.5 mm $CF_{15} = 0.01$
    b. Most approximate ±0.35 mm $CF_{15} = 0.07$
    c. Most approximate ±0.25 mm $CF_{15} = 0.17$
    d. Most approximate ±0.20 mm $CF_{15} = 0.33$
    e. Most approximate ±0.1 mm $CF_{15} = 0.67$
    f. Most approximate ±0.05 mm $CF_{15} = 0.8$
    g. Most approximate ±0.03 mm $CF_{15} = 1.0$

(16) *Ribs present on the part*: The presence of ribs will increase the manufacturing cost of the mold. The quantification of CF_{16} is based on the type and approximate number of ribs present on the part, given as follows:

    a. No ribs $CF_{16} = 0.1$
    b. Shallow (<10 mm in depth) and few (<5) $CF_{16} = 0.12$
    c. Shallow (<10 mm in depth) and many (>5) $CF_{16} = 0.15$
    d. Deep (> 10 mm in depth) and few (<5) $CF_{16} = 0.17$
    e. Deep (> 10 mm in depth) and many (>5) $CF_{16} = 0.4$

(17) *Bosses present on the part*: The presence of bosses will increase the manufacturing cost of the mold. The quantification of CF_{17} is based on the type and approximate number of bosses present on the part, given as follows:

    a. No ribs $CF_{17} = 0.1$
    b. Shallow (<10 mm in depth) and few (<5) $CF_{17} = 0.12$
    c. Shallow (<10 mm in depth) and many (>5) $CF_{17} = 0.15$
    d. Deep (> 10 mm in depth) and few (<5) $CF_{17} = 0.17$
    e. Deep (> 10 mm in depth) and many (>5) $CF_{17} = 0.4$

(18) *Type of ejectors mechanism*: There are four different types of ejector mechanisms, which will result in different costs. The quantification of CF_{18} is given as follows:

    a. Pin ejection $CF_{18} = 0.1$
    b. Blade ejection $CF_{18} = 0.2$
    c. Bar ejection $CF_{18} = 0.35$
    d. Double ejection $CF_{18} = 0.5$

(19) *Number of hydraulic cylinders* ($N_{hydraulic}$): The use of hydraulic cylinders will add extra cost to the mold. The quantification of CF_{19} is given as $CF_{19} = N_{hydraulic}/4$ (The largest number of hydraulic cylinders used in molds is 4).

The above CF types have been identified from a mold making company mainly producing molds for parts in electronic consumer products. The normalization

Early Cost Estimation of Injection Molds

factors used for quantification are based on the previously produced molds in the company. For cost estimation of other types of mold, these factors should be re-examined and the normalization factors re-defined.

Based on the above CF definition, 92 produced molded parts were examined and their CFs extracted. The actual costs of their corresponding molds were calculated based on recorded manufacturing data. These cost values were then multiplied by a constant factor in order to keep the actual ones confidential. As a result, 92 sets of CFs →cost data were collected as shown in Table 8.1.

Table 8.1 The historical data (samples 1-46)

No	CF_1	CF_2	CF_3	CF_4	CF_5	CF_6	CF_7	CF_8	CF_9	CF_{10}	CF_{11}	CF_{12}	CF_{13}	CF_{14}	CF_{15}	CF_{16}	CF_{17}	CF_{18}	CF_{19}	Cost $
1	0.13	0.055	0.0485	0.83	0.20	1.00	0	0.33	0	0.5	0.15	0.20	0.20	0.20	0.67	0.15	0.12	0.10	0.5	169,000
2	0.25	0.005	0.0036	0.18	0.10	0.30	0	0.17	0	0	0.15	0.20	0.10	0.20	0.67	0.10	0.10	0.10	0	23,400
3	0.13	0.05	0.0453	0.4	0.20	1.00	0	0.5	0	0.88	0.40	0.20	0.10	0.10	0.67	0.12	0.10	0.10	0	143,000
4	0.06	0.275	0.086	0.6	0.20	0.50	0	0	0	0	0.15	0.20	0.10	0.20	0.67	0.40	0.10	0.10	0	163,800
5	0.13	0.005	0.0009	0.03	0.10	0.30	0	0	0	0	0.15	0.10	0.10	0.10	0.67	0.10	0.10	0.10	0	17,000
6	0.13	0.1	0.0072	0.16	0.10	0.10	0	0.33	0	0	0.15	0.20	0.10	0.10	0.67	0.12	0.10	0.10	0	37,100
7	0.13	0.1	0.0132	0.27	0.20	0.30	0	0.5	0	0	0.15	0.20	0.20	0.10	0.67	0.17	0.12	0.10	0	81,900
8	0.13	0.2	0.0072	0.16	0.10	0.10	0	0	0	0	0.15	0.20	0.10	0.10	0.67	0.12	0.10	0.35	0	23,400
9	0.06	0.005	0.008	0.23	0.20	0.63	0	0.17	0	0	0.15	0.20	0.20	0.20	0.67	0.12	0.12	0.10	0	25,200
10	0.13	0.01	0.0101	0.22	0.10	0.30	0	0	0	1	0.15	0.20	0.10	0.10	0.67	0.12	0.10	0.10	0	25,400
11	0.06	0.1	0.0117	0.18	0.20	0.10	0	0	0.25	0.38	0.15	0.10	0.10	0.20	0.67	0.15	0.10	0.10	0.25	115,600
12	0.13	0.06	0.0117	0.28	0.20	0.30	0	0.17	1	0	0.15	0.20	0.20	0.20	0.67	0.12	0.12	0.35	0	156,000
13	0.06	0.3375	0.0053	0.33	0.20	0.50	0	0.17	0.25	0.25	0.15	0.10	0.10	0.10	0.67	0.15	0.17	0.10	0.25	202,800
14	0.13	0.08	0.012	0.35	0.20	0.50	0	0.17	1	0	0.15	0.20	0.20	0.20	0.67	0.15	0.12	0.10	0	169,000
15	0.13	0.235	0.017	0.42	0.50	0.63	0	0.67	0	0	0.15	0.20	0.10	0.10	0.67	0.40	0.40	0.10	0	478,400
16	0.06	0.3625	0.0487	0.53	0.20	0.50	0	0.17	0	0	0.15	0.20	0.10	0.10	0.67	0.12	0.40	0.10	0	247,000
17	0.06	0.375	0.035	0.67	0.50	0.50	0	0.17	1	0	0.15	0.10	0.10	0.20	0.67	0.15	0.10	0.10	0.50	257,400
18	0.13	0.0015	0.0009	0.03	0.10	0.30	0	0	0	0.5	0.15	0.20	0.10	0.10	0.67	0.15	0.12	0.10	0	37,100
19	0.25	0.001	0.0002	0.03	0.10	0.10	0	0	0	0	0.15	0.10	0.10	0.10	0.67	0.12	0.12	0.10	0	15,600
20	0.13	0.0025	0.0007	0.04	0.10	0.30	0	0	0	0	0.20	0.20	0.10	0.10	0.67	0.17	0.10	0.10	0	15,600
21	0.06	0.0005	0.001	0.02	0.10	1.00	0	0.08	0	0	0.20	0.10	0.10	0.10	0.80	0.15	0.10	0.10	0	48,100
22	0.13	0.0015	0.0017	0.03	0.50	1.00	0	1	0	0	0.20	0.20	0.10	0.10	0.80	0.15	0.10	0.10	0	39,300
23	0.13	0.03	0.0028	0.17	0.50	0.63	0	0.33	0	0	0.15	0.20	0.20	0.10	0.67	0.17	0.10	0.10	0	114,400
24	0.25	0.03	0.0077	0.01	0.50	0.63	0	0.67	0	0	0.20	0.20	0.10	0.10	0.67	0.12	0.10	0.10	0	101,400
25	0.13	0.015	0.0043	0.02	0.10	0.63	0	0.5	0	0	0.20	0.20	0.10	0.10	0.67	0.15	0.10	0.10	0	78,000
26	0.063	0.35	0.499	0.57	0.50	0.50	0	0.083	0	0.25	0.20	0.10	0.10	0.20	0.67	0.12	0.12	0.10	0.50	247,000
27	0.125	0.005	0.004	0.027	0.10	0.30	0	0.333	0	0.25	0.20	0.50	0.10	0.20	0.67	0.12	0.10	0.10	0	83,000
28	0.063	0.003	9E-04	0.023	0.20	1.00	0	0.167	0	0.625	0.20	0.20	0.10	0.20	0.80	0.10	0.10	0.10	0	98,300
29	0.125	0.006	0.003	0.077	0.10	0.10	0	0.083	0	0	0.20	0.20	0.10	0.10	0.67	0.15	0.10	0.10	0	52,000
30	0.25	0.002	3E-04	0.017	0.10	0.10	0	0	0	0	0.20	0.10	0.10	0.10	0.67	0.10	0.10	0.10	0	16,700
31	0.125	0.005	0.002	0.07	0.10	0.50	0	0	0	0	0.20	0.20	0.10	0.20	0.67	0.40	0.10	0.10	0	69,900
32	0.125	0.005	0.004	0.067	0.10	0.30	0	0	0	0.375	0.20	0.10	0.10	0.20	0.67	0.15	0.10	0.10	0	39,000
33	0.063	0.3	0.053	0.077	0.50	0.50	0	0	0	0	0.20	0.20	0.10	0.10	0.67	0.40	0.10	0.35	0	179,400
34	0.125	0.045	0.009	0.12	0.50	0.50	0	0	0	0	0.10	0.20	0.10	0.10	0.67	0.12	0.12	0.10	0	91,000
35	0.063	0.375	0.007	0.083	0.20	0.63	0	1	0	0	0.15	0.20	0.10	0.10	0.67	0.40	0.10	0.10	0	171,674
36	0.063	0.4	0.037	0.567	0.50	1.00	0	0.167	0	0.25	0.15	0.20	0.10	0.10	0.80	0.15	0.15	0.10	0.25	247,000
37	0.063	0.1	0.035	0.533	0.20	0.10	0	0.083	0.25	0	0.20	0.10	0.20	0.10	0.67	0.17	0.10	0.35	0	63,700
38	0.125	0.004	5E-05	0.05	0.10	0.10	0	0	0	0	0.15	0.10	0.10	0.10	0.67	0.10	0.10	0.10	0	23,000
39	0.125	0.01	0.006	0.2	0.50	0.30	0	0.333	0	0	0.15	0.10	0.30	0.20	0.67	0.15	0.10	0.10	0	41,600
40	0.063	0.1	0.013	0.063	0.20	0.10	0	0.083	0	0	0.15	0.50	0.10	0.20	0.80	0.17	0.10	0.10	0	93,600
41	0.125	0.01	0.003	0.033	0.50	1.00	1	0.333	0	0	0.15	0.10	0.10	0.20	1.00	0.12	0.10	0.10	0	115,000
42	0.25	0.0025	4E-04	0.04	0.50	0.10	0	0	0	0	0.15	0.10	0.10	0.10	1.00	0.12	0.12	0.10	0	110,000
43	0.125	0.08	0.019	0.15	0.20	1.00	0	0	0	0.75	0.15	0.10	0.10	0.20	0.80	0.12	0.12	0.10	0	42,000
44	0.125	0.08	0.019	0.15	0.20	1.00	0	0	0	0.75	0.15	0.10	0.10	0.20	0.80	0.12	0.12	0.10	0	38,000
45	0.125	0.001	2E-04	0.133	0.10	0.10	0	0	0	0.00	0.15	0.10	0.10	0.20	0.80	0.10	0.12	0.10	0	32,000
46	0.125	0.0015	2E-04	0.053	0.10	0.10	0	0	0	0.00	0.15	0.10	0.10	0.20	0.80	0.10	0.10	0.10	0	30,000

Table 8.1 The historical data (samples 47-92) (continued)

No	CF_1	CF_2	CF_3	CF_4	CF_5	CF_6	CF_7	CF_8	CF_9	CF_{10}	CF_{11}	CF_{12}	CF_{13}	CF_{14}	CF_{15}	CF_{16}	CF_{17}	CF_{18}	CF_{19}	Cost $
47	0.125	0.075	0.013	0.0767	0.20	0.75	0	0.333	1.00	0	0.15	0.20	0.10	0.20	0.80	0.12	0.12	0.10	0.25	158,000
48	0.125	0.075	0.013	0.08	0.20	0.75	0	0.333	1.00	0	0.15	0.20	0.10	0.20	1.00	0.12	0.12	0.10	0	156,800
49	0.125	0.075	0.013	0.0767	0.20	0.75	0	0.333	1.00	0.125	0.15	0.20	0.10	0.20	0.80	0.12	0.12	0.10	0	15,800
50	0.0625	0.28	0.025	0.4167	0.20	0.10	0	0.167	0.50	0.02	0.15	0.20	0.10	0.20	1.00	0.17	0.10	0.10	0.50	310,800
51	0.125	0.013	0.0012	0.05	0.10	0.30	0	0	0	0	0.15	0.20	0.10	0.10	0.80	0.12	0.12	0.10	0	52,000
52	0.125	0.0001	0.0003	0.007	0.10	0.10	0	0	0	0	0.20	0.10	0.20	0.20	1.00	0.10	0.10	0.10	0	113,000
53	0.25	0.005	0.0004	0.043	0.10	0.10	0	0	0	0	0.20	0.10	0.20	0.20	1.00	0.10	0.10	0.10	0	23,000
54	0.125	0.008	0.0022	0.04	0.10	0.10	0	0.3	0	0	0.20	0.10	0.20	0.20	1.00	0.12	0.10	0.10	0	38,000
55	0.25	0.004	0.0004	0.083	0.10	0.50	0	0	0	0	0.15	0.10	0.10	0.10	1.00	0.10	0.10	0.10	0	16,500
56	0.125	0.002	0.0001	0.08	0.10	0.30	0	0.2	0	0	0.15	0.10	0.10	0.10	0.80	0.10	0.10	0.10	0	30,000
57	0.25	0.04	0.0025	0.1	0.20	0.50	0	0.3	0	0.5	0.15	0.10	0.10	0.20	0.80	0.10	0.10	0.10	0	45,000
58	0.063	0.325	0.045	0.15	0.20	0.10	0	0	0	0	0.15	0.20	0.10	0.20	0.80	0.10	0.10	0.35	0	179,800
59	0.25	0.01	0.0010	0.107	0.10	0.10	0	0.1	0	0.13	0.15	0.20	0.10	0.20	0.80	0.10	0.12	0.10	0	17000
60	0.125	0.15	0.015	0.367	0.20	0.10	0	0.1	0.25	0.25	0.15	0.20	0.10	0.20	1.00	0.00	0.10	0.10	0.25	156,000
61	0.063	1	0.0833	0.533	0.20	1.00	0	0.3	1	0	0.15	0.20	0.20	0.20	0.67	0.40	0.15	0.10	0.50	429,100
62	0.125	0.3	0.0074	0.083	0.20	0.50	0	0.3	0	1	0.15	0.20	0.10	0.20	0.67	0.17	0.10	0.35	0	70,000
63	0.125	0.05	0.0014	0.067	0.10	0.50	0	0	0	0	0.15	0.10	0.10	0.10	0.67	0.17	0.10	0.10	0	50,000
64	0.063	0.5	0.0181	0.3	0.20	1.00	0	0.2	0	0	0.15	0.20	0.10	0.10	0.67	0.40	0.10	0.10	0	60,000
65	0.063	0.7	0.0533	0.15	0.20	0.63	0	0	0	0	0.15	0.10	0.10	0.10	0.67	0.40	0.10	0.10	0	214,600
66	0.063	0.8	0.0493	0.35	0.20	1.00	0	0.2	0	0	0.15	0.20	0.20	0.20	0.67	0.40	0.10	0.10	0	321,800
67	0.063	0.5	0.0216	0.1	0.20	1.00	0	0.2	0	0	0.15	0.10	0.10	0.10	0.67	0.40	0.12	0.10	0	107,300
68	0.063	1	0.15	0.367	0.10	1.00	0	0.1	0	0	0.15	0.20	0.10	0.10	0.67	0.40	0.12	0.10	0	536,500
69	0.125	0.3	0.0074	0.083	0.20	0.50	0	0.3	0	1	0.15	0.20	0.10	0.20	0.67	0.17	0.12	0.10	0	60,000
70	0.063	0.75	0.0447	0.167	0.20	0.63	0	0.1	0	1	0.15	0.10	0.10	0.10	0.67	0.17	0.12	0.10	0	180,000
71	0.063	0.5	0.04	0.1	0.20	0.63	0	0.1	0	0.75	0.15	0.10	0.10	0.10	0.67	0.40	0.10	0.10	0.25	193,100
72	0.063	0.65	0.0517	0.1	0.20	0.63	0	0.2	0	0.63	0.15	0.10	0.10	0.10	0.67	0.40	0.12	0.10	0.50	214,600
73	0.125	0.01	0.0021	0.093	0.10	0.50	0	0	0	1	0.15	0.10	0.10	0.10	0.67	0.17	0.12	0.10	0	60,000
74	0.125	0.3	0.0074	0.083	0.20	0.50	0	0.3	0	1	0.15	0.20	0.10	0.20	0.67	0.17	0.12	0.10	0	70,000
75	0.125	0.01	0.0021	0.05	0.10	0.50	0	0	0	1	0.15	0.10	0.10	0.10	0.67	0.17	0.10	0.10	0	65,000
76	0.063	0.8	0.0417	0.4	0.10	1.00	0	0.1	0	0	0.15	0.10	0.10	0.20	0.67	0.40	0.40	0.10	0	321,900
77	0.125	0.3	0.0074	0.25	0.20	0.50	0	0.3	0	1	0.15	0.20	0.10	0.20	0.67	0.17	0.17	0.10	0	60,500
78	0.125	0.005	0.0076	0.027	0.10	0.10	0	0	0	0	0.15	0.10	0.10	0.10	0.67	0.12	0.10	0.10	0	35,000
79	0.063	0.008	0.0022	0.26	0.10	0.63	0	0.2	0	0	0.15	0.10	0.10	0.10	0.67	0.10	0.10	0.50	0	40,000
80	0.125	0.004	0.0012	0.033	0.10	0.30	0	0	0	0	0.15	0.10	0.10	0.10	0.67	0.12	0.10	0.10	0	30,000
81	0.125	0.002	0.0001	0.013	0.10	0.10	0	0	0	0	0.15	0.10	0.10	0.10	0.67	0.10	0.10	0.10	0	30,000
82	0.063	0.01	0.02	0.15	0.13	0.63	0	0.1	0	0	0.15	0.10	0.10	0.10	0.67	0.40	0.12	0.10	0	70,000
83	0.063	0.4	0.0583	0.2	0.20	0.63	0	0.3	0	0.5	0.15	0.20	0.10	0.20	0.67	0.40	0.12	0.10	0.25	214,600
84	0.063	0.5	0.0583	0.247	0.20	0.63	0	0.3	0	0	0.15	0.10	0.10	0.20	0.67	0.40	0.12	0.10	0.25	180,000
85	0.063	0.005	0.0008	0.027	0.13	0.63	0	0.3	0	0.13	0.15	0.20	0.10	0.10	0.67	0.17	0.10	0.10	0	40,000
86	0.125	0.065	0.0022	0.117	0.20	0.63	0	0.2	0	0	0.15	0.20	0.10	0.20	0.67	0.40	0.12	0.10	0	90,000
87	0.125	0.003	0.0014	0.027	0.10	0.30	0	0.3	0	0	0.15	0.10	0.10	0.10	0.67	0.12	0.12	0.10	0	30,000
88	0.125	0.003	0.0013	0.027	0.10	0.30	0	0	0	0	0.15	0.10	0.10	0.10	0.67	0.12	0.12	0.10	0	30,000
89	0.125	0.035	0.0040	0.15	0.10	0.63	0	0	0	0.25	0.15	0.20	0.10	0.10	0.67	0.17	0.10	0.10	0	60,000
90	0.063	0.025	0.0043	0.04	0.10	0.63	0	0	0	0.25	0.15	0.10	0.10	0.10	0.67	0.10	0.10	0.10	0	55,000
91	0.25	0.004	0.0001	0.063	0.10	0.10	0	0	0	0	0.30	0.10	0.10	0.10	0.67	0.17	0.10	0.10	0	25,000
92	0.063	0.015	0.0029	0.193	0.10	0.63	0	0	0	0	0.15	0.10	0.10	0.10	0.67	0.15	0.10	0.10	0	60,000

Although the collected samples are only portions of the molds that were produced from the company, they are considered sufficiently representative to cover the most CFs of the whole mold spectrum. For example, the weights of the samples are from 2 g to 2000 g, the projected areas from 5 cm^2 to 14,360 cm^2, and the costs from 20K to 500K (see Fig. 8.2 for the cost distribution of the samples).

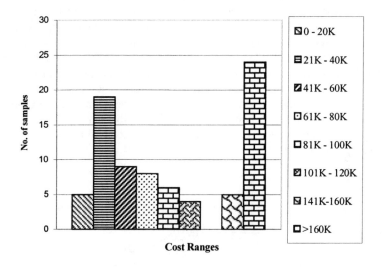

Fig. 8.2 Cost distribution of the samples.

8.4 THE NEURAL NETWORK TRAINING

Generally, there are three steps in training a neural network to learn the relationship between a set of inputs to a set of outputs from a set of input-output samples. The first step is to choose a neural network architecture that includes the number of layers, the number of neurons in each layer, and the transfer functions used for each layer. In the second step, the chosen neural network is trained using the samples until the errors between the actual outputs and the targeted outputs converge. If the errors are within a specified error goal, the training is considered completed. In the third step, the trained neural network is then tested using a set of testing samples that are not involved in the training. If the estimation errors for the testing samples are within the error goal, the training and testing are said to be successful, and the neural network is ready to be used. In general, the whole process is a trial-and-error process that involves some iterations.

8.4.1 The Neural Network Architecture

In the present application, a feed-forward back-propagation neural network is implemented with a program written using the MATLAB Neural Network

Toolbox [14]. The number of neurons in the input layer is the same as the dimension of the cost-related factors, i.e., 19 in this application. The output layer has only one neuron that represents the cost. The neural network can have one or more hidden layers. Based on experience, one hidden layer is sufficient for approximating non-linear relationships [11-12]. Therefore, one hidden layer is used here. The transfer function for the hidden layer is the *tan-sigmoid*, while the linear function, *purelin*, is used for the output layer. The architecture of this neural network is shown in Fig. 8.3 (note that the number of neurons in the hidden layer is to be decided). **CF** is the input vector, **W1** and **W2** are the weight matrices for the interconnections between the input and hidden layers and that between the hidden and output layers, respectively. **B1** and **B2** are the bias vectors for the neurons in the hidden layer and output layer.

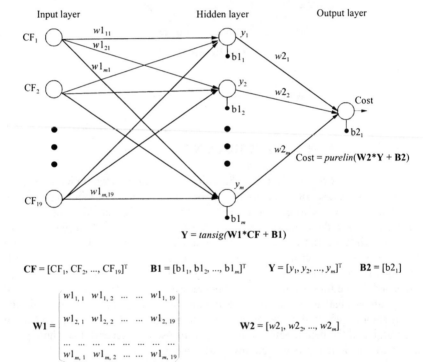

Fig. 8.3 The BP neural network architecture for cost estimation.

Early Cost Estimation of Injection Molds 327

8.4.2 The Training Process

The neural network described above has been trained using the *back-propagation training rule* [14]. Since the number of neurons in the hidden layer is yet to be decided, a trial-and-error procedure is used by starting with 20 neurons and an increment of 10 neurons in each trial, up to 100 neurons. With each specific neural network, the initial weights and biases are generated by random functions with values between -1 and +1. The weights and biases are then adjusted in order to minimize the sum-squared error of the network. This is done by continually changing the values of the weights and biases in the direction of the steepest descent with respect to the error.

Generally, the training will stop when either (i) the sum-squared error is smaller than the error-goal specified by the user, or (ii) the maximum number of epochs is reached. However, the target error-goal is unknown and it was also observed that the sum-squared error tended to converge to a certain value as the number of training epochs increased. Therefore, a very small error-goal was set that virtually could not be reached. On the other hand, a relative changing rate (*RCR*) of the sum-squared error is calculated at any moment when every 500 training epochs have been completed, which is given as follows:

$$RCR_i = \frac{SSE_{(i+1)\times 500}}{SSE_{i\times 500}} \tag{8.2}$$

where

$i = 1, 2, 3, \ldots$
$SSE_{(i+1)\times 500}$ = Sum-squared error at the $(i+1)\times 500$ epochs
$SSE_{i\times 500}$ = Sum-squared error at the $i\times 500$ epochs

Moreover, the maximum number of epochs is set at a very large value. This is to ensure that the training will stop only when the sum-squared error converges to a certain value. At every checking point after 500 epochs, if $0.99 < RCR < 1.01$, the training stops.

In order to train the network quickly and effectively, two improvement measures, namely, *momentum* and *adaptive learning rate* [14], have been incorporated into the training program. The use of *momentum* can reduce back-propagation sensitivity to small details in the error surface and help the network avoid getting stuck in shallow minima. On the other hand, a large learning rate can make the network learn quickly. However, if the learning rate is too large, the solution search keeps jumping over the error minimum without converging. The use of the *adaptive learning rate* is to make the learning rate as a variable that attempts to keep the learning step as large as possible while keeping learning stable.

When the training of a neural network is completed, i.e., the sum-squared error converges according to the *RCR*, it was observed that the estimation errors for most of the training samples reached a very small value (relative error less than

5%). This, however, does not necessarily mean that the neural network has learned well, since data over-fitting could have occurred. A validation process must be applied to the trained neural network to check whether it has generalization capability. This validation process is conducted by applying the trained neural network to samples that have not been involved in the training process. The estimated costs of these samples are compared with their actual costs. If the relative estimation errors of these sample are comparable to those of the training samples, the neural network is said to have generalization capability. Otherwise, the network is a result of over-fitting and cannot be used.

8.4.3 Training and Validation Results

In the present application, the 92 samples shown in Table 8.1 were used as the training samples. Another 10 samples, shown in Table 8.2, were used as the validation samples. There are totally nine neural networks (number of neurons in the hidden layer ranges from 20 to 100) trained, and all of them managed to converge under the *RCR* criterion after several tries. They were then applied to the 10 validation samples and the relative estimation error (*REE*) for each sample is calculated as,

$$REE_i = \frac{Cost_{i,e} - Cost_{i,a}}{Cost_{i,a}} \qquad (8.3)$$

where $Cost_{i,e}$ and $Cost_{i,a}$ are the estimated cost and actual cost of sample *i*, respectively. The resulted *REE* for the ten samples by using the nine different trained neural networks are shown in Table 8.3, in which the last column shows the average *REE* for the ten samples.

Table 8.2 The historical data of validation samples

No	CF$_1$	CF$_2$	CF$_3$	CF$_4$	CF$_5$	CF$_6$	CF$_7$	CF$_8$	CF$_9$	CF$_{10}$	CF$_{11}$	CF$_{12}$	CF$_{13}$	CF$_{14}$	CF$_{15}$	CF$_{16}$	CF$_{17}$	CF$_{18}$	CF$_{19}$	Cost $
1	0.063	0.006	0.0002	0.05	0.13	0.75	0	0.17	0	0	0.15	0.10	0.10	0.10	0.67	0.10	0.10	0.10	0	42,000
2	0.063	0.003	0.0015	0.027	0.10	0.30	0	0	0	0	0.15	0.10	0.10	0.10	0.67	0.15	0.12	0.10	0	35,000
3	0.063	0.5	0.0583	0.247	0.20	0.63	0	0.33	0	0	0.15	0.20	0.10	0.20	0.67	0.40	0.12	0.10	0	190,000
4	0.063	0.4	0.0058	0.2	0.20	0.50	0	0.33	0.5	0	0.15	0.20	0.10	0.20	0.67	0.40	0.12	0.10	0	214,590
5	0.063	0.004	0.0002	0.05	0.13	0.75	0	0.17	0	0	0.15	0.10	0.10	0.10	0.67	0.10	0.10	0.10	0	40,000
6	0.125	0.1	0.0132	0.267	0.20	0.30	0	0.5	0	0	0.15	0.20	0.20	0.10	0.67	0.17	0.12	0.10	0	80,000
7	0.125	0.006	0.0034	0.077	0.10	0.10	0	0.08	0	0.375	0.20	0.20	0.10	0.10	0.67	0.15	0.10	0.10	0	48,700
8	0.125	0.005	0.0021	0.07	0	0.10	0.50	0	0	0	0.20	0.20	0.10	0.20	0.67	0.40	0.10	0.10	0	71,000
9	0.125	0.3	0.0074	0.083	0.20	0.50	0	0.33	0	1	0.15	0.20	0.10	0.20	0.67	0.17	0.12	0.10	0	62,000
10	0.063	0.008	0.0022	0.26	0.10	0.63	0	0.17	0	0	0.15	0.10	0.10	0.10	0.67	0.10	0.10	0.50	0	40,000

Early Cost Estimation of Injection Molds

Table 8.3 The relative estimation errors of the validation samples

No. of neurons in hidden layer	1	2	3	4	5	6	7	8	9	10	Avg. REE
20	2%	25%	43%	-27%	8%	-6%	-1%	-35%	68%	10%	23%
30	7%	31%	43%	-24%	13%	-8%	-6%	-37.%	42%	18%	23%
40	17%	9%	9%	-30%	23%	-4%	-18%	3%	40%	-23%	17%
50	7%	29%	43%	-26%	13%	-1%	-8%	-37%	48%	15%	23%
60	10%	63%	3%	-10%	15%	3%	-30%	-6%	29%	-55%	22%
70	-7%	0%	-4%	0%	-3%	-9%	-14%	-4%	-5%	0%	4.6%
80	-12%	-9%	9%	-9%	-8%	-9%	-10%	7%	58%	-23%	15%
90	7%	-11%	-1%	-16	13%	-6%	9%	6%	21%	-3%	9%
100	10%	-9%	8%	-17%	15%	-6%	-26%	6%	44%	10%	15%

From Table 8.3, it can be seen that the neural network with 70 neurons in the hidden layer produced the best results, while the performance of others are very poor (over-fitting in training is clearly present). This neural network is named as $NN_{0 \to \infty}$, since it covers the molds with the whole range of costs. Among the 10 samples, the average *REE* is 4.6%, and the maximum *REE* (sample 7) is -14%. For early cost estimation of injection molds, this result is considered highly satisfactory.

8.4.4 Neural Networks for Different Cost Ranges

One problem in obtaining the best possible neural network (architecture and parameters) is that when the costs of the samples have a very large range, it is difficult to find a trained neural network to accommodate the samples at the two ends of the cost. For example, all the trained neural networks shown in Table 8.3, except the one with 70 neurons in the hidden layer, produced very large estimation errors on the validation samples. One way to tackle this problem is to build neural networks for different cost ranges. In the present study, it was noticed from Fig. 8.2 that there are 43 samples with costs less than $60K and the other 49 samples with cost more than $60K. Therefore, it was decided to build two neural networks for these two groups of samples. The architectures of these two neural networks is the same as the one described in Section 8.4.1 except that the number of neurons in the hidden layer is to be decided through training and validation. The training procedure and program are the same as described in Section 8.4.2. The validation samples in Table 8.2 are then used to test the trained networks, respectively.

For the 43 samples with costs less than $60K, nine neural networks were trained. Six of the ten samples in Table 8.2 (Nos. 1,2,5,7,9, and 10) have costs less than $60K, and they are used for the test. The resulted REE for the six samples

Table 8.4 Relative estimation errors of the validation samples (<$60K)

No. of neurons in hidden layer	1	2	5	7	9	10	Avg. REE
20	19%	11.4%	25%	-3.5%	-5%	0%	10.6%
30	17%	2.9%	23%	-3.5%	-5%	-2.5%	8.8%
40	4.8%	-11%	10%	-1.4%	-5%	0%	5.4%
50	-4.8%	5.7%	2.5%	8.8%	-5%	2.5%	4.9%
60	2.4%	5.7%	7.5%	8.8%	-3%	-2.5%	5.0%
70	-2.4%	0%	2.5%	8.8%	-5%	0%	3.1%
80	-2.4%	-10%	2.5%	4.7%	-6%	-2.5%	4.8%
90	-2.4%	8.5%	5%	4.7%	-5%	-2.5%	4.7%
100	-9.5%	2.8%	-5%	8.8%	-5%	-2.5%	5.6%

by using the nine different trained neural networks are shown in Table 8.4, in which the last column shows the average *REE* for the six samples. Although the neural network with 70 neurons in the hidden layer produced the best results (average *REE* = 3.1%), the other trained neural networks also produced very good results. The one with 70 hidden neurons is called $NN_{0 \rightarrow 60K}$.

For the 49 samples with costs more than $60K, nine neural networks were also trained. Four of the ten samples in Table 8.2 (Nos. 3, 4, 6, and 8) have costs more than $60K, and they are used to for the test. The resulted *REE* for the four samples by using the nine different trained neural networks are shown in Table 8.5, in which the last column shows the average *REE* for the four samples. The neural network with 80 neurons in the hidden layer produced the best result (Average *REE* = 3.1%). This neural network is called $NN_{60K \rightarrow \infty}$. On the other hand, it was also observed that the other two trained neural networks (with 30 and 100 neurons in the hidden layer) also produced reasonable results. This has further enhanced the idea of training different neural networks for parts with different cost ranges. In this way, the task to find and train a workable neural network for a certain cost range will become easier.

With the three trained neural networks ($NN_{0 \rightarrow \infty}$, $NN_{0 \rightarrow 60K}$, and $NN_{60K \rightarrow \infty}$) at hand, a strategy is needed to decide how to use them for cost estimation of the injection mold for a given new part. Here, the cost estimation procedure is divided into three stages:
(1) Given a new molded part design, the cost-related factors (CFs) are extracted and quantified.
(2) Apply $NN_{0 \rightarrow \infty}$ for rough cost estimation by calculating the cost of the mold (*Cost*_{rough}).

Table 8.5 Relative estimation errors of the validation samples (>$60K)

No. of neurons in hidden layer	3	4	6	8	Avg. REE
20	-11%	-21%	113%	139%	70%
30	0%	-18%	4%	8%	7.5%
40	2%	-48%	28%	42%	30%
50	2%	-48%	28%	42%	30%
60	2%	-14%	39%	8%	15.8%
70	16%	0%	1%	-7%	6%
80	**-5%**	**-2%**	**3%**	**-3%**	**3.1%**
90	22%	-27%	-8%	8%	16%
100	3%	-10%	5%	-6%	5.8%

(3) If $Cost_{rough} < 60K$, apply $NN_{0 \rightarrow 60K}$ to produce the final estimated cost of the mold; otherwise, apply $NN_{60K \rightarrow \infty}$ to produce the final estimated cost of the mold.

8.5 SUMMARY

Traditional product cost estimation approaches have difficulties with early cost estimation due to unavailability of the cost function form and inability to update cost estimation algorithms using actual historical cost data directly. The presented approach of cost function approximation using neural networks has shown the promise of reducing, if not eliminating, these two problems. This approach is applied to early cost estimation of injection molds at the quotation stage. In the development, the cost-related factors (CFs) for making injection molds have been identified based on the molded part design and conceptual mold design only, which fits well to the information availability during the quotation stage. A quantification scheme for the CFs has been established based on all possible states of these CFs. This scheme has been implemented in a mold making company. The CFs are extracted and quantified from a set of previously produced molds and their respective parts (samples), as well as their actual costs. A neural network construction procedure (determination of architecture, training, and validation) has been implemented. Validation results suggest that this approach is able to conduct cost estimation to a satisfactory level. Furthermore, it was proposed and demonstrated that by training different neural networks for different cost ranges, better results can be achieved.

In summary, the advantages of this early cost estimation approach are given as follows:

(1) The CFs cover all the cost aspects related to the part design as well as mold conceptual design. Extracting such factors can be easily done by the quotation engineer or a mold designer.
(2) It is not necessary to know the actual form of cost function. With the previously produced part/mold designs and their cost data, the back-propagation neural network is capable of learning their relationships through function approximation.
(3) The established cost function (in a form of a trained neural network) can be updated by re-training the neural network using new samples from time to time to adjust itself to the new manufacturing environment.

On the other hand, this approach also has certain limitations. First, determination of the number of hidden layers and neurons in each hidden layer is a trial-and-error process. It can be time consuming and there is no guarantee that the finally selected architecture is the best. Secondly, sensitivity study of the CFs on the cost is difficult. Therefore, the current neural network model may not provide the means to guide the designer towards better solutions.

REFERENCES

1. M.M. Andreasson and J. Olesen, "The concept of dispositions", Design, Vol. 1, No. 1, pp. 17-36, 1990.
2. H.H. Jo, H.R. Parsaei, and W.G. Sullivan, "Principles of concurrent engineering," Concurrent Engineering – Contemporary Issues and Modern Design Tools, Chapman & Hall, London, pp. 3-23, 1993.
3. K. Nichols, "Getting engineering changes under control", J of Engineering Design, Vol. 1, No. 1, pp. 5-16, 1990.
4. P. Dewhurst and G. Boothroyd, "Early cost estimating in product design", J of Manufacturing Systems, Vol. 7, No. 3, pp. 183-191, 1988.
5. G. Boothroyd and C. Reynolds, "Approximate cost estimates for typical turned parts", J of Manufacturing Systems, Vol. 8, No. 3, pp. 185-193, 1989.
6. G. Boothroyd and P. Radovanovic, "Estimating the cost of machined components during the conceptual design of a product", Annals of CIRP, Vol. 38, No. 1, pp. 157-160, 1989.
7. C. Poli, J. Escudero, and R. Fernadez, "How part design affects injection-molding tool costs", Machine Design, November 24, 1988, pp. 101-104.
8. L.S. Wierda, "Product cost-estimation by the designer", Engineering Costs and Production Economics, Vol. 13, pp. 189-198, 1988.

9. M.S. Hundal, "Design to cost", Concurrent Engineering – Contemporary Issues and Modern Design Tools, Chapman & Hall, London, pp. 330-351, 1993.
10. N.S. Ong, "Activity-based cost tables to support wire harness design", International J of Production Economics, Vol. 29, pp. 271-289, 1993.
11. Y.F. Zhang, J.Y.H. Fuh, and W.T. Chan, "Feature-based cost estimation for packaging products using neural networks", Computers In Industry, Vol. 32, pp. 95-113, 1996.
12. Y.F. Zhang and J.Y.H. Fuh, "A neural network approach for early cost estimation of packaging products", Computers and Industrial Engineering, Vol. 34, No. 2, pp. 433-450, 1998.
13. T. Khanna, Foundations of Neural Networks, Reading, MA: Addison-Wesley, 1990.
14. The MathWorks Inc., Neural Network Toolbox for Use with MATLAB, 1993.
15. S.K. Lim, A neural network approach for early cost estimation of plastic injection molds, B.Eng. thesis, National University of Singapore, 1996.
16. A. Sharifuddin, Early cost estimation of injection molds, B.Eng. thesis, National University of Singapore, 2000.

9

Case Studies: IMOLD® and IMOLD-Works for Mold Design

9.1 INTELLIGENT MOLD DESIGN AND ASSEMBLY SYSTEMS

Traditionally, computer-aided injection mold design is done directly on a CAD system. Some plug-ins applications were then developed to automate certain portion of the design process. They allow engineers to pick and choose from the best add-on applications available. In this chapter, a knowledge-based mold deign prototype, stand-alone mold design software based on both commercial CAD systems as well as a low-level 3D solid kernel Parasolid will be discussed respectively. The knowledge-based mold design is an early attempt in automating the mold design process but was not fully commercialized due to its programming complexity and limitations in modeling capability. In comparison with UNIX/Linux-based platforms, a Windows-native design environment built with the advantages of user-friendly assembly capabilities, ease-of-use, rapid learning curve, and affordability will be detailed in this chapter.

9.1.1 Knowledge-Based Mold Design Systems

As mentioned previously, designing an injection mold is a very complicated process which requires extensive knowledge from an experienced mold designer. Although CAD systems have replaced traditional drawing boards as design tools, the design process still requires intensive/iterative manipulation of geometrical entities. Furthermore, the design process seldom stops at the initial design, and subsequent modifications may be necessary and are expensive. In such cases, component parameters may need to be recalculated. The complex interrelationship between each mold component causes these design procedures to be tedious and time-consuming. It is therefore natural to develop a knowledge-based expert (KBE) system to automate in part or fully, the injection mold design process. In KBE design systems, the engineering rules used by experienced mold designers need to be collected and stored in a knowledge base, so that the design cycle could be shortened by letting the intelligent system take care of the more repetitive and routine tasks.

The idea to develop a KBE system for plastic injection molds was first attempted by KODAK in their ESKIMO project [1]. Emphasis was placed on the mold base, instead of the mold itself; the prototype for internal use was an integrated system consisting of critical components such as ejector systems, runner and gating systems, sliders and lifters, etc. A KBE system in the design of plastic injection molded parts was also reported by Pratt and Sivakumar [2]. The system enables the designer to present the part-design in terms of hybrid features that associate product and process knowledge with part geometry. Engineering rules about moldability and strength, process and material considerations, and mold design were captured in the system. Other approaches to developing an expert system in design using the Intelligent CAD system can be found in [3,4]. Few fully or semi-automated mold design systems which include a complete mold design functionality have been reported.

Another similar mold design methodology using the KBE system from Intelligent Computer Aided Design (ICAD) was implemented [5]. In this approach, once the engineering knowledge of the product is collected and stored as a product model, design engineers can generate and evaluate new designs quickly and easily by changing the input specifications of the product model, or modify designs by extending or changing the product model. This frees the designer from time-intensive, detailed engineering tasks such as repetitive calculations and allows more time for creative design work. In addition to 3D geometric models, the product design can include various outputs, such as reports on engineering results, data for engineering analysis, etc. that can be downloaded onto a CAD system to form structured bills of materials and manufacturing instructions.

The development environment mentioned above utilizes IDL (ICAD Development Language [6] which sits on top of Allegro LISP platform. Being an object-oriented language, it facilitates the development of a knowledge-based system. Some of the common features of expert-systems such as frames, semantic nets, parent-child relationships are incorporated. The concept of parametric part design was also explored and implemented. The product model takes an input specification, applies the engineering rules, and generates a product. However, in this case, the input specification comes in the forms of both user interactions and the molded product model. Due to the limited facility in the ICAD environment in the manipulation of a solid model, parts of the features in the product molding are pre-processed in the Unigraphics CAD/CAM system (UG-II) [7] before exporting to ICAD. In this approach, engineering rules are captured within the product model

Case Studies

in a systematic format. Since they are captured within the product model itself, they can be easily maintained and modified without affecting other parts. The method of object-oriented programming was initiated by the need to manage large programs efficiently. Each individual part is capable of communicating with other parts for the purpose of generating a complete design. Hence during the design of each product model, particular attention is paid to ensure that it is generic, i.e., it must be able to generate itself and be used by other product models. For example, when a gate needs to be created, the KBE system will simply look for the appropriate product model and create it based on its rules. The knowledge of the entire design methodology is based on a semantic network. The relationships between the features are well defined but not fixed. Depending on the design constraints, a different tree structure (see Fig. 9.1) will be generated. An injection mold design output by this KBE mold design system is shown in Fig. 9.2. However, due to the complexity of programming using KBE languages and limitation in manipulating solid models within the environment, the system was not commercialized for industrial use. The incorporation of commercial CAD/CAM systems has thus simplified the task of developing usable computer-aided mold design and assembly systems as described in the following sections.

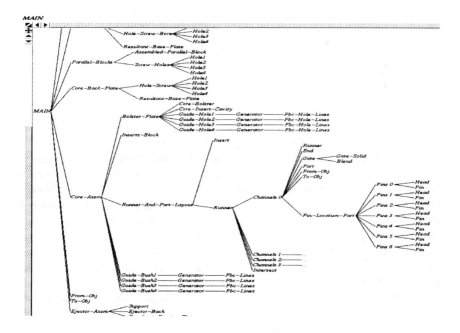

Fig. 9.1 The object-oriented mold assembly tree structure.

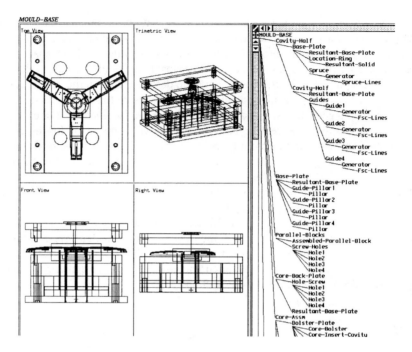

Fig. 9.2 An injection mold design and assembly output from a KBE system [9].

9.1.2 IMOLD® Overview

A comprehensive mold design and assembly system, IMOLD® [8,9,10], was first developed at the National University of Singapore. It is knowledge-based software designed to capture the mold designer's intents and apply the expertise in the specific discipline of injection mold design. The IMOLD® system provides design modules for the complete mold design process, including: parting, cavity layout, gate and runner, mold base, slider, cooling, ejecting, electrode, hot runner, etc. This software can assist a mold designer to significantly shorten the overall mold design lead-time with improved design qualities.

Built on top of the Unigraphics CAD/CAM (UG) system in the early version, IMOLD® (Intelligent Mold Design) system (v2-v5) is a complete 3D solid-based design process application that was developed in conjunction with UG. It enables the mold designer to rapidly create a tool design by providing the necessary tools that are specific to the mold makers. These tools allow a mold designer to concentrate on the practical aspects of mold design rather than the operations of a CAD system. Since IMOLD® is a mold design process application, mold designers are naturally supplied with a consistent method of creating such a design. Starting with a 3D part model as input, it provides a set of functional modules that help

Case Studies 339

the user to complete a mold design process from core and cavity creation to cooling and ejector system design. The output is a complete mold assembly in 3D whose components can be used either collectively or individually for downstream manufacturing applications.

Two notable features of IMOLD® are parametric modeling and automation of routine design tasks [11]. With parametric modeling, the database is concise while a wide range of components with similar shapes can be accommodated easily. Besides, tedious geometric manipulation tasks, such as core and cavity creation and assembly, are made automatic. Moreover, since many industrial design knowledge and practices have been incorporated into the parametric models of the functional parts in an injection mold, the design task has become simply selecting components from the available alternatives.

Usually there are hundreds of parts in a mold assembly. The management of such a large number of parts efficiently becomes an issue in computer-aided injection mold design. By employing the Assembly Navigation Tool in Unigraphics, all the parts in a mold are placed in a hierarchical structure, i.e., an assembly tree, according to the assembly relationships among all the parts. The assembly tree has the overall assembly as its root node. Every branch node represents either a subassembly or a part. The related nodes, namely the parent-child nodes on the same branch and the child nodes sharing the same parent node, are associated by assembly constraints, such as mating and aligning, and/or referring parameters with each other. This assembly tree structure can be used for easy viewing. More importantly, it can be also used as an editing window. Users can simply browse the branches and click the node to modify. Changes in a node are propagated to the other nodes that are associated with it. IMOLD® creates a tree structure assembly whenever a mold is designed for a product model. The structure is created with specifically labeled files to denote the corresponding components of the mold design. All the files of the components have a "pre-name" as a prefix which is specified by the user in the *Data Preparation* module. The tree structure lays out the file of the various components for easy access and identification. Fig. 9.3 shows an example of the Unigraphics Assembly Navigation Tool containing a tool design with a prename of *bmt*.

9.1.3 Development Platforms

IMOLD® system was initially developed on the Unigraphics system running in a UNIX environment using C language. Unigraphics was chosen because it allows users to build a customized system on it using UG/Open Application Programming Interface (API) [13], and it also supports a design with user-defined features due to its parametric nature and the ability to extract design information from the features. API contains a large set of user callable functions/subroutines that access the Unigraphics Graphics Terminal, File Manager, and Database of Unigraphics. It offers an easy and friendly interface between Unigraphics platform and the "add-on" application developer. Unigraphics provides the User Function

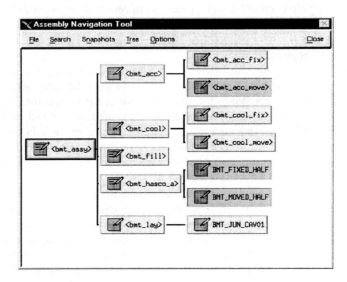

Fig. 9.3 IMOLD® assembly structure [12].

(UFUN) [14] as a software tool to enable an easy interface between Unigraphics and the outside applications.

9.1.4 Functional Modules

IMOLD® comprises of eleven different functional modules and one utility module. This application leads the mold designer through the process of completing virtually every aspect of a mold design. IMOLD® contains comprehensive mold bases and component libraries, but more importantly, it provides extensive tools for quickly defining parting surfaces, inserts, slider and lifter geometries. IMOLD® automatically provides these options in the context of Unigraphics hybrid modeling and assemblies. Eleven of these modules are accessed from the *IMOLD_V4.1* [12] pull-down menu (Fig. 9.4). The twelfth module is the *IMOLD_Tool* option found within the IMOLD®'s application on the Unigraphics menu bar. The main modules are listed as follows:

(1) Data Preparation
The option accesses the Data Preparation module. This module is used to load a product model, set up the mold pull direction, apply a shrinkage factor, and define a containing box for core and cavity creation.

Case Studies

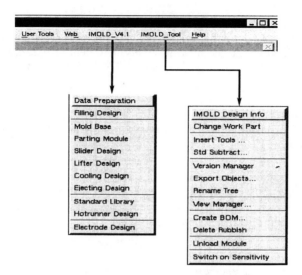

Fig. 9.4 IMOLD® pull-down menu.

(2) Filling Design
The option accesses the Filling Design module (Fig. 9.5). This module is used to determine the number of cavities, gating and runner configurations. Modification and deletion of layout, gates, and runners can also be performed using this module.

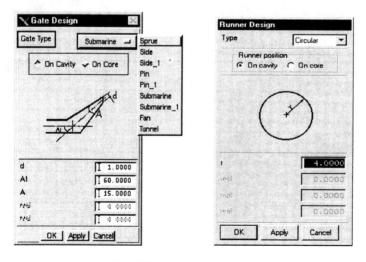

Fig. 9.5 Filling design menu.

(3) Mold Base
The option accesses the Mold Base module (Fig. 9.6). This module is used to load a mold base into the mold assembly. Different brands, series, and standards of mold bases can then be chosen.

(4) Parting
The option accesses the Parting module. This module is used to create parting lines for the creation of cores and cavities. Inserts for the mold can also be designed here.

(5) Slider Design
The option accesses the Slider module (Fig. 9.7). This module is used to design sliders with either the Finger Cams or the Toggle Cams. All slider accessories like guides and wear plates are also added here.

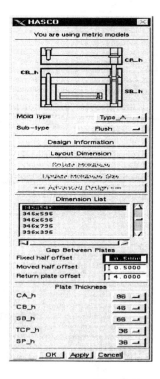
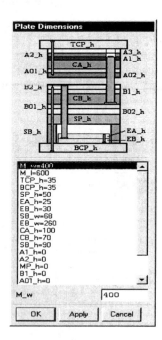

Fig. 9.6 Mold base design menu.

Case Studies

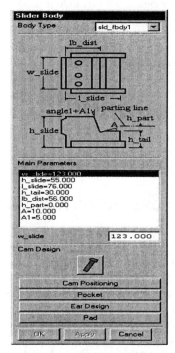 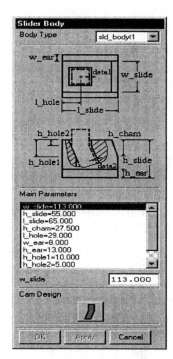

Finger cam slider body   Toggle cam slider body

Fig. 9.7 Slider design menu.

(6) Lifter Design
The option accesses the Lifter module. This module is used to design the lifter head and the body. Several different body styles are provided.

(7) Cooling Design
The option accesses the Cooling Design module. This module is used to design the cooling system. Many different types of cooling patterns and accessories are provided.

(8) Ejecting Design
The option accesses the Ejection Design module. This module is used to design the part ejection. Several types of ejectors are provided and these are all automatically trimmed to the core by IMOLD®.

(9) Standard Library
The option accesses the Standard Part module. This module is used to add any of the IMOLD® supplied standard parts into a mold design.

(10) Hot Runner Design
This is an optional module and it accesses the Hot Runner Design module (Fig. 9.8). The module is used to design the manifold as well as the nozzles and runners used in the hot runner system.

(11) Electrode Design
This is another optional module, which accesses the Electrode Design module. The module is used to design the electrodes for complex part geometry as well as the tool holders for the electrodes.

9.2 A WINDOWS-BASED MOLD DESIGN AND ASSEMBLY SYSTEM

For years, mold design engineers have to deal with two different systems, UNIX and PC. The former is widely used in engineering applications, and the latter is mainly in small- and medium-sized companies. Engineers also need to run corporate office applications such as word processing, spreadsheet, and project management tools, but these are not on their UNIX workstations. So the development of a new mold design application based on Windows platforms is in high demand. One reason for the shift is Microsoft Windows/NT, whose flexibility enables software developers create applications that are affordable and easy to use.

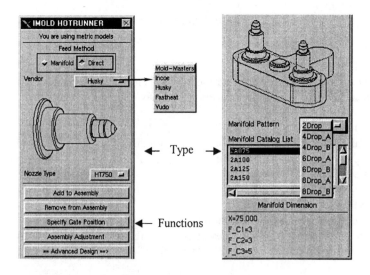

Fig. 9.8 Hot runner design menu.

Case Studies

With the increased availability of sophisticated, low-cost software for Windows, more and more engineers are using PC applications for their daily engineering activities.

In this section, PC-based solid modeling schemes based on *Parasolid* and *SolidWorks* platforms to develop mold applications are introduced. *Parasolid* kernel is selected as the neutral platform due to its wide adoption as a solid modeler embedded in commercial CAD/CAM systems. The architecture and implementation of the Windows-based injection mold design systems are described and each functionality is briefly presented.

9.2.1 3D Windows-Native CAD Systems

High-end 3D solid modeling systems have been on engineers' workstation in large aerospace, consumer products, and automobile companies for years, while many smaller companies are now making the switch from workstations to PCs. One reason for the shift is the flexibility and advancement of Windows-native/NT which has enabled software developers to create applications that are affordable and easy to use. High-end users are finding that mid-range solid modelers, such as *Parasolid*, have met their needs.

The development of a new mold design application based on Windows platforms is in high demand. Application availability, high performance, ease-of-use, standardization, and increased productivity, all at greatly reduced costs nowadays, have led to a shift from UNIX environments to Windows NT. Therefore, any solid modelers running under NT or Windows such as *Parasolid* or *SolidWorks* will become more popular with designers and companies. Using Windows-native environment, designers will be able to adopt the standard Windows functions (e.g., *cut* and *paste*) and interfaces, making the desk-top design tasks more user-friendly and at a lower cost. Another reason for the shift is Microsoft Windows/NT, whose flexibility has also let software developers create applications that are affordable and easy to use.

9.2.1.1 ParaSolid Kernel

Designed as an boundary-representation solid modeler supporting freeform surfaces to serve mechanical CAD/CAM/CAE applications, its extensive functionality is supplied as a library of routines with an object-oriented programming interface. It is essentially a solid modeler to be able to:

- build and manipulate solid objects;
- calculate mass and moments of inertia, and perform interference detection;
- output the objects in various pictorial ways; and
- store the objects in some sort of databases or archives and retrieve them later.

Parasolid [15] provides a robust solid modeling, generalized cellular modeling and integrated surface/sheet modeling capabilities and is designed for easy integration into CAD/CAM/CAE systems to give rapid time to market. *Parasolid* model files are thus very portable. It is therefore a superior platform for developing stand-alone applications. It represents solids (and other types of body) by their boundaries. *Parasolid* has three interfaces to an application: two sit on top of the modeler (side-by-side), and are the means by which the application model manipulates objects and controls the functioning of the modeler. These two are the PK Interface and the Kernel Interface. The other interface lies "beneath" the modeler, and is called by the modeler when it needs to perform data-intensive or system type operations. This downward interface is in three parts: Frustrum, Graphical Output (GO), and Foreign Geometry.

Fig. 9.9 shows the relationships between *Parasolid* and these interfaces. The PK Interface is a library of functions that provide access to *Parasolid*. The Kernel Interface (KI) is the original interface to *Parasolid*, and its functionality is now almost completely replaced by the PK. However, there are still a few areas where KI routines may need to be used. The use of PK functions and KI routines can be mixed within an application as when necessary. The *Frustrum* is a set of functions that must be written by the application programmer. The kernel calls them when

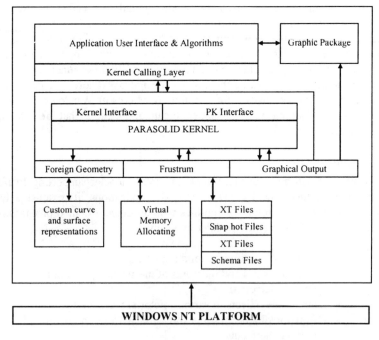

Fig. 9.9 Relationship between *Parasolid* and its applications.

Case Studies

data needs to be saved or retrieved. One of the first things the application programmer needs to do is decide how to manage the storage of data which *Parasolid* outputs through the Frustrum.

Transferring data through the Frustrum usually involves writing to, or reading from, files. The format and location of the files are determined when writing the Frustrum functions. The Graphical Output (GO) functions must also be written by the application programmer. Unlike the Frustrum, the output from these functions is not, in general, data files, but rather instructions to the graphics system for drawing pictures requested from the kernel. Like those in the Frustrum, Graphical Output (GO) functions are called by the kernel. *Parasolid* will call GO functions as a by-product of having been asked to draw something by one of the rendering functions of the PK. The kernel prepares drawing data piecemeal, so that one call to a rendering function will normally generate many calls to the GO functions.

9.2.1.2 SolidWork Platform

The *SolidWorks* Add-In Manager will allow the user to control which third party software is loaded at any time during their *SolidWorks* session. More than one package can be loaded at once, and the settings will be maintained across the *SolidWorks* sessions. *SolidWorks* [16], based on the Windows platform, allows engineers to pick and choose from the best add-on applications available. It was chosen as the platform due to the Windows-native design environment, powerful assembly capabilities, ease-of-use, rapid learning curve, and affordable price. IMOLDWorks (IMOLD® for *SolidWorks*) was developed based on injection mold design procedure by using *SolidWorks* and its API. Fig. 9.10 shows the *SolidWorks* API objects [17].

IMOLD for *SolidWorks* (or simply called IMOLDWorks) [18] is a Windows-native 3D mold design application that runs on Windows platform (see Fig. 9.11). With IMOLDWorks, mold design can be completed much faster than with conventional CAD software by using a single integrated mold assembly database throughout CAD, CAM, mold shop and the mold assembly department. The system provides a designer with an interactive computer-aided design environment, and can both speed up the mold design process and facilitate the standardization, which in turn increases the speed of manufacture. IMOLDWorks was implemented on a Windows/NT through interfacing the Visual C++ codes with the commercial CAD software, *viz. SolidWorks*. The main architecture and the functional modules are shown in Fig. 9.12, which cover the most complete mold design process.

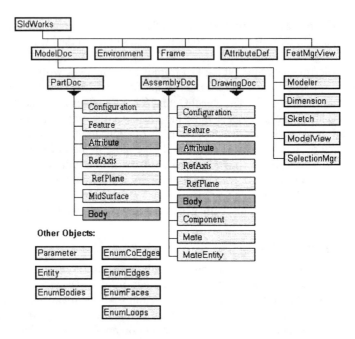

Fig. 9.10 *SolidWorks* API objects.

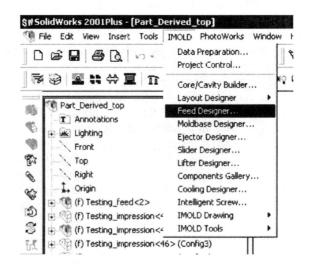

Fig. 9.11 A mold design system running in Windows/NT environment. (Courtesy of IMOLD for *SolidWorks*.)

Case Studies

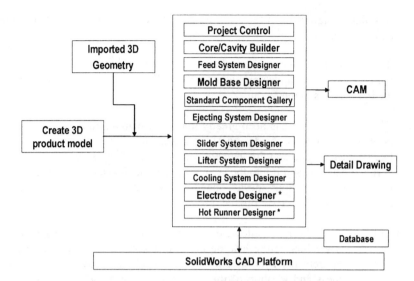

Fig. 9.12 The architecture and functional modules in IMOLDWorks.
(**under development as of v4.0*)

IMOLDWorks provides a vast array of tools that will aide in the development of mold design. The software offers tools in the following areas [19]:

(1) *Data Preparation* is a module which will help designers prepare the product model for design should one of the following considerations apply: if the parting direction of the part is not along the z-axis; and if users do not want to work on the customer's original part. This module will help the designers prepare the part so that should there be any changes in the part eventually, these changes can be reflected in the cores/cavities created in the downstream process.

(2) *Project Control* is the gateway from which design projects are started or continued. It allows users to load current projects or start new ones. With this module, users also define the project code and unit, the material type to use for the project, and the corresponding shrinkage factor. Designers can also specify the various shrinkage factors here should they need directional scaling.

(3) The *Core/Cavity Builder* provides the tools for creating the core and the cavity. It has categorized all these tools into two different processes: Standard and Advanced Parting. Depending on the product, designers can use either or both processes to create the core and the cavity, and these processes en-

sure that designers keep the associativity between the product and the core and cavity inserts.

(4) The *Layout Designer* provides the tools for laying out the impressions in a multi-cavity mold. It also provides tools to edit a layout and translate any of the impressions in the layout.

(5) The *Feed Designer* has been greatly improved with an all-new interface. It provides the tools for the design of the gating and the runner systems. Gates like submarine gate and fan gate are parametrically built so users can easily design a gate and add it into the layout. Linear and *S*-shaped runners are provided for the various design needs. The gates and runners can be automatically subtracted from the inserts.

(6) The *Moldbase Designer* allows designers to preview the mold bases from the various vendors before loading them in. Users can also add in double-ejection requirements with the click of a mouse button and customize each plate thickness and location of the components. Once users have finalized the design of the mold base, the Moldbase Designer will pocket out the components from the plates and remove any unnecessary parts. IMOLD-Works has set up the whole system such that users can even create new mold bases that are unique to their individual companies.

(7) The *Ejector Designer* module lets users add in standard ejector pins from various vendors at locations of users' choice. It also allows for customized ejector pins through its interface and has a trimming facility whereby users can trim all the ejector pins to the core surface at the click of a button and also a function to pocket out the ejectors from the plates.

(8) The *Components Gallery* provides a comprehensive range of standard components that designers can use in their designs. Users can select the standard sizes from the catalog and easily add the components into the design. The Components Gallery ensures that mating conditions are adhered and the components are pocketed out.

(9) The *Slider Designer* has incorporated standard slider bodies and accessories inside and allows the designer to add any of these easily to the side-core, or what is called "slider-head", so that external undercuts can be manufactured in the injection molding cycle. The software, when designing the slider automatically, considers details like the positioning data, stroke, and cam pin angle with reference to the location of the slider. The Slider Designer has both standard sliders and generic sliders in its database.

Case Studies   351

(10) The *Lifter Designer* works in a similar manner as the Slider Designer except that it is used in cases where internal undercuts are present. The module considers details like the positioning data, stroke, and releasing angle with reference to the position of the lifter. Within its database of models, the Lifter Designer has options for standard lifters and generic lifters.

(11) The *Cooling Designer* provides tools to design the cooling circuits easily by specifying the cooling route. Once the circuit has been designed, it can also be modified should designers need to make any changes. In addition, the Cooling Designer also makes considerations for manufacturing by providing functions for drilling and extensions.

(12) IMOLD® Drawing will automate the creation of the core and cavity views of a mold. Section lines can also be defined very easily by picking points on the parent view.

9.2.2 System Implementations

Most of CAD systems provide only the geometric modeling functions that facilitate the modeling and drafting operations, but do not provide mold designers with necessary knowledge to design the molds. Foreseeing this, some "add-ons" software, i.e., IMOLD® and IMOLDWorks, were thus developed on high-level 3D-modeling platforms to facilitate the mold design process. Such an arrangement is advantageous in many ways. The 3D-modeling platform provides the plug-in software with a library of functions, an established user interface and style of programming. As a result, the development time for these "add-ons" are significantly reduced. Fig. 9.13 shows the two add-on applications structure on IMOLD® for Unigraphics (UNIX-based) and IMOLD® for SolidWorks (Windows-based).

A *Parasolid*-based injection mold design system has also been reported by Kong et al. [20]. The Windows-native 3D computer-aided injection mold design system has been implemented on Windows NT using Visual C++, Microsoft Foundation Class (MFC), and the solid modeling kernel-*Parasolid* V11.0. The main advantage is that the Windows-based design system is built on the neural solid kernel, viz. *Parasolid*, and thus it does not require any commercial CAD/CAM system to provide solid modeling and graphical user interfaces (GUI) functions. The system built on top of *Parasolid* kernel is widely used as the internal solid modeler by many CAD/CAM systems and thereby provides a good possibility to input and output to and from other design and manufacturing applications. On the other hand, if the mold design system is built on any commercial CAD/CAM systems, e.g., Unigraphics, Pro/Engineer, *SolidWorks*, AutoCAD, CATIA, I-DEAS, etc., the programming efforts on developing graphical user interfaces (GUIs) and modeling functions will be much less but the cost on licensing these kernels might be much higher. The APIs provided in each commercial CAD software also vary from one to another, and there is no guarantee that the full

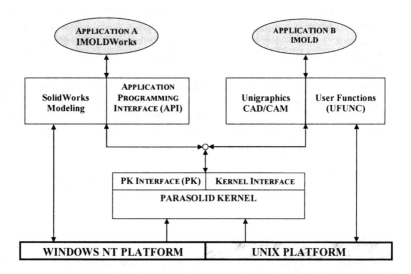

Fig. 9.13 Add-on applications structure for IMOLD[®].

interface and integration can be built based on those APIs provided in the commercial systems. Therefore, using a neutral solid kernel platform such as *Parasolid* will be a good choice in developing a Windows-native mold design system.

9.2.3 Graphical User Interfaces (GUIs)

A GUI was designed in the system for high user-friendliness. Designing an interface allows the users to have a step-by-step idea on how the program precedes the mold design. Fig. 9.14 shows the user interface built for the *Parasolid*-based mold design prototype. The system reported has mainly the parting and mold base building functions that can create a simple mold easily.

As shown in the figure, a 3D modeler sub-system has also been developed, which has the following functions: (1) creation of primitives (block, sphere, cylinder, cone, torus, prism, and other sheet bodies) with user interaction; and file processing functions, e.g., open and save files in *Parasolid* format; transformation of bodies; and Boolean operations on the solid bodies created. This *Parasolid*-based mold design prototype [21] built-in with several modules, such as data prepare, filling design, mold base, and parting design has been tested on industrial parts, and good results have been obtained for the mold generation. The system was written on an object-oriented programming language (Visual C++), which ensures further development and extension. The methodology is mainly addressed for the plastics injection mold design process, but it could be similarly applied to die casting mold design.

Case Studies 353

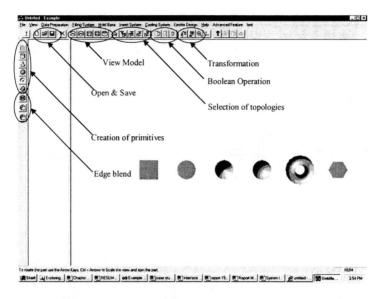

Fig. 9.14 GUIs developed based on *Parasolid* modeler.

Figs 9.15 and 9.16 represent the parting design process of an example part, which is more complicated than those presented in the previous chapters. The example part has three particular features: (i) the existence of free-form surfaces; (ii) stepped parting lines; and (iii) the existence of several through holes, and thus the need to determine the internal parting lines for locating the external parting lines. In Fig. 9.15, the resultant external parting lines are highlighted and displayed together with the extruding directions. The case study is conducted using the designed interface mentioned in the previous section. The core and cavity inserts generated are shown in Fig. 9.16.

Fig. 9.17 is a screen dump from IMOLDWorks, which shows the basic GUIs for mold design use. The functionalities provided in the system appear as icons in Windows. The standard Windows functions can be applied and thus makes the design task more user-friendly. If designers wish to access the *SolidWorks* functions for more advanced modeling features, it is also permitted as IMOLDWorks are embedded inside *SolidWorks* as a truly integrated module. The layout design and core-cavity builder modules are shown in Figs. 9.18 and 9.19, respectively. A complete mold design and assembly displayed within IMOLDWorks is illustrated in Fig. 9.20 that shows the capability of the system in designing a complex industrial mold.

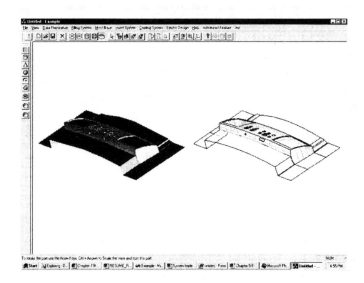

Fig. 9.15 Generation of the parting surfaces (*ParaSolid* platform).

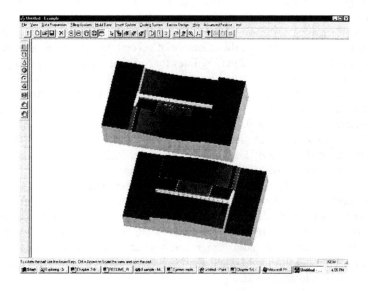

Fig. 9.16 Generation of the core and cavity (*ParaSolid* platform).

Case Studies

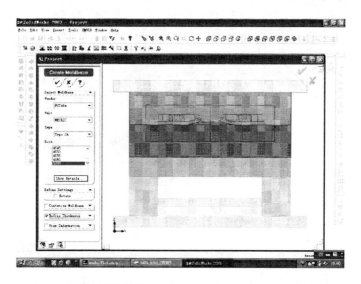

Fig. 9.17 Example of IMOLDWorks user interfaces.

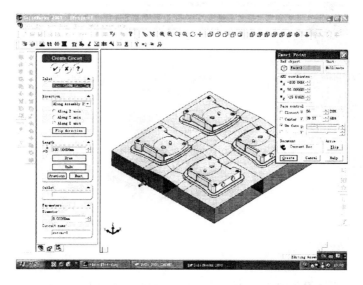

Fig. 9.18 Layout design module (courtesy of IMOLDWorks).

Fig. 9.19 Core and cavity builder. (Courtesy of IMOLDWorks.)

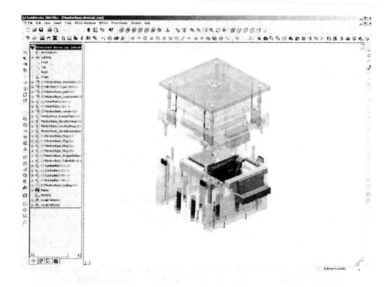

Fig. 9.20 IMOLDWorks, a Windows-based mold design system. (Courtesy of Manusoft Technology Pte Ltd, Singapore.)

9.2.4 Windows-Based Die Casting Die Design Systems

The methodology for Windows-based plastic injection mold design can be similarly applied to the die casting die design. The general process of die-casting die design includes the design of layout, gating system, die-base, parting, ejection system, cooling design, moving core mechanism, etc. These processes are largely similar to the plastic injection molding processes except for the gating system. A typical die-base (also called die set) of die casting die is an assembly consisting of up to fifty component parts. Therefore, it is very tedious and time-consuming to model all the parts one by one, so it is necessary to find a method to automatically generate the die-base in the design system.

In order to facilitate the automatic generation of standard die-base assembly, various parts and die-base assemblies are pre-modeled and stored in a die-base library. All the parts in the assemblies are parametric models, and their dimensions will vary according to the type and dimension of the die-base specified. The automatic generation of die-base from the catalogue-based dimensions requires a database support. Since standard die-bases are commercially available, the establishment of such a database becomes very convenient. A user interface is also necessary for the user to select the desired type and dimension series of the die-base. The user just needs to click on the interface, and the system will automatically import the desired die-base. The Windows-based die casting die design modules can be developed as an add-in Dynamic Linked Library (DLL) file to be used in conjunction with 3D CAD software, such as *SolidWorks*. The implementation of these files can follow the format as shown in Fig. 9.21.

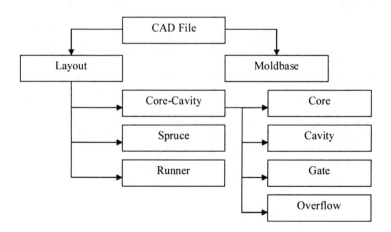

Fig. 9.21 Die casting die design files.

For gate design, a gating feature library is built in advance and adequate gating features are defined and stored in the library before the gating system design continues. During the design process, the geometries of the gating elements are generated automatically by retrieving the gating features from the library and assigning them with estimated geometric parameters. The details of the semi-automated die casting die design have been discussed in Chapter 4.

9.2.5 Illustrative Examples

Essentially, there is no difference in building a Windows-based die casting die system with the plastic injection mold system. The standard APIs provided in any Windows/NT-based CAD systems and the architecture/data-structure adopted for mold design can be also used in die casting die design. It can even build all the functions for die casting die design as a sub-module embedded inside the mold design package as many similar functions such as parting, runners (with overflows), gating, and mold base creation can be shared among these two injection mold design modules. The feeding design menus for runner, gating, filling, overflow, and sprue design are shown in Fig. 9.22 and Fig. 9.23, respectively.

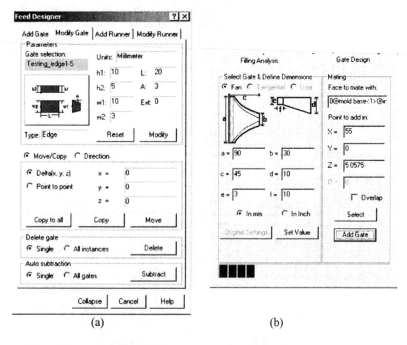

Fig. 9.22 Feeding system design menus: (a) feed designer; (b) gate design.

Case Studies

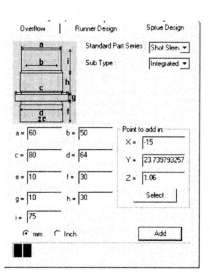

Fig. 9.23 Overflow, runner, and sprue design menu.

The vendor and the size of the mold base can be defined in the mold base design module. The *Moldbase Designer* (Fig. 9.24) offers tools for the selection of a mold base, changing the thickness, and customizing the mold base. In IMOLD-Works, mold bases from popular vendors (e.g., DME, FUTABA, HASCO, HOPPT, LKM, STIEHL, etc.) are parametrically built within IMOLD and made easily available to the designer. Using a single interface, the Moldbase Designer helps users to select and load in standard mold bases. It also shows how modifications, like adding in extra plates, can be made to the mold base. The Moldbase Designer also has capabilities for customizing mold bases in terms of number of plates and plate thickness. In the Moldbase Designer, however, users will see the DME mold bases named as Type-B, N-size, etc. This has been done to facilitate easier selection of the mold bases.

Prior to adding a slider/lifter, the parameters for positioning the slider/lifter must be defined using the *Parameter Selection* tab. The *Components Gallery* (Fig. 9.25) allows the addition of standard components into the design. Users can make changes to the components to either standard sizes or configure it to custom-made sizes. The *Slider/Lifter Design* module (Fig. 9.26) makes the development of the lifter body simple. Users will first determine the location by selecting parameters and then select the type of lifter from the library and add it to the design. The lifter can be modified by users using the edit functions to complete the mold design. To illustrate the applicability to industrial mold design, a complex ejection system and computer monitor mold designed by IMOLDWorks are shown in Figs. 9.27 and 9.28.

360   Chapter 9

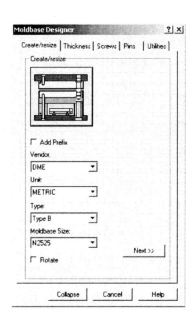

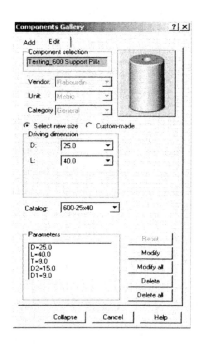

Fig. 9.24 Moldbase Designer menu.   **Fig 9.25** Component library builder.

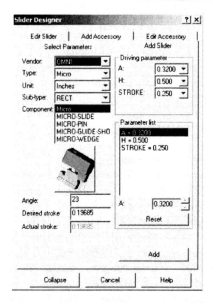

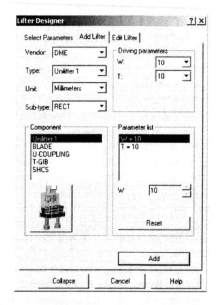

Fig. 9.26 Slider and lifter design modules.

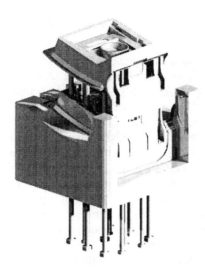

Fig. 9.27 Example 1 – an ejection system design.
(Courtesy of IMOLDWorks, Manusoft Technology Pte Ltd., Singapore.)

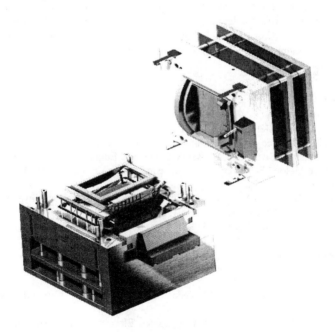

Fig. 9.28 Example 2 – a computer monitor mold designed by IMOLDWorks.
(Courtesy of Manusoft Technology Pte Ltd., Singapore.)

9.3 SUMMARY

In this chapter, several case studies on computer-aided injection mold design prototype and systems have been presented. The methodologies and implementation are given in detail. The knowledge-based mold design systems were the early attempt in automating the heavy experience-dependent tasks, which later led to few key commercialized systems, e.g., *Pro/Mold, IMOLD®, UG Moldwizard*, etc. By the use of those systems, mold designers will be able to cut down their design time of using traditional methods by as much as 70%. Although KBE systems can capture embedded design rules into the systems easily, their limited modeling capabilities cannot provide full functions to create more complicated molds. Due to the increasing computing powder from desk-top machines, Windows-based systems have been well adopted by the industry today. In addition to its relatively low cost, the ease of graphical user interfaces as icons and menus offered by the standard Windows/NT environment has given designers more advantages than traditional UNIX-based systems. It is expected the mid-range CAD systems will dominate the CAD/CAM market in the next decade. Adopting a comprehensive and easy-to use mold design solution under the Windows environment like IMOLD-Works will certainly become the standard for future mold design software. Although lack of rich built-in user interfaces, the neutral solid kernel platform such as *Parosolid* does offer a feasible and economic solution to develop the third-party design application without using any commercial CAD systems that require strong APIs to meet the specific mold design requirements.

REFERENCES

1. R.A. Gammon, ESKIMO: "An Expert System for Kodak Injection Mold Operation" Artificial Intelligence in Engineering Design. Vol 3, pp 81-104, 1991.
2. S.D. Pratt and M. Sivakumar, "A Knowledge-based Engineering system for the Design of Injection Molded Plastic Parts," Advance in Design Automation, ASME, Vol 1,1993, pp 65-71.
3. T. Sakai, E. Usuki and Y. Murakami, "Intelligent CAD system for railway bogie truck design", Proc of INCON'91, IEEE, 1991, pp42-47.
4. D. Cinquegranna, "Intelligent CAD automated plastic injection mold design", Mechanical Engineering, July 1990 issue.
5. K.S. Lee, J.Y.H. Fuh, Y.F. Zhang, Z. Li and A.Y.C. Nee, "IMOLD: An intelligent plastic injection mold design and assembly system", Institute of Engineers Singapore (IES) Journal, 36, no. 4:7-12, 1996.
6. ICAD, User Manuals Vol. 1 to 12, Version 5.0, Concentra Inc., Cambridge, Massachusetts, 1995.
7. UG-II, User Manual Version 10.5, Unigraphics, Electronic Data System, Cypress, CA, 1995.
8. IMOLD[R] was a registered trademark of the National University of Singapore (NUS) and licensed to Manusoft Technology Pte. Ltd., Singapore, a spin-off company from NUS.

Case Studies

9. K.S. Lee, Z. Li, J.Y.H Fuh, Y.F. Zhang, and A.Y.C. Nee, "Knowledge-based injection mold design system", In Proc of the Int'l Conf and Exhibition on Design and Production of Dies and Molds, Istanbul, Turkey, 1997, pp. 45-50.
10. IMOLD home page, http://www.eng.nus.edu.sg/imold, 1998.
11. Y. F. Zhang, J.Y.H. Fuh, K.S. Lee and A.Y.C. Nee, "IMOLD: An Intelligent Mold Design and Assembly System", In Computer Applications in Near Net-Shape Operations, edited by A Y C Nee, S K Ong and Y G Wang, Springer-Verlag, 1999, pp.265-284.
12. IMOLD v4.1 User Manual, Manusoft Plastic Pte. Ltd., Singapore, 1999.
13. UG/open API reference, Vol. 1 to Vol. 4, Unigraphics Solution Co., Maryland Heights, MO, 1997.
14. Unigraphics Essentials User Manuals, Vol. 1, Unigraphics Solution Co., Maryland Heights, MO, 1997.
15. Parasolid On-Line Documentation Web, Parasolid V10.1.123, Unigraphics Solutions Co., Maryland, 1997.
16. 'SolidWorks 2001', SolidWorks Corporation, MA, 2001.
17. SolidWorks 99 API documentation, SolidWorks Corporation, MA, 1999.
18. IMOLDWorks, Manusoft Technology Pte Ltd, Singapore, http://www.imold.com, 2003.
19. IMOLDWorks User's Manual, Manusoft Technology Pte. Ltd., Singapore, 2003.
20. L. Kong, J.Y.H. Fuh, K.S. Lee, Y.F. Zhang and A.Y.C. Nee, X.L. Liu, L.S. Lin, "A Windows-native 3D plastic injection mold design system", J of Materials Processing Technology, 2003, in press.
21. L. Kong, "Development of a Windows-based computer-aided plastic injection mold design system, MEng thesis, National University of Singapore, 2000.

Glossary of Die and Mold Terms

Backing plate	A supporting plate in a mold containing the cavity blocks.
Casting	A part resulting from the solidification of molten metal in a die or mold.
Cavity	The hollow space in a mold where the molten material is formed to the shape.
Cavity (block/insert)	A mold component inserted into the mold block to form an external feature of the molding. It is also referred to as the cavity insert.
Cooling channel/line	A passage through which water flows to cool a die casting die or mold. It is also called a water line.
Core (block/insert)	A part of a mold component that forms an internal feature of the molding. It is usually fixed in the moving half of a mold.
Degating	Separating moldings from the runner or feeding system.
Die	A tool used to impart the shape of a casting. It usually refers two matching steel blocks with cavities that shape the molten metal.
Die casting	A manufacturing process for producing precisely dimensioned parts from non-ferrous metals such as aluminum, magnesium, zinc, copper, etc.
Die-casting	The part produced by the die casting process.

Die casting die	Die casting tool for injection of metal.
Die base/set	An assembly consisting of die component parts.
Draft angle	A slight angle or taper placed on surfaces of a part to allow easy removal of the molding from a mold.
Ejector (pin)	A mechanism (usually a pin) that pushes the solidified molding out of the mold.
Ejector plate	A plate in a mold that carries the ejector pins.
Electrode	A tool used to form the shape of the work-piece geometry by electrical discharging energy for material removal. It comprises the electrode tool and the electrode holder.
Flash	A thin web of materials on a molding which appears at the parting, overflow, or around movable core half due to operating clearances in a mold.
Gate/gating	A small passage (restriction) for molten material connecting the runner to the mold cavity.
Hot runner/ manifold	In contrast to the normal cold runner, a type of runner in which material can be heated and maintained in a molten state and is not ejected with the molded part.
Impression	The mold cavity, may be designated into single impression, or 2-, 4-, 16-, 64-, etc., or multi-impression.
Injection (molding)	A process of forcing molten materials (plastic, metal, or ceramic) into a die or mold.
Insert	A piece of component, usually metal, placed in a mold and around part of which the plastics/metal is molded and becomes an integral part of a mold.
Lifter	The mechanical components (assembly) used to form the internal undercut features that cannot be molded easily in a molding.
Lifter head	The main attachment to the lifter body to form and conform the shape of the inner surface of the molding.
Mold	A mechanical assembly and tooling used in injection molding process to shape (mold) molten material (usually referred to as plastic) into the product needed.
Mold base	A set of mold components that hold the major standard parts provided by mold makers and can be quickly customized according to users' needs.
Molding	The part molded and injected from the injection molding process.
Overflow	A small reservoir added to the cavity (on a parting line) to absorb undesired non-metallic inclusions or impurities during cavity fill.
Parting	A mold design procedure in which the core and cavity inserts are created through the splitting of the mold

Glossary

	block.
Parting line	The intersection boundary (where the two halves of the mold come together) of the surfaces of a part molded by the core and the cavity.
Parting surface	The mating surfaces of the core and cavity in a mold.
Platen	A thick flat steel plate that forms a major part of a molding machine frame.
Plunger	A steel rod that is a clearance fit to the cylinder in which it reciprocates to produce injection.
Runner	A channel joins parting faces of a mold to allow the molten materials to flow from the sprue to the cavity.
Shot	The injection of molten material into a mold, or the solidified metal when it is removed from a die casting die.
Slider	The mechanical components (assembly) used to form the external undercut features in molding.
Sprue	The hole (portion) in a mold or die casting die through which the molten material first enters.
Stripper plate	A steel plate that moves over the male portion of a mold and slides forward to push the molding off the core.
Tie bar	The bars (one of two, three, or four) connecting the stationary platen to the moving platen on a casting/molding machine.
Undercut	A feature on the molding which restricts or inhibits the ejection of the molding from the mold.
Vent/venting	A small channel cut into the parting surface of a mold to let air escape from the cavity as it is filled.
Weld line	A mark on the surface of a molding caused by the separate streams of material flow.

Index

Air traps, 204

CAD/CAM, 1
  die and mold design, 4
  fixture design, 2
  injection molds, 6
  tooling applications, 1
CAE
  analysis procedures, 189
  applications, 187
  challenges in mold design, 214
  cooling analysis, 205
  cooling simulation, 192
  flow simulation, 192
  functionalities, 190
  mold design, 199
  part design, 198
  process design, 202
  product quality assurance, 202
Casting process, 210
Clamping force, 197
Core and cavity
  auto-generation, 78
  builder, 356
  creation, 58
  generation, 80, 166
  mold cavity, 13
  mold core, 13
  side cavities, 13
  side-cores, 13, 110, 112, 117
Cavity
  layout, 82, 139, 147, 170, 200
  multi-cavity, 87
  number, 82, 84, 139, 145
  patterns, 83
  rules, 89

Die casting, 9, 137
  cold chamber, 137
  cooling system, 139
  design, 139, 143, 173, 357
  design process, 139
  design system, 357
  die, 9, 138, 143
  die assembly, 138, 162
  die plate, 167

die-base, 139, 142, 161
die-casting, 137
  gating, 140
  generation, 165
  hot chamber, 137
  runner, 140
  shot sleeve, 160
  spreader, 160
  sprue, 160
Design
  by features, 48
DieWizard, 169

EDM, 254
EDM electrode, 257
  design, 257
  holder, 265
  interference detection, 265
  tool, 258

FA-MOLD, 120
Feature,
  associativity
  creation, 47
  depression, 49
  form, 49, 50, 51
  inter-, 52
  interactive, 48
  intra-, 52
  model, 47
  protrusion, 49
  recognition, 48
  undercut, 49   (*see also* Undercut)
Feed system
  design, 200
Fiber orientation
  prediction, 192
Flow simulation, 192

Graph, 97
  AAG, 98
  EAFEG, 98, 101
  representation, solids, 97
  subgraph, 101

undercut, 103, 114
Gating system, 140
  design, 144, 172
  filling conditions, 153
  filling time, 152
  gate, 140, 157
  gate velocity, 151
  gate-runner, 140, 159
  gating features, 154, 158
  overflow, 141
  parameters, 148
  P-Q^2 diagram, 148
  runner, 141
  shot sleeve, 141, 160

Hybrid variables, 197

IMOLD, 340
  IMOLDWorks (*see* Windows)
  modules, 340
  platforms, 339
Injection mold, 14
  assembly, 32, 34, 54
  automated assembly, 41
  bottom-up approach, 37
  class, 54
  computer-aided, 34
  design, 15, 20, 45, 187, 188
    Parasolid-based, 351
    process, 16
    system, 15
    Windows-based, 46, 344
  development process, 33
  ejector design, 28
  features, 48
  feeding system, 20, 200
  object, 54
  object-oriented, 54, 55
  operation, 54
  product-dependent, 41
  product-independent, 39
  top-down approach, 34, 37
Injection molding, 15
  molding activities, 187

Index

surfaces
  classification, 67
Injection molds, 13
  assembly modeling, 36
  associativity, 50
  cooling, 23, 351
    cooling channels, 205
    cooling system design, 201
  core and skin
    orientations, 197
  early cost estimation, 315
    cost factors, 320
    cost function, 317
    neural network, 319, 325
  ejector, 25, 350
  gate, 20, 140, 157 (*see also*
    Gating system)
  hot runner, 29
  lifters, 110, 359
  machining, 222
    area, 250
    cusp, 249
    cutter selection, 246
      algorithms, 253
      criteria, 248
      system, 270
    finishing, 222
    roughing, 222
    scallop height, 249
    step-over, 252
    time, 250
  modification (*see* Tool path re-
    generation
  mold base, 23, 124, 350, 359
  mold half, 15
  runner, 22
  sliders, 110, 350, 359
  slider and lifter, 28, 126
  sprue, 20
  standard components, 31, 359
  three-plate mold, 16
  two-plate mold, 16

KBE (*see* Knowledge)

Knowledge
  based mold design, 8, 335

Machining interference, 223
  2D interference, 225
  bottleneck, 223
  CC-point, 228
  curvature, 223
  curve offset, 224
  deep profile, 223
  detection, 224, 229, 238
  direct detecting, 230
  global, 232, 239
  local interference, 231, 232, 238
  sharp corner, 265
  surface offset, 227
  z-map, 229
Meld line, 203
Melt-front
  advancement, 197
Mold development
  life cycle, 192
Moldability, 62, 198
Moldability analysis, 188
Molding process
  determination, 188

Part
  design, 188
    verification, 199
  moldability, 198
  quality, 188, 199
  structure optimization, 198
Parting design, 20
  auto-generation, 67
  criteria, 71
  definition, 69
  design, 139
  optimal, 58, 61
    criterion, 61
  parting direction, 13, 58
  parting lines, 13, 64
  parting surfaces, 13, 75
  V-map, 60

Physical variables, 195
Pressure, 195
Process planning, 287
  CAPP, 287
  generative, 288
  hybrid, 293
  generic algorithm, 301
    crossover, 302
    mutation, 302
  process method, 290
  process time, 300
  process type, 290
  simulated annealing, 303
  solution space, 290
Product quality
  assurance, 188

Quality defects, 198

Shear rate, 196
Shrinkage, 139, 203
Structural analysis, 192

Temperature, 196
Tool path
  CL points, affected, 277
  regeneration, 275

Undercut
  blind depression, 101
  EAFEG, 101

feature, 13
  classifications, 91, 116
  definition, 91
  extraction, 91
  recognition, 92, 109, 113
features
  cut-sets, 107
  direction, 94, 96
  draw range, 94
  interacting, 104, 115
  isolated, 102
G-map, 94
graph, 101, 103
hints, 107
hole patching, 79
hybrid method, 106
region, 48
through depression, 101
V-map, 94

Velocity, 196

Warpage, 204
Weld lines, 203
Windows
  -based mold design, 344
  CAD systems, 345
  die casting die design, 357
  GUIs, 352
  IMOLDWorks, 347, 349, 356
  Parasolid, 346, 353
  SolidWorks, 347